FUNDAMENTALS OF LIGHT MICROSCOPY AND ELECTRONIC IMAGING

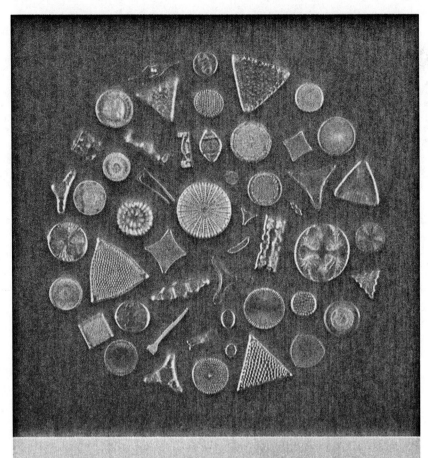
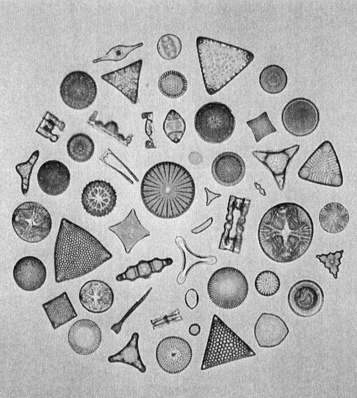

FUNDAMENTALS OF LIGHT MICROSCOPY AND ELECTRONIC IMAGING

Douglas B. Murphy

Department of Cell Biology
Johns Hopkins University School of Medicine

WILEY-LISS

A JOHN WILEY & SONS, INC., PUBLICATION

The cover image is an optical path in the Zeiss Axiophot upright microscope. For details, see the legend to the related Color Plate 1-2. (Courtesy Carl Zeiss, Inc.)

Frontispiece. Diatom exhibition mount, bright-field and dark-field microscopy. (This striking exhibition slide for the light microscope was prepared by Klaus Kemp, Somerset, England.)

This book is printed on acid-free paper. ∞

For ordering and customer service call 1-800-CALL-WILEY.

Library of Congress Cataloging-in-Publication Data:

Murphy, Douglas B.
    Fundamentals of light microscopy and electronic imaging / Douglas B. Murphy.
        p.   cm.
    Includes bibliographical references (p. 357).
    ISBN 0-471-25391-X
    1. Microscopy.   I. Title.

  QH211.M87 2001
  502′.8′2—dc21   2001024021

Printed in the United States of America.

CONTENTS

14. DIGITAL CCD MICROSCOPY   259

15. DIGITAL IMAGE PROCESSING   283

16. IMAGE PROCESSING FOR SCIENTIFIC PUBLICATION   307

PREFACE

Throughout the writing of this book my goal has been how to teach the beginner how to use microscopes. In thinking about a cover, my initial plan was to suggest a silhouette of a microscope under the title "Practical Light Microscopy." However, the needs of the scientific community for a more comprehensive reference and the furious pace of electronic imaging technologies demanded something more. Practitioners of microscopy have long required an instructional text to help align and use a microscope—one that also reviews basic principles of the different optical modes and gives instructions on how to match filters and fluorescent dyes, choose a camera, and acquire and print a microscope image. Advances in science and technology have also profoundly changed the face of light microscopy over the past ten years. Instead of microscope and film camera, the light microscope is now commonly integrated with a CCD camera, computer, software, and printer into electronic imaging systems. Therefore, to use a modern research microscope, it is clear that research scientists need to know not only how to align the microscope optics, but also how to acquire electronic images and perform image processing. Thus, the focus of the book is on the integrated microscope system, with foundations in optical theory but extensions into electronic imaging. Accordingly, the cover shows the conjugate field and aperture planes of the light microscope under the title "Fundamentals of Light Microscopy and Electronic Imaging."

The book covers three areas: optical principles involved in diffraction and image formation in the light microscope; the basic modes of light microscopy; and the components of modern electronic imaging systems and the basic image-processing operations that are required to prepare an image. Each chapter is introduced with theory regarding the topic at hand, followed by descriptions of instrument alignment and image interpretation. As a cell biologist and practitioner of microscopy rather than a physicist or developer of new microscope equipment and methods, the reader will notice that I have focused on how to align and operate microscopes and cameras and have given somewhat abbreviated treatment to the physical theory and principles involved. Nevertheless, the theory is complete enough in its essentials that I hope even experienced microscopists will benefit from many of the descriptions. With the beginner microscopist in mind, each chapter includes practical demonstrations and exercises. The content, though not difficult, is inherently intricate by nature, so the demonstrations are valuable aids in absorbing essential optical principles. They also allow time to pause and reflect on the economy and esthetic beauty of optical laws and principles. If carried out, the demonstrations and exercises also offer opportunities to become acquainted with new biological specimens that the reader may not have confronted or seen before by a new mode of light microscopy. Lists of materials, procedures for specimen preparation, and answers to questions in the problem sets are given in an Appendix. A basic glossary has also been included to aid readers not already familiar with complex terminology. Finally, because the text contains several detailed descriptions of theory and equipment that could be considered ancillary, an effort has been made to subordinate these sections so as to not obscure the major message.

Special thanks are due to many individuals who made this work possible. Foremost I thank profoundly my wife, Christine Murphy, who encouraged me in this work and devoted much time to reading the text and providing much assistance in organizing content, selecting figures, and editing text. I also thank the many students who have taken my microscope courses over the years, who inspired me to write the book and gave valuable advice. In particular, I would like to thank Darren Gray of the Biomedical Engineering Department at Johns Hopkins, who worked with me through every phrase and equation to get the facts straight and to clarify the order of presentation. I would also like to thank and acknowledge the help of many colleagues who provided helpful criticisms and corrections to drafts of the text, including Drs. Bill Earnshaw (University of Edinburgh), Gordon Ellis (University of Pennsylvania), Joe Gall (Carnegie Institution, Department of Embryology), Shinya Inoué (Marine Biological Laboratory), Ernst Keller (Carl Zeiss, Inc.), John Russ (North Carolina State University), Kip Sluder (University of Massachusetts Medical School), and Ken Spring (National Institutes of Health). Finally, I wish to thank many friends and colleagues who provided facts, advice, and much encouragement, including Ken Anderson, Richard Baucom, Andrew Beauto, Marc Benvenuto, Mike Delannoy, Fernando Delaville, Mark Drew, David Elliott, Vickie Frohlich, Juan Garcia, John Heuser, Jan Hinsch, Becky Hohman, Scot Kuo, Tom Lynch, Steven Mattessich, Al McGrath, Michael Mort, Mike Newberry, Mickey Nymick, Chris Palmer, Larry Philips, Clark Riley, Ted Salmon, Dale Schumaker, and Michael Stanley.

I also give special acknowledgment and thanks to Carl Zeiss, Leica Microsystems Nikon Corporation, and Olympus America for providing the color plates that accompany the book.

Finally, I thank Luna Han, Kristin Cooke Fasano and their assistants at John Wiley & Sons for their great patience in receiving the manuscript and managing the production of the book.

Douglas B. Murphy
Baltimore, Maryland

FUNDAMENTALS OF LIGHT MICROSCOPY

OVERVIEW

In this chapter we examine the optical design of the light microscope and review proce-dures for adjusting the microscope and its illumination to obtain the best optical per-formance. The light microscope contains two distinct sets of interlaced focal planes—eight planes in all—between the illuminator and the eye. All of these planes play an important role in image formation. As we will see, some planes are not fixed, but vary in their location depending on the focus position of the objective and condenser lenses. Therefore, an important first step is to adjust the microscope and its illuminator for Koehler illumination, a method introduced by August Koehler in 1893 that gives bright, uniform illumination of the specimen and simultaneously positions the sets of image and diffraction planes at their proper locations. We will refer to these locations frequently throughout the book. Indeed, microscope manufacturers build microscopes so that filters, prisms, and diaphragms are located at precise physical locations in the microscope body, assuming that certain focal planes will be precisely located after the user has adjusted the microscope for Koehler illumination. Finally, we will practice adjusting the microscope for examining a stained histological specimen, review the pro-cedure for determining magnification, and measure the diameters of cells and nuclei in a tissue sample.

OPTICAL COMPONENTS OF THE LIGHT MICROSCOPE

A *compound light microscope* is an optical instrument that uses visible light to produce a magnified image of an object (or specimen) that is projected onto the retina of the eye or onto an imaging device. The word *compound* refers to the fact that two lenses, the objective lens and the eyepiece (or ocular), work together to produce the final magnifi-cation M of the image such that

$$M_{final} = M_{obj} \times M_{oc}.$$

Two microscope components are of critical importance in forming the image: (1) the *objective lens,* which collects light diffracted by the specimen and forms a magnified real image at the real intermediate image plane near the eyepieces or oculars, and (2) the *condenser lens,* which focuses light from the illuminator onto a small area of the specimen. (We define real vs. virtual images and examine the geometrical optics of lenses and magnification in Chapter 4; a real image can be viewed on a screen or exposed on a sheet of film, whereas a virtual image cannot.) The arrangement of these and other components is shown in Figure 1-1. Both the objective and condenser contain multiple lens elements that perform close to their theoretical limits and are therefore expensive. As these optics are handled frequently, they require careful attention. Other components less critical to image formation are no less deserving of care, including the tube and eyepieces, the lamp collector and lamp socket and its cord, filters, polarizers, retarders, and the microscope stage and stand with coarse and fine focus dials.

At this point take time to examine Figure 1-2, which shows how an image becomes magnified and is perceived by the eye. The figure also points out the locations of important focal planes in relation to the objective lens, the ocular, and the eye. The specimen on the microscope stage is examined by the objective lens, which produces a magnified real image of the object in the image plane of the ocular. When looking in the microscope, the ocular acting together with the eye's cornea and lens projects a second real image onto the retina, where it is perceived and interpreted by the brain as a magnified virtual image about 25 cm in front of the eye. For photography, the intermediate image is recorded directly or projected as a real image onto a camera.

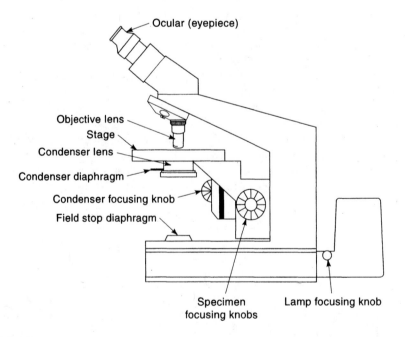

Figure 1-1

The compound light microscope. Note the locations of the specimen focus dials, the condenser focus dial, and the focus dial of the collector lens on the lamp housing. Also note the positions of two variable iris diaphragms: the field stop diaphragm near the illuminator, and the condenser diaphragm at the front aperture of the condenser. Each has an optimum setting in the properly adjusted microscope.

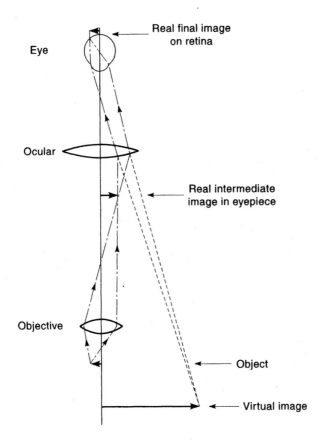

Figure 1-2

Perception of a magnified virtual image of a specimen in the microscope. The objective lens forms a magnified image of the object (called the real intermediate image) in or near the eyepiece; the intermediate image is examined by the eyepiece and eye, which together form a real image on the retina. Because of the perspective, the retina and brain interpret the scene as a magnified virtual image about 25 cm in front of the eye.

Microscopes come in both inverted and upright designs. In both designs the location of the real intermediate image plane at the eyepiece is fixed and the focus dial of the microscope is used to position the image at precisely this location. In most conventional upright microscopes, the objectives are attached to a nosepiece turret on the microscope body, and the focus control moves the specimen stage up and down to bring the image to its proper location in the eyepiece. In inverted designs, the stage itself is fixed to the microscope body, and the focus dials move the objective turret up and down to position the image in the eyepieces.

Note: Inverted Microscope Designs

Inverted microscopes are rapidly gaining in popularity because it is possible to examine living cells in culture dishes filled with medium using standard objectives and avoid the use of sealed flow chambers, which can be awkward. There is also better access to the stage, which can serve as a rigid working platform for microinjection and physiological recording equipment. Inverted designs have their center of mass closer to the lab bench and are therefore less sensitive to vibration. However, there is some risk of physical damage, as objectives may rub against the bottom surface of the stage during rotation of the objective lens turret. Oil immersion objectives are also at risk, because gravity can cause oil to drain down and enter a lens, ruining its optical performance and resulting in costly lens repair. This can be prevented by wrapping a pipe cleaner (the type without the jagged spikes found in a craft store) or by placing a custom fabricated felt washer around the upper part of the lens to catch excess drips of oil. Therefore, despite many advantages, inverted research microscopes require more attention than do standard upright designs.

APERTURE AND IMAGE PLANES IN A FOCUSED, ADJUSTED MICROSCOPE

Principles of geometrical optics show that a microscope has two sets of conjugate focal planes—a set of four *object or field planes* and a set of four *aperture or diffraction planes*—that have fixed, defined locations with respect to the object, optical elements, light source, and the eye or camera. The planes are called *conjugate,* because all of the planes of a given set are seen simultaneously when looking in the microscope. The field planes are observed in normal viewing mode using the eyepieces. This mode is also called the orthosocopic mode, and the object image is called the orthoscopic image. Viewing the aperture or diffraction planes requires using an eyepiece telescope or Bertrand lens, which is focused on the back aperture of the objective lens (see Note and Fig. 1-3). This mode of viewing is called the aperture, diffraction, or conoscopic mode, and the image of the dif-

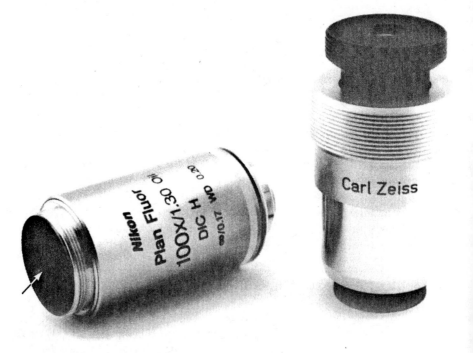

Figure 1-3

The back aperture of an objective lens and a focusable eyepiece telescope.

fraction plane viewed at this location is called the conoscopic image. In this text we refer to the two viewing modes as the *normal* and *aperture* viewing modes and do not use the terms *orthoscopic* and *conoscopic,* although they are common in other texts.

Note: Using an Eyepiece Telescope to View the Objective Back Aperture

An *aperture* is a hole or opening in an opaque mask designed to eliminate stray light from entering the light path, and most field and aperture planes of a microscope contain apertures. A fixed circular aperture is found at or near the rear focal plane of the objective lens. (The precise location of the back focal plane is a function of the focal length of the lens; for objectives with short focal lengths, the focal plane is located inside the lens barrel.) The aperture mask is plainly visible at the back surface of the objective lens (Fig. 1-3). We refer to this site frequently in the text.

The *eyepiece telescope* (sometimes called a phase or centering telescope) is a special focusable eyepiece that is used in place of an ocular to view the back aperture of the objective lens and other aperture planes that are conjugate to it. To use the telescope, remove the eyepiece, insert the eyepiece telescope, and focus it on the circular edge of the objective back aperture. Some microscopes contain a built-in focusable telescope lens called a *Bertrand lens* that can be conveniently rotated into and out of the light path as required.

The identities of the sets of conjugate focal planes are listed here, and their locations in the microscope under conditions of Koehler illumination are shown in Figure 1-4. The terms *front aperture* and *back aperture* refer to the openings at the front and back focal planes of a lens from the perspective of a light ray traveling from the lamp to the retina. Knowledge of the location of these planes is essential for adjusting the microscope and for understanding the principles involved in image formation. Indeed, the entire design of a microscope is based on these planes and the user's need to have access to them. Taken in order of sequence beginning with the light source, they are as follows:

**Field Planes
(normal view through the eyepieces)**

- lamp (field) diaphragm
- object or field plane
- real intermediate image plane (eyepiece field stop)
- retina or camera face plate

Aperture Planes (aperture view through the eyepiece telescope)

- lamp filament
- front aperture of condenser (condenser diaphragm)
- back aperture of objective lens
- exit pupil of eyepiece (coincident with the pupil of the eye)

The *exit pupil* of the eyepiece, which occupies the location of one of the aperture planes, is the disk of light that appears to hang in space a few millimeters above the back lens of the eyepiece; it is simply the image of the back aperture of the objective lens. Normally we are unaware that we are viewing four conjugate field planes when looking through the eyepieces of a microscope. As an example of the simultaneous visibility of conjugate focal planes, consider that the image of a piece of dirt on a focused specimen could lie in any one of the four field planes of the microscope: floaters near the retina,

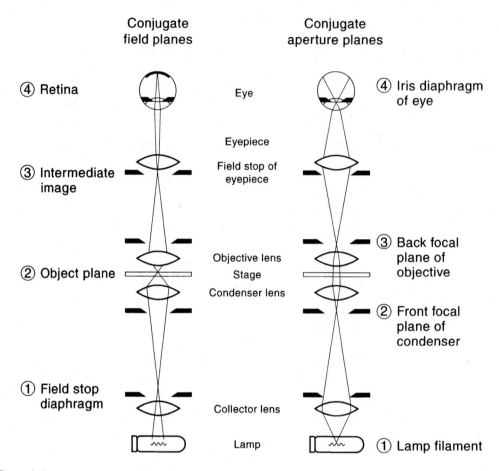

Figure 1-4
The locations of conjugate focal planes in a light microscope adjusted for Koehler illumination. Note the locations of four conjugate field planes (left) and four conjugate aperture planes (right) indicated by the crossover points of rays in the diagrams. The left-hand diagram shows that the specimen or object plane is conjugate with the real intermediate image plane in the eyepiece, the retina of the eye, and the field stop diaphragm between the lamp and the condenser. The right-hand drawing shows that the lamp filament is conjugate with aperture planes at the front focal plane of the condenser, the back focal plane of the objective, and the pupil of the eye.

dirt on an eyepiece reticule, dirt on the specimen itself, or dirt on the glass plate covering the field diaphragm. With knowledge of the locations of the conjugate field planes, the location of the dirt can be determined quickly by rotating the eyepiece, moving the microscope slide, or wiping the cover plate of the field diaphragm.

Before proceeding, take the time to identify the locations of the field and aperture planes on your microscope in the laboratory.

KOEHLER ILLUMINATION

Illumination is a critical determinant of optical performance in light microscopy. Apart from the intensity and wavelength range of the light source, it is important that the light emitted from different locations on the filament be focused at the front aperture of the condenser. The size of the illuminated field at the specimen is adjusted so that it matches the specimen field diameter of the objective lens being employed. Because each source point contributes equally to illumination in the specimen plane, variations in intensity in the image are attributed to the object and not to irregular illumination from the light source. The method of illumination introduced by August Koehler fulfills these requirements and is the standard method used in light microscopy (Fig. 1-5). Under the conditions set forth by Koehler, a *collector lens* on the lamp housing is adjusted so that it focuses an image of the lamp filament at the front focal plane of the condenser while completely filling the aperture; illumination of the specimen plane is bright and even. Achieving this condition also requires focusing the condenser using the condenser focus dial, an adjustment that brings two sets of conjugate focal planes into precise physical locations in the microscope, which is a requirement for a wide range of image contrasting techniques that are discussed in Chapters 7 through 12. The main advantages of Koehler illumination in image formation are:

- *Bright and even illumination in the specimen plane and in the conjugate image plane.* Even when illumination is provided by an irregular light source such as a lamp filament, illumination of the object is remarkably uniform across an extended area. Under these conditions of illumination, a given point in the specimen is illuminated by every point in the light source, and, conversely, a given point in the light source illuminates every point in the specimen.

- *Positioning of two different sets of conjugate focal planes at specific locations along the optic axis of the microscope.* This is a strict requirement for maximal spatial resolution and optimal image formation for a variety of optical modes. As we will see, focusing the stage and condenser positions the focal planes correctly, while adjusting the field and condenser diaphragms controls resolution and contrast. Once properly adjusted, it is easier to locate and correct faults such as dirt and bubbles that can degrade optical performance.

ADJUSTING THE MICROSCOPE FOR KOEHLER ILLUMINATION

Take a minute to review Figure 1-4 to familiarize yourself with the locations of the two sets of focal planes: one set of four field planes and one set of four aperture planes. You will need an eyepiece telescope or Bertrand lens to examine the aperture planes and to make certain adjustments. In the absence of a telescope lens, you may simply remove an eyepiece and look straight down the optic axis at the objective aperture; however, without a telescope the aperture looks small and is difficult to see. The adjustment procedure is given in detail as follows. You will need to check your alignment every time you change a lens to examine a specimen at a different magnification.

Note: Summary of Steps for Koehler Illumination

1. Check that the lamp is focused on the front aperture of the condenser.

2. Focus the specimen.

3. Focus the condenser to see the field stop diaphragm.

4. Adjust the condenser diaphragm using the eyepiece telescope.

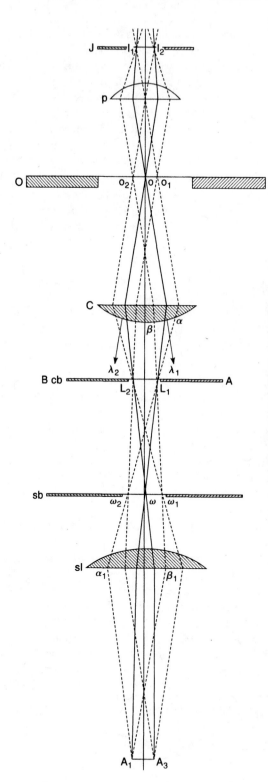

August Kohler
1866-1948

- *Preliminaries.* Place a specimen slide, such as a stained histological specimen, on the stage of the microscope. Adjust the condenser height with the condenser focusing knob so that the front lens element of the condenser comes within ~1–2 mm of the specimen slide. Do the same for the objective lens. Be sure all diaphragms are open so that there is enough light (including the illuminator's field diaphragm, the condenser's front aperture diaphragm, and in some cases a diaphragm in the objective itself). Adjust the lamp power supply so that the illumination is bright but comfortable when viewing the specimen through the eyepieces.

- *Check that the lamp fills the front aperture of the condenser.* Inspect the front aperture of the condenser by eye and ascertain that the illumination fills most of the aperture. It helps to hold a lens tissue against the aperture to check the area of illumination. Using an eyepiece telescope or Bertrand lens, examine the back aperture of the objective and its conjugate planes, the front aperture of the condenser, and the lamp filament. Be sure the lamp filament is centered, using the adjustment screws on the lamp housing if necessary, and confirm that the lamp filament is focused in the plane of the condenser diaphragm. This correction is made by adjusting the focus dial of the collector lens on the lamp housing. Once these adjustments are made, it is usually not necessary to repeat the inspection every time the microscope is used. Instructions for centering the lamp filament or arc are given in Chapter 3. Lamp alignment should be rechecked after the other steps have been completed.

- *Focus the specimen.* Bring a low-power objective to within 1 mm of the specimen, and looking in the microscope, carefully focus the specimen using the microscope's coarse and fine focus dials. It is helpful to position the specimen with the stage controls so that a region of high contrast is centered on the optic axis before attempting to focus. It is also useful to use a low magnification dry objective (10–25×, used without immersion oil) first, since the *working distance*—that is, the distance between the coverslip and the objective—is 2–5 mm for a low-power lens. This reduces the risk of plunging the objective into the specimen slide and causing damage. Since the lenses on most microscopes are *parfocal*, higher magnification objectives will already be in focus or close to focus when rotated into position.

Figure 1-5

August Koehler introduced a new method of illumination that greatly improved image quality and revolutionized light microscope design. Koehler introduced the system in 1893 while he was a university student and instructor at the Zoological Institute in Giessen, Germany, where he performed photomicrography for taxonomic studies on limpets. Using the traditional methods of critical illumination, the glowing mantle of a gas lamp was focused directly on the specimen with the condenser, but the images were unevenly illuminated and dim, making them unsuitable for photography using slow-speed emulsions. Koehler's solution was to reinvent the illumination scheme. He introduced a collector lens for the lamp and used it to focus the image of the lamp on the front aperture of the condenser. A luminous field stop (the field diaphragm) was then focused on the specimen with the condenser focus control. The method provided bright, even illumination, and fixed the positions of the focal planes of the microscope optics. In later years, phase contrast microscopy, fluorescence microscopy with epi-illumination, differential interference contrast microscopy, and confocal optical systems would all utilize and be critically dependent on the action of the collector lens, the field diaphragm, and the presence of fixed conjugate focal planes that are inherent to Koehler's method of illumination. The interested reader should refer to the special centenary publication on Koehler by the Royal Microscopical Society (see Koehler, 1893).

- *Focus and center the condenser.* With the specimen in focus, close down (stop down) the *field diaphragm*, and then, while examining the specimen through the eyepieces, focus the angular outline of the diaphragm using the condenser's focusing knob. If there is no light, turn up the power supply and bring the condenser closer to the microscope slide. If light is seen but seems to be far off axis, switch to a low-power lens and move the condenser positioning knobs slowly to bring the center of the illumination into the center of the field of view. Focus the image of the field diaphragm and center it using the condenser's two centration adjustment screws. The field diaphragm is then opened just enough to accommodate the object or the field of a given detector. This helps reduce scattered or stray light and improves image contrast. The condenser is now properly adjusted. We are nearly there! The conjugate focal planes that define Koehler illumination are now at their proper locations in the microscope.

- *Adjust the condenser diaphragm while viewing the objective back aperture with an eyepiece telescope or Bertrand lens.* Finally, the condenser diaphragm (and the built-in objective diaphragm, if the objective has one) is adjusted to obtain the best resolution and contrast, but is not closed so far as to degrade the resolution. In viewing the condenser front aperture using a telescope, the small bright disk of light seen in the telescope represents the objective's back aperture plus the superimposed image of the condenser's front aperture diaphragm. As you close down the condenser diaphragm, you will see its edges enter the aperture opening and limit the objective aperture's diameter. Focus the telescope so that the edges of the diaphragm are seen clearly. Stop when ~3/4 of the maximum diameter of the aperture remains illuminated, and use this setting as a starting position for subsequent examination of the specimen. As pointed out in the next chapter, the setting of this aperture is crucial, because it determines the resolution of the microscope, affects the contrast of the image, and establishes the depth of field. It is usually impossible to optimize for resolution and contrast at the same time, so the 3/4 open position indicated here is a good starting position. The final setting depends on the inherent contrast of the specimen.

- *Adjust the lamp brightness.* Image brightness is controlled by regulating the lamp voltage, or if the voltage is nonadjustable, by placing neutral density filters in the light path near the illuminator in specially designed filter holders. *The aperture diaphragm should never be closed down as a way to reduce light intensity,* because this action reduces the resolving power and may blur fine details in the image. We will return to this point in Chapter 6.

The procedure for adjusting the microscope for Koehler illumination seems invariably to stymie most newcomers. With so many different focusing dials, diaphragm adjustments, viewing modes, eyepiece changes, image planes, filter placements, and lamp settings to worry about, this is perhaps to be expected. To get you on your way, try to remember this simple two-step guide: *Focus on a specimen and then focus and center the condenser.* Post this reminder near your microscope. If you do nothing else, you will have properly adjusted the image and aperture planes of the microscope, and the rest will come quickly enough after practicing the procedure a few times. Although the adjustments sound complex, they are simple to perform, and their significance for optical performance cannot be overstated. The advantages of Koehler illumination for a number of optical contrasting techniques will be revealed in the next several chapters.

Note: Focusing Oil Immersion Objectives

The *working distance*—that is, the distance between the front lens element and the surface of the coverslip—of an oil immersion lens is so small ($\sim$60 μm for some oil immersion lenses) that the two optical surfaces nearly touch each other when the specimen is in focus. Due to such close tolerances, it is unavoidable that the lens and coverslip will occasionally make contact, but this is usually of little consequence. The outermost lens elements are mounted in a spring-loaded cap, so the lens can be compressed a bit by the specimen slide without damaging the optics. The lens surface is also recessed and not coplanar with the surface of the metal lens cap, which prevents accidental scratching and abrasion.

Begin focusing by bringing the lens in contact with the drop of oil on the coverslip. The drop of oil expands as the lens is brought toward focus, and at contact (essentially the desired focus position) the oil drop stops expanding. If overfocused, the microscope slide is pushed up off the stage by a small amount on an inverted microscope; on an upright microscope the spring-loaded element of the objective compresses a bit. Retract the lens to the true focal position and then examine the specimen. In normal viewing mode it should only be necessary to change the focus by a very small amount to find the specimen. It can help to move the specimen stage controls with the other hand to identify the shadows or fluorescence of a conspicuous object, which may serve as a guide for final focus adjustment. Notice that if focus movements are too extreme, there is a risk that the objective (on an upright microscope) or the condenser (on an inverted microscope) might break the microscope slide, or worse, induce permanent strain in the optics. Focusing with oil immersion optics always requires extra care and patience.

Before observing the specimen, examine the back focal plane of the objective with an eyepiece telescope to check for lint and oil bubbles. An insufficient amount of oil between the lens and coverslip can cause the entire back aperture to be misshapen; if this is the case, focusing the telescope will bring the edge of the oil drop into sharp focus. These faults should be removed or corrected, as they will significantly degrade optical performance. Finally, when using immersion oil, never mix oils from different companies since slight differences in refractive index will cause pronounced blurring.

PRECAUTIONS FOR HANDLING OPTICAL EQUIPMENT

- *Never strain, twist, or drop objectives or other optical components.* Optics for polarization microscopy are especially susceptible to failure due to mishandling.
- *Never force the focus controls of the objective or condenser, and always watch lens surfaces as they approach the specimen.* This is especially important for high-power oil immersion lenses.
- *Never touch optical surfaces.* In some cases, just touching an optical surface can remove unprotected coatings and ruin filters that cost hundreds of dollars. Carefully follow the procedures for cleaning lenses and optical devices.

Exercise: Calibration of Magnification

Examine a histological specimen to practice proper focusing of the condenser and setting of the field stop and condenser diaphragms. A 1 μm thick section of pancreas or other tissue stained with hematoxylin and eosin is ideal. A typical histological specimen is a section of a tissue or organ that has been chemically fixed, embedded in epoxy resin or paraffin, sectioned, and stained with dyes specific for nucleic acids, proteins, carbohydrates, and so forth. In hematoxylin and eosin (H&E) staining, hematoxylin stains the nucleus and cell RNA a dark blue or purple color, while eosin stains proteins (and the pancreatic secretory granules) a bright orange-pink. When the specimen is illuminated with monochromatic light, the contrast perceived by the eye is largely due to these stains. For this reason, a stained histological specimen is called an *amplitude specimen* and is suitable for examination under the microscope using bright field optics. A suitable magnification is 10–40×.

Equipment and Procedure

Three items are required: a focusable eyepiece, an eyepiece reticule, and a stage micrometer (Fig. 1-6). The eyepiece reticule is a round glass disk usually containing a 10 mm scale divided into 0.1 mm (100 μm) units. The reticule is mounted in an eyepiece and is then calibrated using a precision stage micrometer to obtain a conversion factor (μm/reticule unit), which is used to determine the magnification obtained for each objective lens. The reason for using this calibration procedure is that the nominal magnification of an objective lens (found engraved on the lens barrel) is only correct to within ± 5%. If precision is not of great concern, an approximate magnification can be obtained using the eyepiece reticule alone. In this case, simply measure the number of micrometers from the eyepiece reticule and divide by the nominal magnification of the objective. For a specimen covering 2 reticule units (200 μm), for example: 200 μm/10× = 20 μm.

The full procedure, using the stage micrometer, is performed as follows:

- To mount the eyepiece reticule, unscrew the lower barrel of the focusing eyepiece and place the reticule on the stop ring with the scale facing upward. The stop ring marks the position of the real intermediate image plane. Make sure the reticule size matches the internal diameter of the eyepiece and rests on the field stop. Carefully reassemble the eyepiece and return it to the binocular head. Next focus the reticule scale using the focus dial on the eyepiece and then focus on a specimen with the microscope focus dial. The images of the specimen and reticule are conjugate and should be simultaneously in sharp focus.

- Examine the stage micrometer slide, rotating the eyepiece so that the micrometer and reticule scales are lined up and partly overlapping. The stage micrometer consists of a 1 or 2 mm scale divided into 10 μm units, giving 100 units/mm. The micrometer slide is usually marked 1/100 mm. The conversion factor we need to determine is simply the number of μm/reticule unit. This conversion factor can be calculated more accurately by counting the number of micrometers contained in *several* reticule units in the eyepiece. The procedure

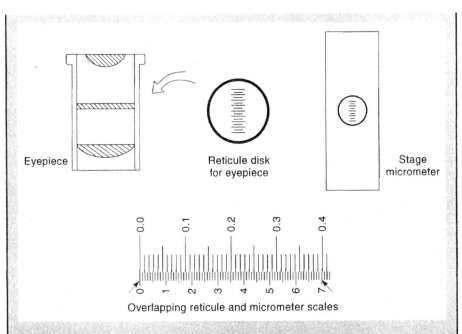

Figure 1-6
The eyepiece reticule and stage micrometer used for determining magnification. The typical eyepiece reticule is divided into 1/100 cm (100 μm unit divisions), and the stage micrometer into 1/100 mm (10 μm unit divisions). The appearance of the two overlapping scales is shown at the bottom of the figure.

    must be repeated for each objective lens, but only needs to be performed one time for each lens.

- Returning to the specimen slide, the number of eyepiece reticule units spanning the diameter of a structure is determined and multiplied by the conversion factor to obtain the distance in micrometers.

Exercise

1. Calibrate the magnification of the objective lens/eyepiece system using the stage micrometer and an eyepiece reticule. First determine how many micrometers are in each reticule unit.

2. Determine the mean diameter and standard deviation of a typical cell, a nucleus, and a cell organelle (secretory granule), where the sample size, n, is 10. Examination of cell organelles requires a magnification of 40–100X.

3. Why is it wrong to adjust the brightness of the image using either of the two diaphragms? How else (in fact, how should you) adjust the light intensity and produce an image of suitable brightness for viewing or photography?

LIGHT AND COLOR

OVERVIEW

In this chapter we review the nature and action of light as a probe to examine objects in the light microscope. Knowledge of the wave nature of light is essential for understanding the physical basis of color, polarization, diffraction, image formation, and many other topics covered in this book. The eye-brain visual system is responsible for the detection of light including the perception of color and differences in light intensity that we recognize as contrast. The eye is also a remarkably designed detector in an optical sense—the spacing of photoreceptor cells in the retina perfectly matches the requirement for resolving the finest image details formed by its lens (Fig. 2-1). Knowledge of the properties of light is important in selecting filters and objectives, interpreting colors, performing low-light imaging, and many other tasks.

LIGHT AS A PROBE OF MATTER

It is useful to think of light as a probe that can be used to determine the structure of objects viewed under a microscope. Generally, probes must have size dimensions that are similar to or smaller than the structures being examined. Fingers are excellent probes for determining the size and shape of keys on a computer keyboard, but fail in resolving wiring patterns on a computer's integrated circuit chip. Similarly, waves of light are effective in resolving object details with dimensions similar to the wavelength of light, but generally do not do well in resolving molecular and atomic structures that are much smaller than the wavelength. For example, details as small as 0.2 μm can be resolved visually in a microscope with an oil immersion objective. *As an approximation, the resolution limit of the light microscope with an oil immersion objective is about one-half of the wavelength of the light employed.*

    Visible light, the agent used as the analytic probe in light microscopy, is a form of energy called electromagnetic radiation. This energy is contained in discrete units or quanta called photons that have the properties of both particles and waves. Photons as electromagnetic waves exhibit oscillating electric and magnetic fields, designated E and

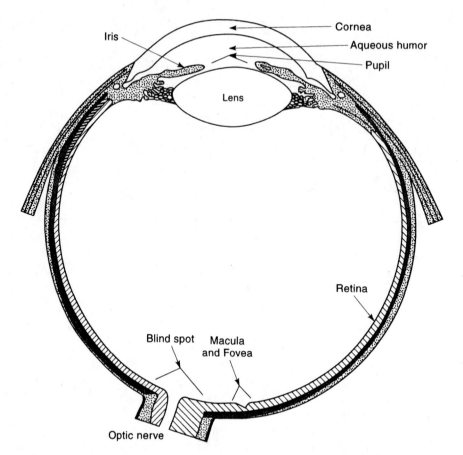

Figure 2-1

Structure of the human eye. The cornea and lens of the eye work together with the eyepiece to focus a real magnified image on the retina. The aperture plane of the eye-microscope system is located in front of the lens in the pupil of the eye, which functions as a variable diaphragm. A large number of rod cells covers the surface of the retina. The 3 mm macula, or yellow spot, contains a 0.5 mm diameter fovea, a depressed pit that contains the majority of the retina's cone cells that are responsible for color vision. The blind spot contains no photoreceptor cells and marks the site of exit of the optic nerve.

B, respectively, whose amplitudes and directions are represented by vectors that oscillate in phase as sinusoidal waves in two mutually perpendicular planes (Fig. 2-2). Photons are associated with a particular energy (ergs), which determines their wavelength (nm) and vibrational frequency (cycles/s). It is important to realize that the electromagnetic waves we perceive as light (400–750 nm, or about 10^{-7} m) comprise just a small portion of the entire electromagnetic spectrum, which ranges from 10^4 m (radio waves) to 10^{-10} m (γ-rays) (Fig. 2-3). The figure also compares the sizes of cells, molecules, and atoms with the wavelengths of different radiations. See Hecht (1998) and Longhurst (1967) for interesting discussions on the nature of light.

Although it is frustrating that light cannot be defined in terms of a single physical entity, it can be described through mathematical relationships that depict its dual particle- and wavelike properties. The properties of energy, frequency, and wavelength are

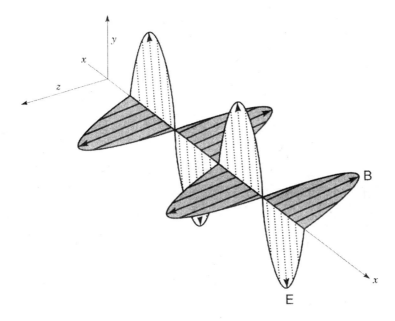

Figure 2-2

Light as an electromagnetic wave. The wave exhibits electric (E) and magnetic (B) fields whose amplitudes oscillate as a sine function over dimensions of space or time. The amplitudes of the electric and magnetic components at a particular instant or location are described as vectors that vibrate in two planes perpendicular to each other and perpendicular to the direction of propagation. However, at any given time or distance the E and B vectors are equal in amplitude and phase. For convenience it is common to show only the electric field vector (E vector) of a wave in graphs and diagrams and not specify it as such.

related through the following equations, which can be used to determine the amount of energy associated with a photon of a specific wavelength:

$$c = \nu\lambda,$$

$$E = h\nu,$$

and combining,

$$E = hc/\lambda,$$

where c is the speed of light (3×10^{10} cm/s), ν is the frequency (cycles/s), λ is the wavelength (cm), E is energy (ergs), and h is Plank's constant (6.62×10^{-27} erg-seconds). The first equation defines the velocity of light as the product of its frequency and wavelength. We will encounter conditions where velocity and wavelength vary, such as when photons enter a glass lens. The second equation relates frequency and energy, which becomes important when we must choose a wavelength for examining live cells. The third equation relates the energy of a photon to its wavelength. Since $E \sim 1/\lambda$, 400 nm blue wavelengths are twice as energetic as 800 nm infrared wavelengths.

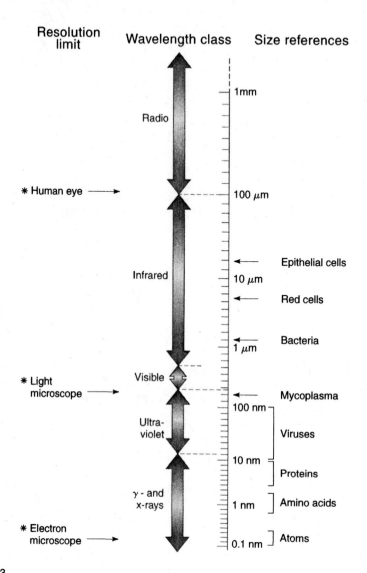

Figure 2-3

The electromagnetic spectrum. The figure shows a logarithmic distance scale (range, 1 mm to 0.1 nm). One side shows the wavelength ranges of common classes of electromagnetic radiation; for reference, the other side indicates the sizes of various cells and macromolecules. Thus, a red blood cell (7.5 μm) is 15 times larger than a wavelength of visible green light (500 nm). The resolution limits of the eye, light microscope, and electron microscope are also indicated. For the eye, the resolution limit (0.1 mm) is taken as the smallest interval in an alternating pattern of black and white bars on a sheet of paper held 25 cm in front of the eye under conditions of bright illumination. Notice that the range of visible wavelengths spans just a small portion of the spectrum.

LIGHT AS PARTICLES AND WAVES

For the most part, we will be referring to the wave nature of light and the propagation of electromagnetic radiation as the movement of planar wavefronts of a specific wavelength through space. The propagation vector is linear in a homogeneous medium such as air or glass or in a vacuum. The relatively narrow spectrum of photon energies (and corresponding frequencies) we experience as light is capable of exciting the visual pigments in the rod and cone cells in the retina and corresponds to wavelengths ranging from 400 nm (violet) to 750 nm (red). As shown in Figure 2-4, we depict light in various ways depending on which features we wish to emphasize:

- As *quanta* (photons) of electromagnetic radiation, where photons are detected as individual quanta of energy (as photoelectrons) on the surfaces of quantitative measuring devices such as charge-coupled device (CCD) cameras or photomultiplier tubes.

- As *waves*, where the propagation of a photon is depicted graphically as a pair of electric (E) and magnetic (B) fields that oscillate in phase and in two mutually perpendicular planes as functions of a sine wave. The vectors representing these fields vibrate in two planes that are both mutually perpendicular to each other and perpendicular to the direction of propagation. For convenience it is common to show only the wave's electric field vector (E vector) in graphs and diagrams and not specify it as such. When shown as a sine wave on a plot with x, y coordinates, the amplitude of a wave on the y-axis represents the strength of the electric or magnetic field, whereas the x-axis depicts the time or distance of travel of the wave or its phase relative to some other reference wave. At any given time or distance, the E and B field vectors are equal in amplitude and phase. Looking down the x-axis (the propagation axis), the plane of the E vector may vibrate in any orientation through 360° of rotation about the axis. The angular tilt of the E vector along its propagation axis and

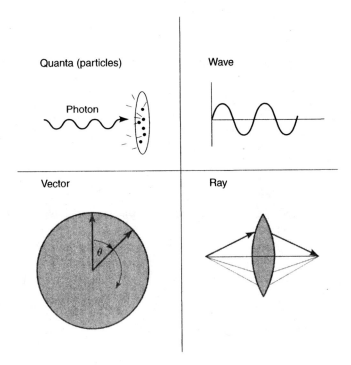

Figure 2-4
Light as quanta, waves, vectors, and rays.

relative to some fixed reference is called the *azimuthal angle* of orientation. Commonly, the sine waves seen in drawings refer to the average amplitude and phase of a beam of light (a light train consisting of a stream of photons), not to the properties of a single electromagnetic wave.

- As *vectors,* where the vector length represents the amplitude, and the vector angle represents the advance or retardation of the wave relative to an imaginary reference. The vector angle is defined with respect to a perpendicular drawn through the focus of a circle, where 360° of rotation corresponds to one wavelength (2π radians).

- As *rays* or *beams,* where the linear path of a ray (a light train or stream of photons) in a homogeneous medium is shown as a straight line. This representation is commonly used in geometrical optics and ray tracing to show the pathways of rays passing through lenses of an imaging system.

THE QUALITY OF LIGHT

As an analytic probe used in light microscopy, we also describe the kind or quality of light according to the degree of uniformity of rays comprising an illuminating beam (Fig. 2-5). The kinds of light most frequently referred to in this text include:

- *Monochromatic*—waves having the same wavelength or vibrational frequency (the same color).

- *Polarized*—waves whose E vectors vibrate in planes that are parallel to one another. The E vectors of rays of sunlight reflected off a sheet of glass are plane parallel and are said to be linearly polarized.

- *Coherent*—waves of a given wavelength that maintain the same phase relationship while traveling through space and time (laser light is coherent, monochromatic, and polarized).

- *Collimated*—waves having coaxial paths of propagation through space—that is, without convergence or divergence, but not necessarily having the same wavelength, phase, or state of polarization. The surface wavefront at any point along a cross-section of a beam of collimated light is planar and perpendicular to the axis of propagation.

Light interacts with matter in a variety of ways. Light incident on an object might be absorbed, transmitted, reflected, or diffracted, and such objects are said to be opaque, transparent, reflective, or scattering. Light may be absorbed and then re-emitted as visible light or as heat, or it may be transformed into some other kind of energy such as chemical energy. Objects or molecules that absorb light transiently and quickly re-emit it as longer wavelength light are described as being phosphorescent or fluorescent depending on the time required for re-emission. Absorbed light energy might also be re-radiated slowly at long infrared wavelengths and may be perceived as heat. Light absorbed by cells may be damaging if the energy is sufficient to break covalent bonds within molecules or drive adverse chemical reactions including those that form cyto-toxic free radicals. Finally, a beam of light may be bent or deviated while passing through a transparent object such as a glass lens having a different refractive index (*refraction*), or may be bent uniformly around the edges of large opaque objects (*dif-*

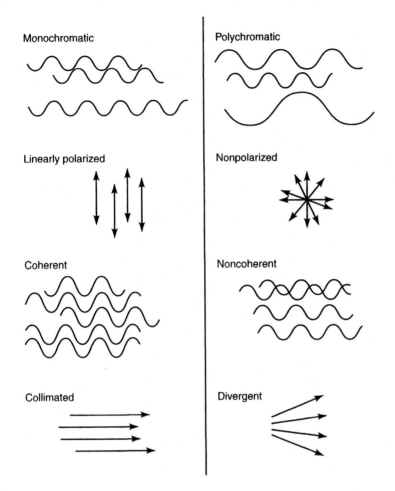

Figure 2-5
Eight waveforms depicting variations in the quality of light.

fraction), or even scattered by small particles and structures having dimensions similar to the wavelength of light itself (also known as diffraction). The diffraction of light by small structural elements in a specimen is the principal process governing image formation in the light microscope.

PROPERTIES OF LIGHT PERCEIVED BY THE EYE

The eye-brain system perceives differences in light intensity and wavelength (color), but does not see differences in the phase of light or its state of polarization. Thus, laser light, which is both coherent and polarized, cannot be distinguished from random light having the same wavelength (color). We will restrict our discussion here to the perception of light intensity, since the perception of color is treated separately in the following section.

The brightness of a light wave is described physically and optically in terms of the *amplitude (A)* of its E vector, as depicted in a graph of its sine function. Indeed, the amplitudes of sine waves are shown in many figures in the text. However, the nervous activity of photoreceptor cells in the retina is proportional to the light *intensity (I)*, where intensity is defined as the rate of flow of light energy per unit area and per unit time across a detector surface. Amplitude (energy) and intensity (energy flux) are related such that the intensity of a wave is proportional to the square of its amplitude, where

$$I \propto A^2.$$

For an object to be perceived, the light intensity corresponding to the object must be different from nearby flanking intensities and thereby exhibit contrast, where *contrast (C)* is defined as the ratio of intensities,

$$C = \Delta I / I_b,$$

ΔI is the difference in intensity between an object and its background, and I_b is the intensity of the background. If $I_{obj} \sim I_b$, as it is for many transparent microscope specimens, $C = 0$, and the object is invisible. More specifically, visibility requires that the object exceed a certain *contrast threshold*. In bright light, the contrast threshold required for visual detection may be as little as 2–5%, but should be many times that value for objects to be seen clearly. In dim lighting, the contrast threshold may be 200–300%, depending on the size of the object. The term *contrast* always refers to the ratio of two intensities and is a term commonly used throughout the text.

PHYSICAL BASIS FOR VISUAL PERCEPTION AND COLOR

As we will emphasize later, the eye sees differences in light intensity (contrast) and perceives different wavelengths as colors, but cannot discern differences in phase displacements between waves or detect differences in the state of polarization. The range of wavelengths perceived as color extends from 400 nm (violet) to 750 nm (red), while peak sensitivity in bright light occurs at 555 nm (yellow-green). The curves in Figure 2-6 show the response of the eye to light of different wavelengths for both dim light (night or rod vision) and bright light (day or cone vision) conditions. The eye itself is actually a *logarithmic* detector that allows us to see both bright and dim objects simultaneously in the same visual scene. Thus, the apparent difference in intensity between two objects I_1 and I_2 is perceived as the logarithm of the ratio of the intensities, that is, as $\log_{10}(I_1/I_2)$. It is interesting that this relationship is inherent to the scale used by Hipparchus (160–127 B.C.) to describe the magnitudes of stars in 6 steps with 5 equal intervals of brightness. Still using the scale today, we say that an intensity difference of 100 is covered by 5 steps of Hipparchus' stellar magnitude such that $2.512 \log_{10} 100 = 5$. Thus, each step of the scale is 2.512 times as much as the preceding step, and $2.512^5 = 100$, demonstrating that what we perceive as equal steps in intensity is really the log of the ratio of intensities. The sensitivity of the eye in bright light conditions covers about 3 orders of magnitude within a field of view; however, if we allow time for physiological adaptation and consider both dim and bright lighting conditions, the sensitivity range of the eye is found to cover an incredible 10 orders of magnitude overall.

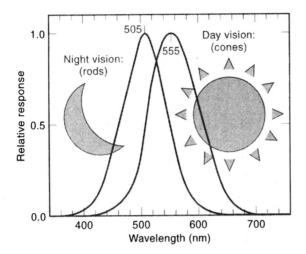

Figure 2-6

The spectral response of the eye in night and day vision. The two curves have been normalized to their peak sensitivity, which is designated 1.0; however, night (rod) vision is 40 times more sensitive than day (cone) vision. Rhodopsin contained in rod cells and color receptor pigments in cone cells have action spectra with different maxima and spectral ranges.

The shape and distribution of the light-sensitive rod and cone cells in the retina are adapted for maximum sensitivity and resolution in accordance with the physical parameters of light and the optics of the eye-lens system. Namely, the outer segments of cone cells, the cells responsible for color perception in the fovea, are packed together in the plane of the retina with an intercellular spacing of 1.0–1.5 μm, about one-half the radius of the minimum spot diameter (3 μm) of a focused point of light on the retina. The small 1.5 μm cone cell diameter allows the eye to resolve structural details down to the theoretical limit calculated for the eye-lens system. For an object held 25 cm in front of the eye, this corresponds to spacings of ~0.1 mm. It appears nature has allowed the light receptor cells to utilize the physics of light and the principles of lens optics as efficiently as possible!

Rod cell photoreceptors comprise 95% of the photoreceptors in the retina and are active in dim light, but provide no color sense. Examine Figure 2-1 showing the structure of the eye and Figure 2-7 showing the distribution of rod cells in the retina. Rods contain the light-sensitive protein, *rhodopsin,* not the photovisual pigments required for color vision, and the dim light vision they provide is called *scotopic vision.* Rhodopsin, a photosensitive protein, is conjugated to a chromophore, 11-cis-retinal, a carotenoid that photoisomerizes from a cis to trans state upon stimulation and is responsible for electrical activity of the rod cell membranes. The peak sensitivity of the rod photoreceptor cells (510 nm) is in the blue-green region of the visual spectrum. Rod cell vision is approximately 40 times more sensitive to stimulation by light than the cone cell receptors that mediate color vision. Bright light rapidly bleaches rhodopsin, causing temporary blindness in dim lighting conditions, but rhodopsin isomerizes gradually over a 20–30 min period, after which rod receptor function is largely restored. Full recovery may require several hours or even days—ask any visual astronomer or microscopist! To avoid photobleaching your rhodopsin pigments and to maintain high visual sensitivity for dim specimens (common with polarized light or fluorescence optics), you

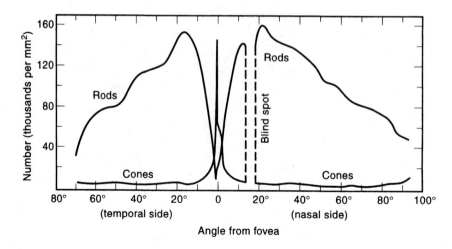

Figure 2-7

Distribution of rod and cone cells in the retina. The number of cells per mm² is plotted vs. the angle from the fovea as seen from the lens. The fovea is distinct in having a paucity of rods and an abundance of cones. The blind spot lacks photoreceptor cells.

should make observations in a darkened room. Red light illumination in the otherwise darkened microscope room is also commonly employed, because red wavelengths bleach the rhodopsin inefficiently (see Fig. 2-8 for differences in absorption spectra of visual pigments), yet allow you to see to operate equipment and take notes.

Cone cell photoreceptors comprise only 5% of the retinal photoreceptor cells and are contained nearly exclusively in the small central fovea of the retina, a 0.5 mm diameter spot that is responsible for color perception and visual acuity. Vision dominated by the function of cones under bright light conditions is called photopic vision. Cone cells contain red-, green-, or blue-sensitive pigment proteins that are also conjugated to 11-cis-retinal. The color photovisual pigments are highly homologous to each other and share about 40% amino acid sequence homology with rod cell rhodopsin (Nathans, 1984). Absorption spectra for purified rhodopsin and the three color pigments are shown in Figure 2-8.

POSITIVE AND NEGATIVE COLORS

As discussed in this section, color can be described as the addition or subtraction of specific wavelengths of light. Light is perceived as white when all three cone cell types (red, green, and blue) are stimulated equally as occurs when viewing a nonabsorbing white sheet of paper in sunlight. It was found over a century ago by James Clerk Maxwell (1831–1879) that color vision can be approximated by a simple tristimulus system involving red, green, and blue color stimulation. By varying the relative intensities of the three colors, all of the colors of the visual spectrum can be created, ranging from violet to red. Positive colors are created by combining different color wavelengths. A fine example of mixing wavelengths to create positive colors can be made using three slide projectors, each equipped with monochromatic red, green, and blue cellophane filters (the kind used for RGB color analysis) from a scientific supply house. The filters are mounted in slide holders and covered with an opaque aluminum foil mask containing a

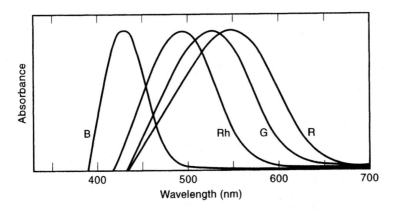

Figure 2-8

Absorption difference spectra of the four human visual pigments. The four pigment proteins were cloned, purified, and characterized with respect to their absorption spectra in vitro. Photobleaching difference spectra were obtained by subtracting an absorption spectrum measured after light exposure from one measured prior to light exposure. The pigments show maxima in the expected red, green, and blue regions of the visual spectrum. The values are close to those measured for rod and cone cells in vivo and confirm Maxwell's theory for RGB-based color vision over a century ago. (Courtesy of Jeremy Nathans, Johns Hopkins University.)

1 cm diameter hole in the center. Three color disks can be projected on a screen and made to overlap as shown in Figure 2-9. Try it and experience why mixing magenta and green light gives white. *Negative colors,* in contrast, are generated by the subtraction (absorption) of light of a specific wavelength from light composed of a mixture of wavelengths. A pigment that looks red, for example, absorbs blue and green wavelengths, but reflects red, so it is red by default. To appreciate this point, it is informative to examine colored objects (paints and pigments) with a handheld *spectroscope* under bright white light illumination. It is fascinating that yellow, cyan-blue, and magenta pigments are composed, respectively, of equal mixtures of red and green, green and blue, and blue and red wavelengths.

Thus, perception of the color yellow can arise in two ways: (1) by simultaneous stimulation of the red and green cone cells by a monochromatic yellow (580 nm) light source—the red and green photovisual pigments exhibit broad excitation spectra that

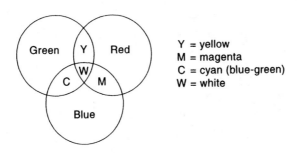

Figure 2-9

Addition colors of the red-green-blue tricolor system. This color display can be produced by projecting the colors of RGB color filters with three separate slide projectors on a screen as described in the text.

overlap significantly in the 580 nm band region and are both stimulated almost equally; or (2) by stimulating the red and green cones separately with a mixture of distinct red and green wavelengths, each wavelength selectively stimulating red and green cone cells in the retina. In either case, the color yellow is defined as the simultaneous stimulation of both red and green visual pigments. Perception of other colors requires stimulation of one, two, or all three cone cell types to varying degrees. The mixing of different colored paints to produce new colors, which is our common experience in producing colors, is actually a subtractive process. Consider why mixing yellow and blue paints produces a green color: Yellow pigment (reflects red and green, but absorbs blue) and blue pigment (reflects blue and green, but absorbs red) gives green because green is the only wavelength not absorbed by the mixture of yellow and blue pigments. Thus, combining blue and yellow wavelengths of light gives white, but mixing blue and yellow pigments gives green! Removal of a specific wavelength from the visual spectrum is also the mechanism for producing interference colors, and is discussed in Chapter 9. A useful overview on the perception of color when viewing natural phenomena is given by Minnaert (1954).

Exercise: Complementary colors

A complementary color is a color that gives white light when mixed with its complement. Thus, yellow and cyan-blue are complementary colors as are the color-pairs green with magenta and red with cyan. Our perception of complementary colors is due to the red, green, and blue photovisual pigments in the cone cells of the retina. Note that mixing wavelengths of different colors is distinct from mixing colored pigments.

Combining Red, Green, and Blue Light. To experience the relationships among complementary colors, prepare 3 slide projectors each containing a red, blue and green color filter sandwiched together with masks containing a 1 cm diameter hole, and project three disks of red, green and blue color on a projection screen. Focus each projector so the edges of the circular masks are sharp. Move the projectors to partially overlap the colors so that it is possible to see that red plus green gives yellow, red plus blue gives magenta, and blue plus green gives cyan. The overlap of red, green and blue gives white. Thus, when all 3 color types of cone cells in the retina are saturated, we see white light.

Mixing Colored Pigments. As is known, mixing yellow and blue pigments gives green. The reason for this is that blue and yellow pigments reflect green light; all other wavelengths are absorbed by the blue and yellow dyes in the mixture. To demonstrate this, prepare 2 beakers with 500 mL water and add 8 drops of blue and yellow food coloring separately to each beaker. Display the beakers on a light box. The generation of green by mixing the yellow and blue pigmented solutions is different from the mixing of blue and yellow light, which gives white light, as was demonstrated above.

Removing Colors from a White Light Source. The relationship between complementary colors and subtraction colors can be demonstrated using a bright white light source, a slit, a diffraction grating and a large diameter magnifying glass to form the image of the slit on a projection screen. Set up the optical bench apparatus with a bright xenon light source, a 1 mm wide slit made from razor blades, an IR blocking filter, and a holographic diffraction grating as described in Figure 2-10. Intercept the dispersed color spectrum with a white card and exam-

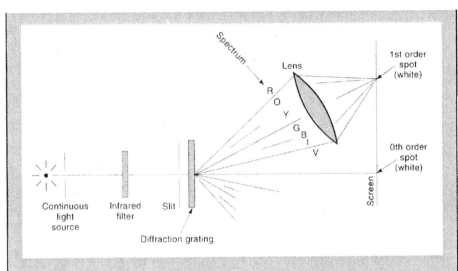

Figure 2-10
Optical bench setup for demonstrating complementary colors.

ine the spectral colors. With the help of a partner, examine the colors on the card with the spectroscope. Scan the spectroscope back and forth across the spectrum and confirm that each color is pure, monochromatic, and located in the spectrum according to its wavelength. Next intercept the spectrum between the grating and the projection screen with a 4–6 inch diameter magnifying glass and focus the image of the slit on a projection screen. Notice that the color of the focused image of the slit is white. It is clear that the individual spectral colors have been recombined to give white light. Next insert an opaque 1 cm wide white paper strip into the light path to remove a band of wavelengths such as red and orange from the spectrum. Note the corresponding color of the slit in the image plane. The color of the slit on the screen and the color of the blocked light are called complementary colors. What are the complementary colors for red, green, and blue?

Examine the Colors of After-images. Another way to examine complementary colors is to produce "after images" on the retina. Stare at a collection of large brightly colored objects for 30–60 seconds in a brightly illuminated area. Do not let your eyes wander, but focus steadily on a central spot in the visual field. Then shift your vision to stare at a blank white sheet of paper or brightly illuminated white wall. Do you see an after-image composed of complementary colors of the objects? The complementary colors are seen because the cones stimulated by certain colors become depleted and temporarily exhausted after prolonged viewing, so unstimulated cones in the same area of the retina provide the principal stimulus when you look at a white page. Cyan-blue, magenta-red, and yellow are the three complementary colors to red, green, and blue.

ILLUMINATORS, FILTERS, AND THE ISOLATION OF SPECIFIC WAVELENGTHS

OVERVIEW

To obtain optimal imaging performance in the light microscope, the specimen must be properly illuminated. This requires proper selection of wavelength and intensity and correct alignment and focus of the lamp. The first objective is met by matching the particular application to the proper combination of illuminator and filters. Since research microscopes may be equipped with a variety of lamps, including quartz halogen lamps and other tungsten filament lamps, mercury, xenon, and metal halide arc lamps, we discuss the energy and spectral output of various illuminators. Filters that adjust light intensity and provide wavelengths of a particular color are also discussed. For example, if the microscope is equipped with a constant wattage power supply, the intensity must be controlled using neutral density filters; similarly, colored glass filters and interference filters are used to isolate specific color bandwidths. It is the combination of illuminator and filters that determines the quality of light directed to the condenser for illuminating the specimen. While all forms of light microscopy require selecting illuminators and filters, knowledge of their action and function becomes especially important in fluorescence and confocal fluorescence microscopy. We close the chapter by discussing how illuminators and filters must be carefully considered when examining living cells.

ILLUMINATORS AND THEIR SPECTRA

Successful imaging requires delivery to the condenser of a focused beam of light that is bright, evenly distributed, constant in amplitude, and versatile with respect to the range of wavelengths, convenience, and cost. Alignment and focus of the illuminator are therefore essential and are the first steps in adjusting the illumination pathway in the microscope. A number of incandescent filament lamps and arc lamps are available to meet the needs of various applications. The spectra of the principal lamps used in microscopy are shown in Figures 3-1 and 3-2 and are summarized here.

Incandescent lamps with tungsten wire filaments and inert argon gas are frequently used for bright field and phase contrast optics and are bright enough for certain applications requiring polarized light. Tungsten and quartz halogen lamps are convenient and

29

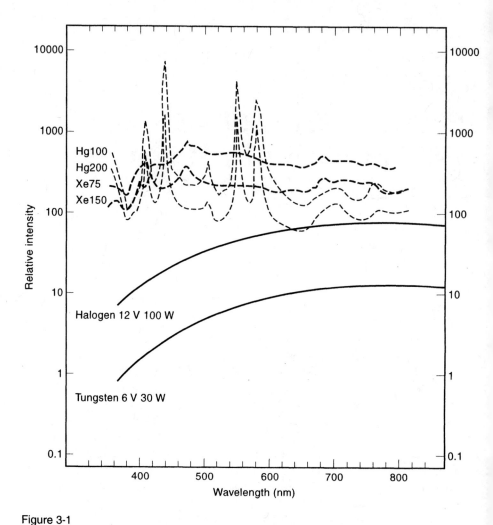

Figure 3-1
The spectra of various illuminators. Tungsten filament lamps give continuous emission, but their output is reduced at shorter wavelengths; mercury and xenon arc lamps are brighter, although mercury contains prominent emission lines in the visible range. Notice that over much of its range, the intensity of a 75 W xenon bulb (XBO) is several times greater than that of a 100 W mercury lamp (HBO). Although higher-wattage arc lamps generate a lower luminous flux (lumens/mm²/s), they cover a much larger area and their total luminous output is considerably greater.

inexpensive, easy to replace, and provide bright, even illumination when used together with a ground glass filter; hence their popularity in nearly all forms of light microscopy. These lamps produce a continuous spectrum of light across the visual range, with peak output occurring in the red and infrared (IR) range and blue and ultraviolet (UV) output being the weakest. Excitation of the filament is regulated by a continuously variable power supply. As voltage and excitation are increased, brightness increases and the spectrum shifts to increasingly higher-energy photons with shorter wavelengths. Therefore, color balance of the light from an incandescent lamp varies depending on the voltage applied to the lamp. When producing color micrographs, a specific voltage is selected in order to obtain a consistent and predictable spectrum of wavelengths. Spe-

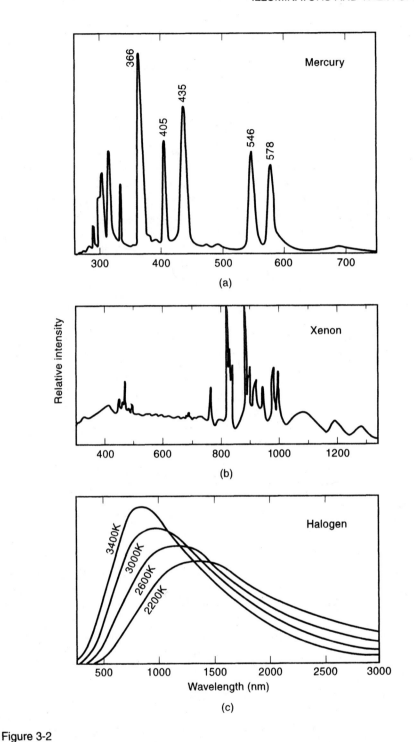

Figure 3-2

Detailed spectra of arc and tungsten filament lamps. (a) Mercury arc lamp. A continuous spectrum in the visible range is superimposed with bright emission lines, of which the most useful are at 366, 405, 435, 546, and 578 nm. Significant emission occurs in the UV (below 400 nm) and IR portions of the spectrum (not shown). (b) Xenon arc lamp. In the visible range, this lamp produces a continuous spectrum without major emission lines. Emission is significant in the UV and very large in the infrared. (c) Quartz halogen tungsten filament lamp. Much of the radiation is in the infrared range. Intensity increases and the peak of radiation intensity shifts to lower visible wavelengths as voltage is increased.

cial tungsten film and photographic filters are available to correct for the red-rich spectral output of these lamps.

Ion arc lamps are 10–100 times brighter than incandescent lamps and can provide brilliant monochromatic illumination when combined with an appropriate filter, but the increase in brightness comes with some inconveniences in mechanical alignment, shorter lifetime, and higher cost. Two types are commonly used: 75 W xenon and 100 W mercury arc lamps. Both lamps produce continuous spectra across the entire visible range from 400–750 nm and extending into the ultraviolet and infrared. In fact, only about a fifth of the output of these arc lamps is in the visible portion of the spectrum, the remainder being in the ultraviolet and infrared, so special blocking filters must be used when examining living cells, which are sensitive to UV and IR radiation.

Arc lamps tend to flicker due to subtle variations in power. This can be annoying, especially during time lapse recording, but stabilized power supplies are now available that minimize this problem. It is also common practice to avoid turning the lamp on and off frequently, as this poses a risk to nearby electronic equipment and shortens the life of the bulb. If there is a 20–30 min pause in the use of the microscope, it is better to leave the lamp on than to turn it off and reignite it. There are, however, new lamp designs that use a variable transformer to control light intensity. As the bulb is turned down, a heating mechanism keeps the lamp hot so that brightness increases immediately when the power is turned up again. The lifetime of mercury and xenon lamps is generally rated at 200 hours; however, the UV output of a mercury lamp weakens (sometimes considerably) with prolonged use, since metal vapors from the electrodes become deposited on the glass envelope. In addition, the arc becomes unstable and begins to flicker. Although arc lamps are expensive, the actual cost works out to be about 50 cents per hour, so it is advisable to replace them after their nominal lifetime has expired even if they are still working.

The *mercury arc lamp* is distinct in emitting several prominent emission lines, some of which are up to 100 times brighter than the continuous background: 254 (far UV), 366 (near UV), 405 (violet), 435 (deep blue), 546 (yellow-green), 578, 579 (yellow doublet band), plus several lines in the IR. The 546 nm green line of the mercury arc lamp is a universal reference for calibrating wavelengths in a number of optical devices and is a favorite among biologists for examining living cells. UV emission accounts for about half of the output of the mercury lamp, so care must be taken to protect the eyes and living cells that are illuminated by it. When changing and aligning a new lamp, avoid staring at the unattenuated beam; when examining live cells, use a green bandpass filter plus a UV-blocking filter such as a Schott GG420 glass filter. Since mercury lamps also emit in the IR, heat-cut filters are desirable to block these wavelengths as well.

The spectrum of the *xenon arc lamp* is largely continuous and lacks prominent emission lines. Its advantage is bright, uniform output across the entire range of visual wavelengths. At blue-green and red wavelengths it is significantly brighter than a 100 W mercury lamp, making it advantageous for certain applications in fluorescence microscopy. Since about half of the light emitted from a xenon lamp is in the IR, special IR-blocking filters, such as a Schott BG38 or BG39 glass filter and/or an IR-reflecting mirror, are used to attenuate and block these wavelengths and protect cells from excess heat. The detectors of electronic cameras, particularly those of CCD cameras, are also particularly sensitive to infrared light, which can fog the image. Although the intensity of a 75 W xenon lamp is high, the distance between the lamp electrodes is small—only 0.75 mm—which can make it difficult to obtain an even distribution of the light across the front aperture of the condenser and therefore across the specimen in the object plane.

Metal halide lamps, which have a spectral output similar to mercury, are becoming popular because they are bright (150 W), have a long bulb life (1000 hr), and have a large electrode gap (5 mm). Nikon Corporation promoted their use in combination with a liquid fiber bundle for delivering bright, homogeneous illumination for video microscopy.

Demonstration: Spectra of Common Light Sources

Please note: *Never look directly at unattenuated mercury or xenon beams, because they are, respectively, extremely UV- and IR-rich and potentially damaging to the eye!*

It is useful to become familiar with the spectra of common illuminators by indirect inspection of their light with a spectroscope or a diffraction grating. There are several ways to do this:

- For display and discussion in a group, set up the I-beam/optical bench and project the diffraction pattern of a diffraction grating on a projection screen as shown in Figure 2-10.

- For individual study, wrap a piece of aluminum foil containing a narrow slit aperture over the mouth of the illuminator and examine the slit at a distance of several feet with a handheld diffraction grating held close to the eye. A transparent holographic grating works best. To make a slit, cut a 1 cm long slit in the foil using a razor blade, while holding the foil placed against a sheet of stiff cardboard.

- An inexpensive handheld spectroscope based on a sine-wave (holographic) diffraction grating is available from Learning Technologies, Cambridge, Massachusetts. Direct the unfiltered beam of the illuminator onto a white card or projection screen positioned several feet away and follow the instructions for using the spectroscope. The advantage of the spectroscope is that it permits you to determine the wavelengths of colors and emission lines from a built-in wavelength scale. You should perform these observations in a darkened room.

Examine the spectrum of a tungsten lamp or quartz halogen lamp first. The continuous, smooth nature of the spectrum and the relative falloff in brightness at the blue end of the spectrum are characteristic. Examine the spectrum with the power supply adjusted at the minimum and maximum permissible settings to see the shift in the peak spectral output. As power increases, the intensity of shorter bluer wavelengths increases. (The peak emission wavelength in the infrared also decreases, but this cannot be seen visually.)

Next, examine the spectrum of the xenon arc and notice the uniform intensity across the entire visible range. Careful inspection will show that the spectrum is not perfectly smooth, but rather has weak emission lines in the visible range near 470 nm (blue) and also at the red end of the spectrum near 680 nm. Fifty percent of the output of this lamp is in the IR, where prominent, though invisible, emission lines occur at >800 nm.

Finally, inspect the mercury arc spectrum with its continuous spectrum and superimposed prominent emission lines. Half of the output of this lamp is in the UV, with one of the most prominent (but invisible) emission lines being located at 366 nm. This wavelength is commonly used for photoactivation of caged fluorescent compounds, stimulation of UV-excitable fluorescent dyes, and conversion of colchicine to its photoisomer, lumicolchicine. This line and the 405 nm violet line can be visualized using the optical bench setup by holding a piece of fluorescent white paper in the proper location in the spectrum in a darkened room. A suitable piece of paper can be found using a handheld near-UV black light to select for sheets that exhibit brilliant bluish white fluorescence due to the presence of phenolic compounds in the paper. The 405 nm violet line and the 366 nm near-UV line suddenly leap into view when the white card is inserted into the blue end of the spectrum.

ILLUMINATOR ALIGNMENT AND BULB REPLACEMENT

Microscope illuminators consist of a lamp housing with a lamp and concave reflector, a focusable collector lens, an electrical socket for holding the bulb, and an external power supply. The socket and power cord, in particular, deserve attention. Oxidized metal surfaces of the socket electrodes and the copper conducting wire in an arc lamp should be cleaned with an emery cloth each time the lamp is changed to assure good electrical contact. The socket's power cord should never be crimped (as occurs when the illuminator is shoved against a wall) as this action loosens wires, which can lead to inconvenient and expensive repair. The bulb, rear reflector, and front collector lens should be kept clean of all dirt, lint, and finger oils. At the time of changing a bulb and after the collector lens and metal housing have been removed, move the illuminator's adjustment screws with a screwdriver to observe their effect on the location of the bulb and the reflector. Some arc lamp housings only contain adjustment screws for the rear reflector, whereas others contain additional screws for adjusting the bulb itself. Arc lamp illuminators should be maintained in an upright position during use to preserve the life of the bulb. *Never ignite an arc lamp when it is outside its protective metal housing!*

After a bulb is changed and aligned, the image of the arc or filament should be focused in the front aperture plane of the condenser using the illuminator's collector lens focusing dial, which is located on the illuminator housing. On some microscopes it may be possible to place a lens tissue across the front of the condenser aperture or to stop down the condenser diaphragm in order to see the image of the focused filament. Alternatively, the focused image of the filament or arc may be viewed at its conjugate location at the objective back aperture using an eyepiece telescope or Bertrand lens. In this case, the light should be turned down to a minimum or attenuated with appropriate neutral density filters. To see the image clearly, it may be necessary to remove a ground glass diffusing screen, whose function is to remove the pattern of the filament from the image of the specimen.

Alignment of a new bulb is especially critical for mercury or xenon arc lamps, such as those mounted in epi-illuminators used for fluorescence microscopy, because the arc in the lamp is small (~1 mm) and the arc's image at the condenser aperture must be positioned on the optic axis of the microscope. Light from an arc lamp can be safely examined after attenuation with a fluorescence filter set plus additional neutral density filters. It is easier on the eye to examine the green excitation light provided by a rhodamine filter set. A similar procedure can be applied for arc lamps used in transillumination mode.

Demonstration: Aligning a 100 W Mercury Arc Lamp in an Epi-illuminator

- Always turn off the power supply and allow the lamp to cool completely before changing a failed bulb. Since arc lamps are under moderately high pressure when they are hot, an applied mechanical force can cause them to explode. After replacing a bulb, secure the lamp socket to the lamp housing and fasten the housing to the microscope before reigniting the lamp.

- Place neutral density filters in the light path sufficient to block ~97% of the light and place a rhodamine fluorescence filter cube into position so that the 546 nm green line of the arc is reflected onto the specimen plane. Insert additional UV- and IR-blocking filters into the path to protect the eyes.

- Tape a white card or paper on the microscope stage, focus an objective lens until you see an intense, focused dot on the card, and mark the location with a pen. The dot defines the position of the optic axis of the microscope (Fig. 3-3).

- Without disturbing the card, rotate the objective turret to an empty position and observe an intense, extended spot of light on the card. Focus the collector lens of the lamp until the bright primary image of the arc is sharply defined. If the arc's image does not coincide with the dot, you will need to adjust the bulb using the adjustment screws on the illuminator housing.

- There should also be a dimmer reflection image of the arc, which is focused and aligned using the reflector's adjustment screws on the lamp housing. Position the reflection image so that it is on top of or next to the primary image of the arc.

- Slowly defocus the lamp collector lens and ascertain that the beam expands uniformly around the black dot on the card. This is an important test of

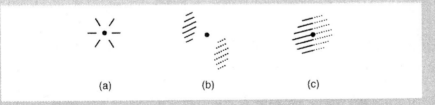

(a)   (b)   (c)

Figure 3-3

Alignment of an arc lamp. (a) The optic axis is marked on a white card with a pen as described in the text. (b) When the collector lens is properly adjusted, the direct image and reflection image of the arc are seen as two separate patches of light on the card. (c) Adjustment screws on the illuminator housing are moved to align the two spots on the optic axis.

alignment. If the arc's image does not expand symmetrically, you must make an additional round of adjustments. Sometimes it helps to expand the image slightly with the collector lens before making additional adjustments.

- While looking in the microscope at a focused fluorescent specimen, adjust the collector lens of the illuminator until the image is bright and uniform from one edge of the field to the other and across all diameters of the field. At this position the arc is focused on the objective's back aperture. In epi-illumination, the objective lens functions as both a condenser and an objective.

For incandescent filament lamps the procedure is easier. After remounting the bulb, turn on the power supply and examine the lamp filament in the microscope using a telescope eyepiece or Bertrand lens. Focus the telescope to view the edge of the condenser diaphragm at the front aperture of the condenser. In the absence of a diffuser screen, the filament should be sharply focused; if not, adjust the lamp's collector lens. Notice the pattern of vertical lines representing the filament. Center the image of the filament, then center the image of the reflection so that the vertical lines of the primary and reflection images are interdigitated (see Fig. 3-4). Some illuminators do not contain adjustable collector lenses for low-power lamps. If the filament image seems off-center, try remounting the bulb.

"FIRST ON—LAST OFF": ESSENTIAL RULE FOR ARC LAMP POWER SUPPLIES

It is very important to understand the potential hazard of turning on and off an arc lamp power supply located near functioning electronic equipment. Arc lamps should be turned on and allowed to stabilize for a minute or two *before* turning on the other pieces of nearby electronic equipment. Although the power supply and cable are generally well shielded, a momentary 20,000–50,000 V surge passing between the DC power supply and the arc lamp generates magnetic fields that are strong enough to damage sensitive integrated circuits in nearby VCRs, electronic cameras, and computers. Upon turning

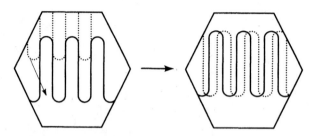

Figure 3-4

Alignment of a filament lamp. In the absence of a diffuser screen, the image of the lamp can be seen in the back aperture of the objective lens using an eyepiece telescope or Bertrand lens. The reflection image of the filament is adjusted to interdigitate with the filament loops seen in the direct image.

the lamp off, the reverse procedure should be used: First, turn off the accessory electronics, then turn off the arc lamp. It is advisable to post a simple warning on or near the arc lamp power supply that states *First on—Last off* to act as a reminder to yourself and other microscope users. Second, when an arc lamp fails, remember that some power supplies try to continually reignite the lamp with a series of high-voltage pulses that can be heard as a rapid series of clicks. When heard, the supply should be shut off immediately; otherwise, electromagnetic fields might damage peripheral equipment as well as the power supply itself. There is always the risk that this may happen when you are out of the room and away from the microscope. To protect against this, and especially if time lapse studies are performed with no one at the microscope, be sure the power supply is protected with an *automatic trigger override switch.*

FILTERS FOR ADJUSTING THE INTENSITY AND WAVELENGTH OF ILLUMINATION

Selecting and adjusting the lamp for a particular application is important, but the actual control of the wavelength and intensity of illumination in a microscope requires the use of filters, so it is important to understand the fundamentals of their action and performance. The microscopist needs to know how to interpret the transmission spectra of filters, select the best among several possible filters for a given application, and explain differences in image quality, fluorescence quality, and cell behavior obtained with different filter combinations. This is particularly true in fluorescence microscopy, where filters must match the excitation and emission requirements of fluorescent dyes. Fortunately, the task is manageable, and filtering light to select a band of wavelengths from the illuminating beam presents many challenges and rewards.

Neutral density filters regulate light intensity, whereas *colored glass filters* and *interference filters* are used to isolate specific colors or bands of wavelengths. There are two classes of filters that regulate the transmission wavelength: edge filters and bandpass filters (Fig. 3-5). Edge filters are classified as being either *long-pass* (transmit long wavelengths, block short ones) or *short-pass* (transmit short wavelengths, block long ones), whereas bandpass filters transmit a band of wavelengths while blocking wavelengths above and below the specified range of transmission. Optical performance is defined in terms of the efficiency of transmission and blockage (% transmission), and by the steepness of the so-called cut-on and cut-off boundaries between adjacent domains of blocked and transmitted wavelengths. Edge filters are described by referring to the wavelength giving 50% of peak transmission; *bandpass* filters are described by citing the full width at half maximum transmission (FWHM), and by specifying the peak and central transmitted wavelengths. FWHM is the range of the transmitted band of wavelengths in nanometers and is measured as the distance between the edges of the bandpass peak where the transmission is 50% of its maximum value. For high-performance filters these boundaries are remarkably sharp, appearing on transmission spectra as nearly vertical lines.

In part, the resurgence of light microscopy as an analytic tool in research has been driven by the technologies used for depositing thin films of dielectric materials and metals on planar glass substrates. Companies now manufacture interference filters with transmission and blocking efficiencies approaching 100% and with bandwidths as narrow as 1 nm anywhere along the UV-visible-IR spectrum—a truly remarkable accomplishment. This achievement has stimulated development of new research applications

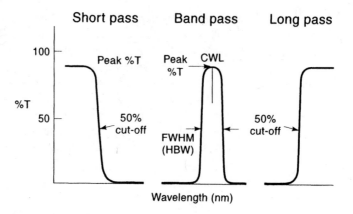

Figure 3-5

Filters for isolating the wavelength of illumination. Short-pass and long-pass filters, sometimes called edge filters, block or transmit wavelengths at specific cut-off wavelengths. Bandpass filters exhibit broadband or shortband transmission centered on a particular band of wavelengths. Filter performance is defined by the central wavelength (CWL) and by the full width at half maximum (FWHM). Another term for FWHM is halfbandwidth (HBW). A bandpass filter can also be created from two overlapping short-pass and long-pass filters.

involving laser-based illumination, new fluorescent dyes and molecules, and ratio imaging at multiple specific wavelengths.

Neutral Density Filters

Neutral density filters are used in microscopy to attenuate uniformly the intensity of light over the entire range of visible wavelengths. They are commonly employed in differential interference contrast (DIC), polarization, and fluorescence microscopy with high-intensity arc lamps that cannot be regulated with an adjustable power supply. In these circumstances, neutral density filters must be used. As discussed in Chapter 6, it is impermissible to reduce the intensity of illumination by closing down the condenser diaphragm, as this action affects resolution and contrast. A light-absorbing filter is the only solution.

Neutral density (ND) filters have a neutral gray color like smoked glass and are usually calibrated in units of *absorbance* or optical density (OD), where

$$OD = \log_{10}(1/T),$$

and T is the transmittance (intensity of transmitted light/intensity of incident light). Thus, a 0.1 OD neutral density filter gives 79% transmission and blocks 21% of the incident light. Other manufacturers indicate the transmittance directly. ND filters can be stacked in a series, in which case the total density of the combination is equal to the sum of the individual filter densities.

ND filters are either absorbing or reflecting. Absorbing filters contain rare earth elements throughout the glass, so there is no reflective coating that can be scratched off, and their orientation in the beam is not critical. Reflecting ND filters contain an evaporated coating of metal on one of the surfaces, so care must be taken not to scratch them. These filters must be inserted into the beam with the reflective surface facing the light source. They can, however, be cheaper and thinner, and are the filter of choice for use with lasers.

Colored Glass Filters

Colored glass filters are used for applications not requiring precise definition of transmitted wavelengths. They are commonly used to isolate a broad band of colors or as long-pass filters to block short wavelengths and transmit long ones. Colored glass filters contain rare earth transition elements, or colloidal colorants such as selenide, or other substances to give reasonably sharp transmission-absorption transitions at a wide range of wavelength values across the visual spectrum. Since colored glass filters work by absorbing quanta of nontransmitted wavelengths, they can be heat sensitive and subject to altered transmission properties or even breakage after prolonged use. However, as the absorbent atoms are contained throughout the glass and are not deposited on its surface, colored glass filters offer major advantages: They are less resistant to physical abrasion and chemical attack from agents contained in fingerprints and other sources, and their optical performance is not sensitive to the angle of incidence of incoming rays of light. Colored glass filters are also less expensive than interference filters and are generally more stable and long-lived.

Interference Filters

Interference filters often have steeper cut-in and cut-off transmission boundaries than colored glass filters and therefore are frequently encountered in fluorescence microscopy where sharply defined bandwidths are required. Interference filters are optically planar sheets of glass coated with dielectric substances in multiple layers, each $\lambda/2$ or $\lambda/4$ thick, which act by selectively reinforcing and blocking the transmission of specific wavelengths through constructive and destructive interference (Fig. 3-6). Bandpass filters transmit a limited range of wavelengths that experience reinforcement through constructive interference between transmitted and multiple reflected rays; wavelengths that do not reinforce each other destructively interfere and are eventually back-reflected out of the filter.

Interference filters contain layers of *dielectric substances*, electrically nonconductive materials of specific refractive index, typically optically transparent metal salts such as zinc sulfide, sodium aluminum fluoride (cryolite), magnesium fluoride, and other substances. In some designs semitransparent layers of metals are included as well. The interface between two dielectric materials of different refractive index partially reflects incident light backward and forward through the filter, and is essential for constructive interference and reinforcement. The wavelength that is reinforced and transmitted depends on the thickness and refractive index (the optical path) of the dielectric layers. The coatings are built up in units called cavities, with 1 cavity containing 4 or 5 alternating layers of dielectric salts separated by a spacer layer (Fig. 3-7). The steepness of the transmission boundary and the definition of filter performance are increased by increasing the number of cavities. An 18-cavity filter may contain up to 90 separate dielectric layers. The deposition of salts is performed by evaporation of materials in a computer-controlled high-vacuum evaporator equipped with detectors for optical interference, which are used to monitor the thicknesses of the various layers. The final layer is covered

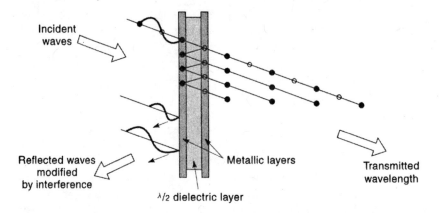

Figure 3-6

The action of an interference filter. An interference filter selectively transmits waves centered at a particular wavelength. For the filter shown, two thin films of metal cover a layer of dielectric material with an optical path of exactly $\lambda/2$ for a particular wavelength. The angle of the incident beam is usually perpendicular to the face of the filter, but is shown obliquely to reveal the behavior of transmitted waves. Since transmitted and multiply reflected waves of the designated wavelength are in phase, principles of constructive interference allow reinforcement and transmission through the filter. Shorter and longer wavelengths experience destructive interference and exit the filter as a back-reflection.

with glass or overcoated with a scuff-resistant protective coating of silicon monoxide or silicon dioxide (referred to as quartz) to guard against abrasion. The availability of computers and programs to model the behavior of multiple dielectric cavities has stimulated a revolution in thin film technology, allowing significant improvements in the selection of specific wavelengths and in the intensity and contrast of fluorescence images. The technology has also stimulated research for the production of new fluorescent dyes, optical variants of green fluorescent protein (GFP), fluorescent crystals, and other substances for use with a variety of illuminators and lasers. Because filter production is technology dependent and labor intensive, interference filters remain relatively expensive.

Interference bandpass filters for visible wavelengths frequently transmit wavelengths in the UV and IR range that may not be included in transmission spectra and documentation provided by the manufacturer. For microscopy involving live cell applications, it is safest to obtain the extended transmission spectra of all filters used and to employ efficient UV- and IR-blocking filters, particularly when UV- and IR-rich mercury or xenon arc lamps are used. Even with fixed fluorescent cells, an IR-blocking filter, such as a BG38 or BG39 glass filter, is frequently used as a heat-cut filter to protect optics and to prevent image fogging on IR-sensitive CCD cameras.

Interference filters gradually deteriorate upon exposure to light, heat, humidity, and especially exposure to abrasion, fingerprints, and harsh chemicals. Gently remove fingerprints and deposits from interference filters with a lens tissue and a neutral lens cleaner. Care must be taken not to rub interference coatings too hard, as this might scratch the surface, making the filter less efficient and shortening its life. Filters containing semitransparent metal coatings are usually mounted with the shiniest (silvery, nearly colorless) side of the filter facing the light source. The bindings at filter edges are usually inscribed with arrows to aid in the orientation and placement of filters in the light path of the microscope.

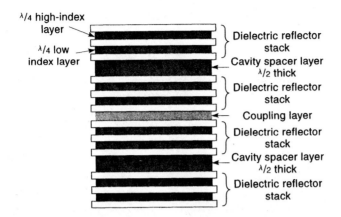

Figure 3-7

Structure of an all-dielectric interference filter. The revolution in thin film technology continues to drive the development of high-performance interference filters. The 2-cavity filter shown contains alternate layers of high- and low-refractive index dielectric materials, each λ/4 and λ/2 thick, with 5 such layers defining a cavity. Computers control the deposition of layers of dielectric materials in a vacuum evaporator while film thickness is determined by interference optics. Dozens of such layers are deposited during a single run in the evaporator. Three cavities are the industry standard, but 18-cavity filters with 90 separate layers that can produce bandwidths of less than 1 nm are now routinely produced.

EFFECTS OF LIGHT ON LIVING CELLS

Since the energy per quantum is related to wavelength ($E = hc/\lambda$), short wavelengths are more energetic than long ones. UV wavelengths flanking the blue end of the visual spectrum (200–400 nm) are particularly damaging to cells, because photons of ultraviolet light are energetic enough to break covalent bonds, thereby creating reactive free radicals that chemically alter and denature macromolecules such as proteins, nucleic acids, lipids, and small metabolites. Damage to membrane proteins, such as ion channels and gates, is a particular concern. Photons of infrared radiation (750–1000 nm) are less energetic than those corresponding to visible wavelengths, but are strongly absorbed by carbon bonds in macromolecules such as DNA and by water, leading to accumulation of kinetic energy (heat) and denaturation of molecules. Visible light itself is unique because it is absorbed relatively poorly by living cells, particularly at green and yellow wavelengths. For the most part, cellular absorption of visible light is considerably less than for the flanking UV and IR wavelengths. Since green light is relatively nontoxic and marks the peak sensitivity for human color vision, the 546 nm green line of the mercury arc lamp is commonly used for monochromatic illumination of living cells.

It is apparent that live cells must be protected from unwanted UV and IR radiation. IR- and UV-blocking filters, such as Schott filters BG38 (for IR) and GG420 (for UV), are especially useful, since the spectra generated by mercury and xenon arc lamps used in microscopy are rich in UV and IR radiation (for mercury, 30% UV, 40% IR, 30% visible; for xenon, 5% UV, 70% IR, and 25% visible). Phototoxicity in the microscope is recognized by the cessation of cell motility and the arrest of organelle movement; within 3 seconds of exposure to the full spectrum of a 100 W mercury arc, amoebae retract filopodia and freeze, their cytoplasm appearing to have gelled. Upon further exposure, cells respond by blebbing and swelling, and particles contained in dilated vesicles begin to exhibit vibrational movements (Brownian movements) that are not as obvious in the living state. It is useful to observe cells without protective filters to become familiar with these effects.

Cells may require additional chemical protection from the buildup of high concentrations of damaging free radicals. For well chamber slides the simplest measures are to increase the volume of medium in the chamber to dilute the radicals or to use anti–free radical reagents such as 10 mM ascorbate or succinate in the culture medium to neutralize free radicals as they form. Alternatively, the rate of free radical formation can be slowed by reducing the concentration of dissolved oxygen in the culture medium using a mixture of oxygen-scavenging enzymes, such as catalase, glucose oxidase, and D-glucose, or supplementing the medium with Oxyrase (Oxyrase, Inc., Mansfield, Ohio), which is a commercial preparation of respiratory particles of bacterial cell membranes that contains oxygen-removing enzymes. To maintain low oxygen concentrations, cells must be enclosed in specially designed flow cells. If well chambers are used, the medium should be covered with a layer of lightweight nontoxic mineral oil to prevent recharging of the medium with atmospheric oxygen. The presence of these agents is usually harmless to vertebrate somatic cells, since cells in most tissues exist in a low-oxygen environment.

LENSES AND GEOMETRICAL OPTICS

OVERVIEW

In this chapter we discuss some essential principles of geometrical optics, the action of lenses on light as revealed by ray tracing and explained by principles of refraction and reflection (Fig. 4-1). With the help of a few simple rules and from studying examples, we can understand the process of magnification, the properties of real and virtual images, the aberrations of lenses, and other phenomena. We also examine the designs and specifications of condenser and objective lenses, review the nomenclature inscribed on the barrel of an objective lens that specifies its optical properties and conditions for use, and give some practical advice on the cleaning of optical components.

IMAGE FORMATION BY A SIMPLE LENS

To understand microscope optics, we begin by describing some of the basic properties of a thin, simple lens. A thin lens has a thickness that is essentially negligible, and by simple we mean consisting of a single lens element with two refracting surfaces. The *principal plane* and *focal plane* of a lens are defined as those planes, within the lens and in the focused image, respectively, where rays or extensions of rays intersect and physically reunite. Thus, for a simple positive lens receiving a collimated beam of light, the plane in the lens in which extensions of incident and emergent rays intersect is called the principal plane, and the plane in which rays intersect to form an image is the focal plane. The *focal length* of a lens is the distance between the principal plane and the focal plane. Lenses can be either positive or negative (Fig. 4-2). A *positive lens* converges parallel incident rays and forms a real image; such a lens is thicker in the middle than at the periphery and has at least one convex surface. (See the Note for definitions of real and virtual images.) Positive lenses magnify when held in front of the eye. A *negative lens* causes parallel incident rays to diverge; negative lenses are thinner in the middle than at the periphery, and have at least one concave surface. Negative lenses do not form a real image, and when held in front of the eye, they reduce or demagnify.

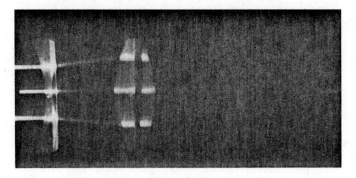

Figure 4-1
Geometrical optics of a positive lens. (From Hecht, 1998.)

For any given lens there are two principal planes: one each for the front and back surface of the lens. For the special case of a thin biconvex lens, the two principal planes are coincident in the middle of the lens. Microscope objectives contain multiple lens elements, some of which may be united with transparent sealing compound to make a complex *thick lens*. The principal planes of thick lenses are physically separated, but their locations can be determined by ray tracing. Most lens elements used in microscope optics are ground and polished with spherical curvatures.

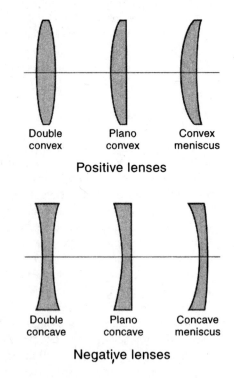

Figure 4-2
Examples of positive and negative lenses.

Note: Real and Virtual Images

Images can be defined as regions where rays, or the extensions of rays, become convergent as the result of refraction by a lens or reflection by a mirror. If the rays intersect and physically reunite, the image is said to be real. A *real image* can be seen on a viewing screen or recorded on a piece of film when a screen or film is placed in the image plane. If rays diverge, but the imaginary backward extensions of the rays become convergent and intersect, the image is said to be virtual. The plane occupied by a *virtual image* cannot be observed on a viewing screen or recorded on film. To be perceived, a real image must be formed on the retina of the eye. In the case of viewing an image in a microscope, a real image is formed on the retina but is perceived as a virtual image located some 25 cm in front of the eye. Lens configurations giving real and virtual images are described in this chapter.

The geometric parameters of a *simple thin lens* are described in Figure 4-3, where the vertical line represents the combined principal planes of a thin biconvex lens of focal length f. The object, an arrow on the left-hand side of the figure, is examined by the lens and imaged as a magnified real image (magnified inverted arrow) in the image plane on the right. The *focal length* is shown as the distance f from the principal plane of the lens to its *focal point F*, the front and rear focal lengths having the same value. The optic axis is shown by a horizontal line passing through the center of the lens and perpendicular to its principal plane. The *object distance a* (distance from the object to the principal plane of the lens) and *image distance b* (distance from the image to the principal plane of the lens) are also indicated.

The *focal length* of any simple lens can be determined by aiming the lens at a bright "infinitely distant" light source (> 30 times the focal length) such as a lamp across the room or a scene outdoors; by focusing the image on a sheet of paper held behind the lens, the focal length is readily determined (Fig. 4-4). We will now examine the basic rules that determine the action of a simple convex lens.

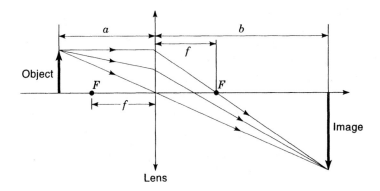

Figure 4-3

Geometrical optics of a simple lens. The focal length *f*, focal point *F*, object-lens distance *a*, and lens-image distance *b* are indicated.

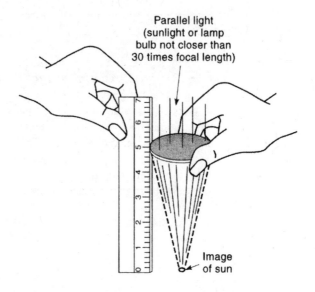

Figure 4-4

Determining the focal length of a simple lens. The image of a distant source is projected by the lens on a viewing surface; the focal length is the distance between the focal plane, and the lens as measured with a ruler.

RULES OF RAY TRACING FOR A SIMPLE LENS

The three rules governing ray tracing for a simple lens are depicted in Figure 4-5 and are listed as follows:

1. A light ray passing through the center of a lens is not deviated.
2. A light ray parallel with the optic axis will, after refraction, pass through the rear focal point.
3. A ray passing through the front focal point will be refracted in a direction parallel to the axis.

Notice that the intersection of any two of the three key rays just described identifies the location of the image plane.

OBJECT-IMAGE MATH

The well-known *lens equation* describes the relationship between focal length f and object and image distances, a and b:

$$1/f = 1/a + 1/b,$$

or

$$b = af/(a - f).$$

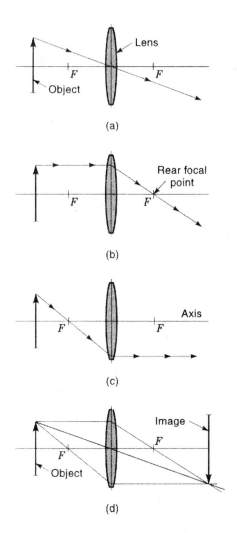

Figure 4-5
Principles governing ray tracing for a thin lens. (a) A ray, incident at any angle, that passes through the center of the lens remains undeviated from its original course on the other side of the lens. (b) Any ray traveling parallel to the optic axis and refracted by the lens always passes through the rear focal point F. (c) A ray passing through the front focal point of a lens at any angle is refracted and follows a path parallel to the optic axis. (d) The intersection of any two of the three rays described defines the location of the image plane.

Further, the magnification factor M of an image is described as:

$$M = b/a,$$

or

$$M = f/(a - f).$$

From these relationships the action of a lens on image location and magnification can be deduced. Depending on the location of the object relative to the focal point of the lens,

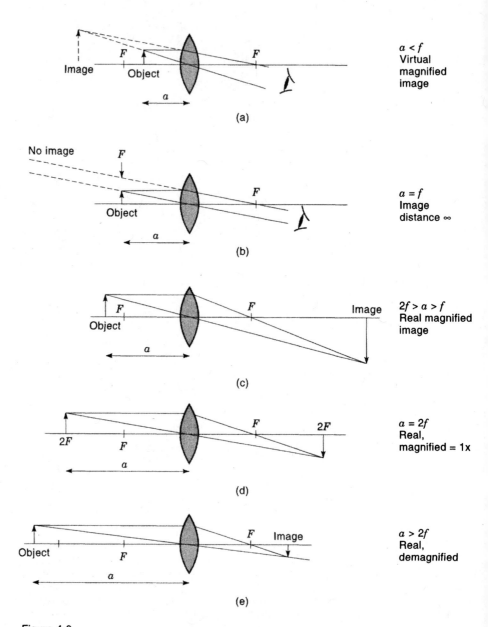

Figure 4-6

Object-image relationships. (a) $a < f$; (b) $a = f$; (c) $2f > a > f$; (d) $a = 2f$; (e) $a > 2f$.

the image may be real or virtual, and either magnified or demagnified (Fig. 4-6; Spencer, 1982). It is good practice to work through the following relationships using any handheld magnifier or simple biconvex lens such as the front lens of a binocular (10–50 mm is convenient), whose focal length has been determined by the procedure just described. The object should be self-luminous, moderately bright, and well defined. A 5–10 mm diameter hole cut in a large opaque cardboard sheet placed in front of a lamp works well as an object. Work in a partially darkened room. It is useful to remember the principal conditions describing object-image relationships for a simple positive lens:

- $a < f$: No real image exists that can be projected on a screen. If the eye is placed behind the lens, a virtual image is perceived on the far side of the lens.
- $a = f$: The image distance b is infinite, so no image exists that can be projected on a screen. We used this condition previously to determine the focal length of a lens, only in reverse: Parallel beams of light from an "infinitely distant" object converge at the focal length of the lens. This is the case for image formation in a telescope.

For the condition that $a > f$, a real image is always formed. The unique domains for this condition are as follows:

- $2f > a > f$: A real *magnified* image is formed. This arrangement is used for producing the first real image in a microscope.
- $a = 2f$: This is a specialized case. Under this condition, $b = 2f$ also. A real image is formed, but there is *no magnification* and $M = 1$.
- $a > 2f$: A real *demagnified* image is formed and $M < 1$.

In the case of a microscope objective lens focused on a specimen, the image is both real and magnified, meaning that the object is located at a distance a between $1f$ and $2f$ ($2f > a > f$) (Fig. 4-7). Since the focused objective is very near the specimen, we deduce that the focal length of the objective must be very short, only a few millimeters. In the course of using the focusing dials of a microscope, the image comes into sharp focus when the correct object distance a has been obtained, and we obtain the correct adjustment without even thinking about object and image distances. In practice, focusing a microscope positions the image (the real intermediate image plane) at a fixed location in the front aperture of the eyepiece; when the microscope is defocused, there is still a real image nearby, but it is not in the proper location for the ocular and eye to form a focused image on the retina. Finally, notice that the image distance b is many centimeters long. The ratio b/a (the magnification M) usually ranges from $<10–100$. Thus, when a microscope with finite focus objectives is focused on a specimen, the specimen lies just outside

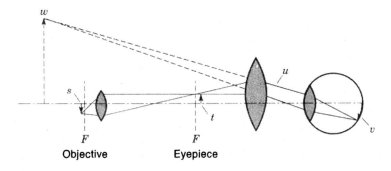

Figure 4-7

Location of real and virtual images in a light microscope marked s through w. Note that the specimen at *s* lies just outside the focus of the objective, resulting in a real, magnified image at *t* in the eyepiece. The primary image at *t* lies just inside the focus of the eyepiece, resulting in diverging rays at *u*. The cornea and lens of the eye form a real image of the object on the retina at *v*, which because of the diverging angle at *u* perceives the object as a magnified virtual image at *w*.

the front focal point of the objective, while the intermediate image is located at a distance 10–100 times the focal length of the objective in the eyepiece. For more detailed discussions on the topic, refer to Pluta (1988) or Hecht (1998).

Modern microscopes with *infinity focus objective lenses* follow the same optical principles already described for generating a magnified real image, only the optical design is somewhat different. For an objective with infinity focus design, the specimen is located at the focus of the lens, and parallel bundles of rays emerging from the back aperture of the lens are focused to infinity and do not form an image; it is the job of the *tube lens* in the microscope body to receive the rays and form the real intermediate image at the eyepiece. The advantage of this design is that it allows greater flexibility for microscope design while preserving the image contrast and resolution provided by the objective. Items such as waveplates, compensators, DIC prisms, reflectors, and fluorescence filter sets can be placed anywhere in the "infinity space" between the back of the objective and the tube lens. As long as these items have plane-parallel elements, their location in the infinity space region of the imaging path is not critical. If we consider the combination of objective plus tube lens as the effective objective lens, then the same optical rules pertain for generating a real magnified image and we observe that the relationship $2f > a > f$ is still valid. Sketches showing the infinity space region and tube lens in upright and inverted microscopes are shown in Color Plates 4-1 and 4-2.

The function of the eyepiece or ocular is to magnify the primary image another 10-fold, and together with the lens components of the eye, to produce a real magnified image of the intermediate image on the retina. Thus, the object of the eyepiece is the intermediate image made by the objective lens. Note that in the case of the ocular, $0 < a < f$, so the object distance is less than one focal length, resulting in a virtual image that cannot be focused on a screen or recorded on film with a camera. However, when the eye is placed behind the eyepiece to examine the image, the ocular-eye combination produces a real secondary image on the retina, which the brain perceives as a magnified virtual image located about 25 cm in front of the eye. The visual perception of virtual images is common in optical systems. For example, we also "see" virtual images when we employ a handheld magnifying glass to inspect small objects or when we look into a mirror.

THE PRINCIPAL ABERRATIONS OF LENSES

Simple lenses of the type already discussed have spherical surfaces, but a spherical lens is associated with many intrinsic optical faults called *aberrations* that distort the image in various ways. Of these faults, the major aberrations are chromatic aberration, spherical aberration, coma, astigmatism, curvature of field, and distortion (Fig. 4-8). Corrective measures include use of compound lens designs, use of glass elements with different refractive indexes and color dispersion, incorporation of aspherical lens curvatures, and other methods. The tube lens performs an additional important function in removing residual aberrations of the objective lens. In some microscopes the eyepieces also help perform this function. Objective lenses are designed to correct for aberrations, but can never completely remove them. It is common that a solution for correcting one fault worsens other faults, so the lens designer must prioritize goals for optical performance and then work toward the best compromise in correcting other aberrations. For these reasons, objective lenses vary considerably in their design, optical performance, and cost.

On-axis aberrations

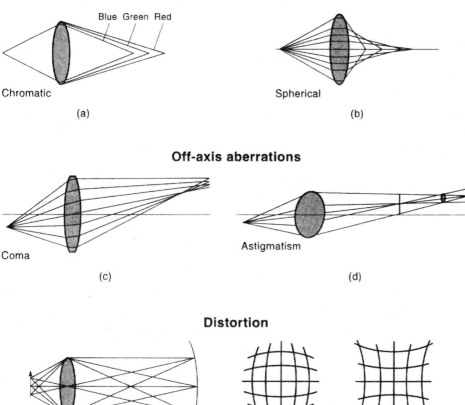

Chromatic

(a)

Spherical

(b)

Off-axis aberrations

Coma

(c)

Astigmatism

(d)

Distortion

Field curvature

Barrel distortion

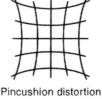

Pincushion distortion

(e)

Figure 4-8
Aberrations of a simple lens. (a) Chromatic aberration: Parallel incident rays of different wavelength are focused at different locations. (b) Spherical aberration: Incident rays parallel to the optic axis and reaching the center and the periphery of the lens are focused at different locations. (c) Coma: Off-axis rays passing through the center and periphery of the lens are focused at different locations. (d) Astigmatism: An off-axis aberration causes waves passing through the vertical and horizontal diameters to focus an object point as a streak. (e) Distortion and field curvature: The image plane is curved and not planar. So-called barrel and pincushion distortions produce images that are not high in fidelity compared to the object.

Chromatic aberration occurs because a lens refracts light differently depending on the wavelength. Blue light is bent inward toward the optic axis more than red light. The result is disastrous: Blue wavelengths are focused in an image plane closer to the lens than the image plane for red wavelengths. Even at the best focus, point sources are surrounded by color halos, the color changing depending on the focus of the objective, the image never becoming sharp. Since each wavelength is focused at a different distance from the lens, there is also a difference in magnification for different colors (chromatic magnification difference). The solution is to make compound lenses made of glasses having different color-dispersing properties. For example, glass types known as crown and flint are paired together to make an achromatic doublet lens that focuses blue and red wavelengths in the same image plane.

Spherical aberration is the undesirable consequence of having lenses figured with spherical surfaces, the only practical approach for lens manufacture. Parallel rays incident at central and peripheral locations on the lens are focused at different axial locations, so there is not a well-defined image plane and a point source of light at best focus appears as a spot surrounded by a bright halo or series of rings. For an extended object, the entire image is blurred, especially at the periphery. One common solution is to use a combination of positive and negative lenses of different thicknesses in a compound lens design. Lenses corrected for spherical aberration are intended for use under a specific set of working conditions. These include the coverslip thickness, the assumption that the focal plane is at or near the coverslip surface, the refractive index of the medium between the lens and coverslip, the wavelength of illumination, and others. Thus, users employing well-corrected lenses can unknowingly induce spherical aberration by using coverslips having the wrong thickness or refractive index. Special lenses are available with adjustable correction collars so that spherical aberration can be minimized for specimens distant from the coverslip, or when it is desirable to be able to use various immersion media (Brenner, 1994).

Coma refers to a streak of light with the shape of a comet's tail that appears to emanate from a focused spot at the periphery of an image. Coma occurs for object points that are off the optic axis—that is, when object rays hit the lens obliquely. It is the most prominent off-axis aberration. Rays passing through the edge of the lens are focused closer to the optic axis than are rays that pass through the center of the lens, causing a point object to look like a comet with the tail extending toward the periphery of the field. Coma is greater for lenses with wider apertures. Correction for this aberration is made to accommodate the diameter of the object field for a given lens.

Astigmatism, like coma, is an off-axis aberration. Rays from an object point passing through the horizontal and vertical diameters of a lens are focused as a short streak at two different focal planes. The streaks appear as ellipses drawn out in horizontal and vertical directions at either side of best focus, where the focused image of a point appears as an extended circular patch. Off-axis astigmatism increases with increasing displacement of the object from the optic axis. Astigmatism is also caused by asymmetric lens curvature due to mistakes in manufacture or improper mounting of a lens in its barrel.

Curvature of field is another serious off-axis aberration. Field curvature indicates that the image plane is not flat, but has the shape of a concave spherical surface as seen from the objective. Different zones of the image can be brought into focus, but the whole image cannot be focused simultaneously on a flat surface as would be required for photography. Field curvature is corrected by the design of the objective and additionally by the tube or relay lens and sometimes the oculars.

Distortion is an aberration that causes the focus position of the object image to shift laterally in the image plane with increasing displacement of the object from the optic axis. The consequence of distortion is a nonlinear magnification in the image from the center to the periphery of the field. Depending on whether the gradient in magnification is increasing or decreasing, the aberration is termed pincushion or barrel distortion after the distorted appearance of a specimen with straight lines such as a grid or reticule with a pattern of squares or rectangles. Corrections are made as described for field curvature.

DESIGNS AND SPECIFICATIONS OF OBJECTIVE LENSES

Achromats are red-blue corrected (meaning for wavelengths at 656 and 486 nm). Spherical correction is for midspectrum yellow-green light at 540 nm. These objectives give satisfactory performance in white light and excellent performance in monochromatic light, and are quite suitable for low magnification work at 30–40× and lower. They are also much less expensive than more highly corrected lens designs (Fig. 4-9).

Fluorite or *semiapochromat* lenses contain elements made of fluorite or fluorspar (CaF_2) or synthetic lanthanum fluoride—materials giving low color dispersion (Fig. 4-9). Corrections for color dispersion and curvature of field are easily applied. The combination of good color correction, extremely high transparency (including to near UV light) and high contrast makes them favorites for immunofluorescence microscopy, polarization and differential interference contrast (DIC) microscopy, and other forms of light microscopy. The maximum obtainable numerical aperture (NA) is about 1.3.

Apochromats are expensive, highly color-corrected designs suitable for color photography using white light (Fig. 4-9). These lenses are red-, green-, blue-, and dark blue–corrected for color, and are corrected for spherical aberration at green and blue wavelengths. This design tends to suffer some curvature of field, but is corrected in *plan-apochromatic* lenses. The high degree of color correction makes them desirable for fluorescence microscopy, since various fluorescence wavelengths emitted from a multiple-stained specimen are accurately focused in the same image plane. It is also possible to obtain very large NAs (up to 1.4) with this lens design, making them desirable for low light applications such as dim fluorescent specimens. Newer designs are now transparent to near UV light, making them suitable for fluorescence microscopy involving UV-excitable dyes.

A summary of the characteristics of some commonly used objective lenses is provided in Table 4-1.

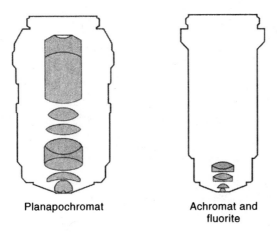

Planapochromat   Achromat and
                                     fluorite

Figure 4-9

Objective lens designs. Two popular lenses for fluorescence microscopy are shown. Apochromatic lenses may contain 12 or more lens elements to give bright, flat images with excellent color correction across the visual spectrum. Fluorite lenses have fewer lens components and produce sharp, bright images. These lenses exhibit excellent color correction and transmit UV light.

TABLE 4-1.   Characteristics of Selected Objective Lenses[a]

M	Type	Medium (n)	WD (mm)	NA	d_{min} (μm)	DOF (μm)	B
5	Achromat	1	9.9	0.12	2.80	38.19	0.1
10	Achromat	1	4.4	0.25	1.34	8.80	0.4
20	Achromat	1	0.53	0.45	0.75	2.72	1.0
25	Fluorite	1.515	0.21	0.8	0.42	1.30	6.6
40	Fluorite	1	0.5	0.75	0.45	0.98	2.0
40	Fluorite	1.515	0.2	1.3	0.26	0.49	17.9
60	Apochromat	1	0.15	0.95	0.35	0.61	2.3
60	Apochromat	1.515	0.09	1.4	0.24	0.43	10.7
100	Apochromat	1.515	0.09	1.4	0.24	0.43	3.8

[a]The magnification (M), type of lens design, refractive index (n) of the intervening medium (air or immersion oil), working distance (WD), numerical aperture (NA), minimum resolvable distance (d), depth of field (DOF), and brightness (B) are indicated. Terms are calculated as: wave-optical depth of field, $n\lambda/NA^2$; brightness in epi-illumination mode, $10^4\ NA^4/M^2$. Resolution and depth of field are discussed in Chapter 6.

Special Lens Designs

Other performance characteristics such as working distance, immersion design, and UV transparency are optimized in special lens designs:

- *Long working distance lenses* allow focusing through thick substrates (microscope slides, culture dishes) or permit the introduction of devices such as micropipettes between the specimen and the lens. The *working distance* is the distance between the surface of the front lens element of the lens and the surface of the coverslip. In contrast, conventional oil immersion lenses have short working distances that may be as small as 60 μm.

- *Multi-immersion* and *water immersion lenses* are used for examination of specimens covered with medium in a well chamber, and in some cases, can be placed directly on the surface of a specimen such as a slice of tissue. These lenses usually contain a focusable lens element to correct for spherical aberration.

- *UV lenses* made of quartz and other UV-transparent materials support imaging in the near-UV and visible range (240–700 nm).

Markings on the Barrel of a Lens

The engraved markings on the barrel of the objective describe the lens type, magnification, numerical aperture, required coverslip thickness if applicable, and type of immersion medium (Fig. 4-10).

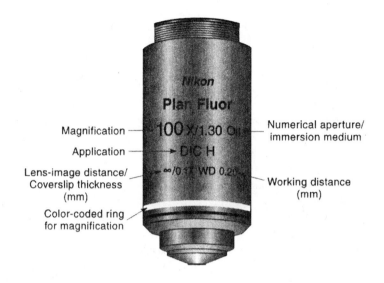

Mag.	1X	2X	4X	10X	20X	40X	50X	60X	100X
Color code	Black	Gray	Red	Yellow	Green	Light blue	Light blue	Dark blue	White

Figure 4-10

Key for interpreting the markings on the barrel of an objective lens. Markings on the lens barrel indicate the type of lens and correction, initial magnification, immersion medium, numerical aperture, lens-image distance, and required coverglass thickness. For quick reference, the color-coded ring, near the thread, denotes the initial magnification, while the color-coded ring near the front lens denotes the type of immersion medium (black-immersion oil, white-water, orange-glycerin, yellow-methylene iodide, red-multi-immersion).

Image Brightness

Notice in Table 4-1 that the ratio of numerical aperture to magnification determines the light-gathering power of a lens and hence the image *brightness B*. *B* is defined through the relationships

$$B \propto (NA/M)^2 \text{ (transillumination mode)}$$

and

$$B \propto (NA^4/M^2) \text{ (epi-illumination mode)}$$

where M is the magnification, and NA is the numerical aperture, a geometric parameter related to the light-gathering power of an objective lens. Numerical aperture as a primary determinant of the spatial resolution of an objective is discussed in Chapters 5 and 6. The values for magnification and NA are indicated on the lens barrel of the objective. A 60×/1.4 NA apochromatic objective lens gives among the brightest images, and because its image is color corrected across the entire visual spectrum and is flat and substantially free of common aberrations, it is popular in fluorescence microscopy. The 40×/1.3 NA fluorite lens is significantly brighter, but is less well corrected.

CONDENSERS

Imaging performance by a microscope depends not only on the objective lens but also on the light delivery system, which includes the illuminator and its collector lens, and of particular importance, the condenser. High-performance condensers are corrected for chromatic and spherical aberrations and curvature of the focal plane (field curvature). However, most of these aberrations are still apparent when using the Abbe condenser, a very common condenser that is based on a two-lens design (Fig. 4-11). The three-lens aplanatic condenser (indicating correction for spherical aberration and field curvature) is superior, but still exhibits chromatic aberration. The highly corrected achromatic-aplanatic condenser has five lenses including two achromatic doublet lenses, provides NAs up to 1.4, and is essential for imaging fine details using immersion-type objectives. These condensers are corrected for chromatic aberration at red and blue wavelengths, spherical aberration at 550 nm, and field curvature. Such a condenser can be used dry for numerical apertures up to ~0.9, but requires immersion medium for higher NA values, although the condenser is commonly used dry even with oil immersion objectives. Note, however, that for maximal resolution, the NA of the condenser must be equal to the NA of the objective, which requires that both the condenser and the objective should be oiled.

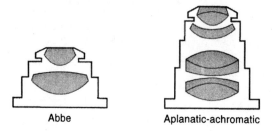

Abbe   Aplanatic-achromatic

Figure 4-11

Two common microscope condensers. The Abbe condenser contains two achromatic doublet lenses and gives very good performance for dry lenses of low to medium power. The achromatic-aplanatic condenser is useful for lenses with NA > 0.5, and is essential for oil immersion lenses with high numerical apertures. For low NA performance, the top element of this condenser can be removed. This condenser focuses light in a flat focal plane and is highly corrected for the significant lens aberrations.

OCULARS

Oculars or eyepieces are needed to magnify and view the image produced by the objective. To make optimal use of the resolution afforded by the objective, an overall magnification equal to 500–1000 times the NA of the objective lens is required. More magnification than this gives "empty" magnification, and the image appears highly magnified but blurry. For most applications, 10× eyepieces work well. When higher magnifications are required for a specific objective, a magnifying booster lens (in Zeiss microscopes, an Optovar lens magnification system) can be rotated into the optical path. Alternatively, a different set of higher-magnification eyepieces can be employed. If the eyepiece is examined when the microscope is focused and the lamp turned on, a bright disk can be seen floating in space a few millimeters outside the eyepiece. The disk is called the *exit pupil* or *Ramsden disk* and represents the aperture at the back focal plane of the objective lens. When viewing a focused specimen, the exit pupil of the eyepiece will be found to be coincident with the *entrance pupil* of the eye, an adjustment that occurs automatically as we focus the specimen.

Eyepiece specifications are engraved on the lens barrel to indicate their magnifying power and field of view. An eyepiece engraved 10×/20 indicates a 10× magnification and 20 mm diameter field of view. The field of view marking also provides a quick reference for determining the diameter of the field in the specimen plane as seen with a given objective. For example, when combined with a 100× objective lens, this eyepiece would give 20 mm/100, or 200 μm for the diameter of the visible object field. Other special design features of the ocular are designated by letter codes, the most common of which indicate high eyepoint (distance between ocular surface and Ramsden disk) for glasses wearers, additional corrections for color and flatness of field, and wide field or wide angle of view. Eyepieces also come with focusable and nonfocusable eye lenses. At least one focusable eyepiece should be included on the microscope to allow parfocal adjustment of the optics so that the same focal plane examined by the eye will be in focus on a camera mounted on the microscope. Oculars are based around a general design containing two achromatic doublet lenses (the field and eye lenses) and a *field stop*, a raised ridge or flange along the inside wall of the ocular that marks the site of the intermediate image plane. In oculars of so-called Huygenian design, the field stop and image plane are located between the field and eyepiece lenses; in Ramsden designs, the focal plane and field stop are located in front of the field lens below the eyepiece. To use an eyepiece reticule, the eyepiece is unscrewed and the reticule is placed in the image plane and rests on the flange comprising the field stop.

MICROSCOPE SLIDES AND COVERSLIPS

Many objectives are designed to be used with standard (1.1 mm thick) glass slides and coverslips of a certain thickness, usually 0.17 mm, which corresponds to thickness grade 1.5. Other coverslip thicknesses induce spherical aberration and give poorer performance, especially when used with high, dry lenses above 40×. For lenses with an NA < 0.4, coverslip thickness is not particularly important. *Remember the thickness of your slides and coverslips counts!* Refer to the following chart when ordering coverslips:

Grade Number	Thickness (mm)
0	0.083–0.13
1	0.13–0.16
1.5	0.16–0.19 (standard)
2	0.19–0.25

THE CARE AND CLEANING OF OPTICS

Maintenance and care are required to protect an expensive optical instrument and to guarantee that optimal high-contrast images will be obtained from it. Neglect, such as not removing immersion oil, forgetting to cover open ports and apertures, or accidentally twisting or dropping an objective lens can ruin optical performance. Even if the microscope is left unused but unprotected on the lab bench, image quality can deteriorate rapidly due to the accumulation of dust from the air. James (1976) and Inoué and Spring (1997) provide detailed descriptions on the effect of dirt on the microscope image and the cleaning of optical surfaces. Following are a few tips for maintaining the performance and image quality of your microscope.

Dust

Keep the microscope protected with a plastic or cloth cover. Wipe dust off the microscope body and stage with a damp cloth. Keep the objective lens turret spotless and free of dust, immersion oil, spilled culture medium, and salt solutions. Hardened thread grease, or additionally on inverted microscopes, dried immersion oil, buffers, and media can weld objectives onto the rotating objective turret, making them difficult to remove or exchange. If an objective is frozen fast, place a drop of water (if salts) or oil-penetrating agent (if oil) at the lens-turret interface to loosen the material before trying to remove the objective. Keep all openings covered with caps so that dust cannot enter the microscope and soil inaccessible lenses, mirrors,and prisms in the microscope body. Make use of the plastic caps designed to cover all objective ports, eyepiece sleeves, and camera ports that are unoccupied and empty. The internal optical pathway should always be completely protected from airborne dust.

Immersion Oil

When finished with an observing session, gently wipe off and clean away excess oil with a high-quality lens tissue and then clean the lens surface with an agent designed for cleaning microscope optics. Immersion oil is a slow-acting solvent that can weaken the cementing compounds that act as a seal between the front lens element and the metal lens cap of the objective. Residual oil should be removed with a lens tissue wetted with a mild lens cleaner such as the solution sold by Edmund Scientific Company (Barrington, New Jersey). Commercial glass cleaners such as Sparkle and Windex are also effective in removing immersion oil, but these generally contain an acid or base that has the potential to erode the antireflection coating on the front lens element and should be used only if a neutral cleaner is not available. For more tenacious deposits, try, in order, ethanol or ethyl ether. However, do not use toluene or benzene, as these solvents used over time will eventually dissolve the front lens sealing compounds. Generally, it is advisable to remove immersion oil and contaminating liquids with the objective lens still mounted on the microscope, as this will avoid mishandling or dropping, the worst accidents that can befall a valuable lens.

Scratches and Abrasions

Never wipe the surfaces of objectives with papers or cloths that are not certified to be free of microscopic abrasives. All objectives contain an exposed optical surface that must be protected from abrasion. Strands of wood fibers in coarse paper, or worse, the stick end of a cotton swab applicator, are strong enough to place dozens of permanent scratch marks (sleeks) on the front lens element with a single swipe. Once present, scratches cannot be removed, and their effect (even if hardly visible) is to scatter light and permanently reduce image contrast. Further, most lenses contain an antireflection coating composed of layers of a dielectric material; each layer is just a few atoms thick. Although antireflection surfaces are protected with a layer of silicon monoxide, you should use only high-quality lens tissue and apply only a minimum of force to wipe off drops of excess oil.

Mechanical Force

Never apply strong physical force to an objective lens or other optical component. To bring another objective into position, move the rotating turret; do not grab and pull on a lens to bring it into position. Also, never remove a stuck objective with a vice-grips or a pipe wrench! If the threads of an objective become stuck to the rotating turret from dried culture medium, oil, or corrosion, apply a drop of water or lens cleaner or penetrating oil to loosen the objective and avoid using force. Likewise, never drop an objective onto the lab bench or floor. Also, do not allow an objective to strike exposed edges of the microscope stage or the condenser (easy to do on some inverted microscope designs). Impacts of this kind cause two forms of irreparable damage: (1) They can crack the compounds that seal the top lens element and metal lens cap, thus allowing immersion oil to penetrate into the lens and coat internal lens elements, causing permanent damage. (2) They can induce permanent stresses in the glass lens components of an objective and may severely degrade its performance in sensitive forms of light microscopy that use polarized light.

Exercise: Constructing and Testing an Optical Bench Microscope

Microscope Construction and Testing

- Determine the focal lengths of the lenses using the method described in the text and label them with pieces of lab tape applied to their edges. Use three 50 mm lenses for the objective, ocular, and illuminator's collector lens to construct the optical bench microscope.

- Mount in order of sequence on the optical bench: a tungsten lamp illuminator, 50 mm collector lens, specimen holder, objective lens, and ocular.

- Handwrite the letter a with a fine marker pen on a sheet of lens tissue or on a microscope slide, and tape it to the specimen holder, centering the letter on the optic axis.

- Position the collector lens about 75 mm away from the lamp filament. Position the specimen about 20 cm away from the collector lens. The light from the illuminator should be focused into a 1–2 cm diameter spot centered on the object (the letter a).

- Using the lens equation $1/f = 1/a + 1/b$, calculate the object-lens distance that gives an image-lens distance of ~30 cm, and mount the objective lens at the calculated position.

- Place a sheet of paper (a paper screen) at the 30 cm location to confirm that the intermediate image of the letter *a* is both real and magnified. Notice that the real intermediate image is both inverted and upside-down. Confirm that it is necessary to position the lens between 1 and 2 focal lengths away from the object to obtain a real intermediate image of the object.

- Mount the ocular at a position that gives a magnified image of the letter when looking down the axis of the microscope through the ocular. The eye might have to be positioned several inches behind the ocular to see the specimen clearly. Note the image–ocular lens distance.

- Place a paper screen in the plane of the virtual image to confirm that it is indeed virtual—that is, no image is produced on the screen. You have now created a compound light microscope!

Answer the following questions about your microscope:

1. Is the object located precisely at the focal length of the objective lens?
2. Is the real intermediate image located precisely at the focal length of the ocular?
3. Explain why the eye-brain perceives a magnified virtual image, while the retina receives a real image from the ocular lens.

Lens Aberrations

Prepare a specimen consisting of a 5 cm² piece of aluminum foil with a dozen small pinholes contained within a 5 mm diameter circle, and mount it on the optical bench microscope. The ideal specimen has some pinholes on or close to the axis and other pinholes near the periphery of the field of view. What lens aberrations do you observe while examining the pinhole images? Here are some tips:

- *Chromatic aberration:* Move the objective lens back and forth through focus and notice how the fringes around the points of light change color depending on the lens position.

- *Spherical aberration:* Pinholes at the periphery of the visual field look blurry. The blurriness can be reduced by creating a 5 mm hole in an opaque mask and placing it in the back focal plane of the objective lens.

- *Coma and astigmatism:* At best focus, peripheral pinholes look like streaks with comet tails that radiate from the center of the field (coma). As the objective is moved back and forth through the plane of best focus, the streaks become drawn out into elliptical shapes that change their orientation by 90° on each side of focus (astigmatism).

- *Distortion and curvature of field:* At best focus, the focal plane is curved like the shape of a bowl, so only one zone of a certain radius can be focused at one time (curvature of field). To view pincushion or barrel distortion, replace the pinhole specimen with a fine mesh copper grid used for electron microscopy and examine the pattern of the square mesh of the grid on the viewing screen (distortion).

DIFFRACTION AND INTERFERENCE IN IMAGE FORMATION

OVERVIEW

This chapter deals with diffraction and interference in the light microscope—the key principles that determine how a microscope forms an image. Having just concluded a section on geometrical optics where image locations and foci are treated as points, lines, and planes, it is surprising to learn that in the microscope the image of a point produced by a lens is actually an extended spot surrounded by a series of rings and that a focal plane is contained in a three-dimensional slab of finite thickness. These properties are due to the diffraction of light (see Fig. 5-1). In the microscope, light from the illuminator is diffracted (literally broken up in the sense of being scattered or spread) by the specimen, collected by the objective lens, and focused in the image plane, where waves constructively and destructively interfere to form a contrast image. The scattering of light (diffraction) and its recombination (interference) are phenomena of physical optics or wave optics. We study these processes, because they demonstrate how light, carrying information from an object, is able to create an image in the focal plane of a lens. With a working knowledge of diffraction, we understand why adjusting the condenser aperture and using oil immersion techniques affect spatial resolution. Diffraction theory also teaches us that there is an upper limit beyond which a lens cannot resolve fine spatial features in an object. In studying diffraction, we see that complex optical phenomena can be understood in mathematically precise and simple terms, and we come to appreciate the microscope as a sophisticated optical instrument. Readers interested in the physical optics of diffraction and interference of light can refer to the excellent texts by Hecht (1998) and Pluta (1988).

DEFINING DIFFRACTION AND INTERFERENCE

Diffraction is the spreading of light that occurs when a beam of light interacts with an object. Depending on the circumstances and type of specimen, diffracted light is perceived in different ways. For example, when a beam of light is directed at the edge of an object, light appears to bend around the object into its geometric shadow, a region not directly illuminated by the beam (Fig. 5-2a). The situation reminds us of the behavior of

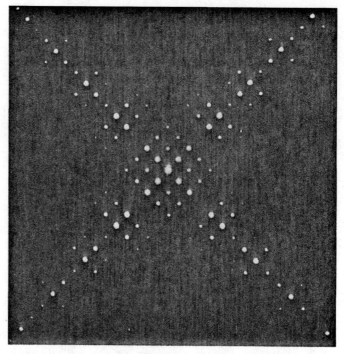

Figure 5-1

Diffraction image of a copper mesh grid. A 400-mesh grid was illuminated with a laser pointer, and its diffraction pattern was photographed on a projection screen. Multiple diffraction spots are observed due to the relatively large spacing between grid bars and the coherent laser light source.

water waves incident on a log floating in a pond. The waves wrap around the ends of the log into the geometrical shadow; instead of reflecting away from the ends of the log, they seem to grab hold of the ends and swing themselves around into the sheltered zone. The redirected component of diffracted light is easily observed when a tree or person is backlighted by a strong light source under conditions where the background behind the object is still dark; the bright line outlining the silhouette of the object is diffracted light. Of particular interest to us is the image of a point source of light in the microscope, since images are composed of a myriad of overlapping points. As we will see, waves emanating from a point in the object plane become diffracted at the margins of the objective lens (or at the edges of a circular aperture at the back focal plane of the lens), causing the image of the point to look like a spot. Thus, the image of a point in a microscope is not a point at all, but a diffraction pattern with a disk of finite diameter. Because of diffraction, an object's image never perfectly represents the real object and there is a lower limit below which an optical system cannot resolve structural details.

Diffraction is also observed when a beam of light illuminates a microscope slide covered with fine dust or scratches (Fig. 5-2b). The spreading of an automobile's headlight beams on a foggy night is another good example of the phenomenon. In these cases, diffraction is defined as the scattering of light by small particles having physical dimensions similar to the wavelength of light. The amount of scattering and angle of

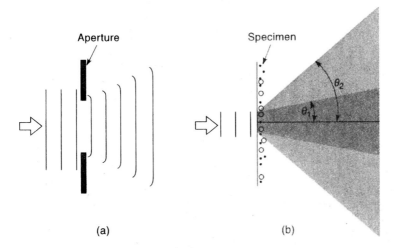

Figure 5-2

Diffraction at an aperture and at a substrate containing fine particles. (a) The electric field of a planar wavefront becomes disturbed by diffraction upon passage through an aperture. The waves appear to grab hold of the aperture and swing around into its geometric shadow. The amplitude profile of a transmitted wavefront is also no longer uniform and remains permanently altered after passage through the aperture (not shown). (b) A substrate containing a layer of a mixture of fine particles (0.2 and 2 μm diameter) diffracts an incident planar wavefront into scattered beams that diverge at different angles. The angle of spreading (θ) is inversely proportional to the size of the particles.

spreading of the beam depend on the size and density of the light-diffracting particles. In the case of illuminated specimens in the microscope, there are therefore two primary sites of diffraction: one at the specimen itself and another in the aperture of the objective lens. Image formation by a lens is critically dependent on these events. If light passes through an object but does not become absorbed or diffracted, it remains invisible. It is the spreading action or diffraction of light at the specimen that allows objects to become visible, and the theory of image formation is based on this principle.

Just as diffraction describes the scattering of light by an object into divergent waves, *interference* describes the recombination and summation of two or more superimposed waves, the process responsible for creating the real intermediate image of an object in the focal plane of a microscope. In a real sense, diffraction and interference are manifestations of the same process. The traditional way of describing interference is to show the recombination of waves graphically in a plot depicting their amplitude, wavelength, and relative phase displacement (Fig. 5-3). The addition of two initial waves produces a resultant wave whose amplitude may be increased (*constructive interference*) or diminished (*destructive interference*).

The term *interference* is frequently taken to mean that waves annihilate each other if they are out of phase with each other by $\lambda/2$, and indeed a graphical depiction of interfering waves reinforces this idea. However, since it can be demonstrated that all of the photon energy delivered to a diffracting object such as a diffraction grating can be completely recovered (e.g., in a resultant diffraction pattern), it is clear that photons do not self-annihilate; rather their energies become redistributed (channeled) during diffraction

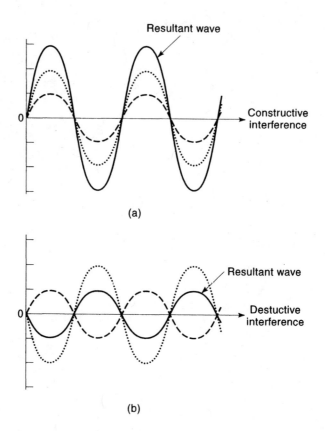

(a)

(b)

Figure 5-3

Two coincident waves can interfere if their E vectors vibrate in the same plane at their point of intersection. Two waves are shown that vibrate in the plane of the page. In these examples, both waves (dotted and dashed curves) have the same wavelength, but vary in amplitude. The amplitude of a resultant wave (solid curve) is the arithmetic sum of the amplitudes of the two original waves. (a) Constructive interference occurs for two waves having the same phase. (b) Destructive interference occurs for waves shifted in phase; if the amplitudes of the waves are the same and the relative phase shift is λ/2, the wave is eliminated.

and interference in directions that permit constructive interference. Likewise, interference filters do not destroy the light they do not pass; they merely reflect or absorb it. It is therefore best to think of diffraction and interference as phenomena involving the redistribution of light waves and photon energy. Accordingly, wave constructions of the kind shown in Figure 5-3 are best thought of as devices that help us calculate the energy traveling in a certain direction or reaching a certain point. The mechanism by which light energy becomes redistributed is still not understood.

THE DIFFRACTION IMAGE OF A POINT SOURCE OF LIGHT

The image of a self-luminous point object in a microscope or any other image-generating instrument is a diffraction pattern created by the action of interference in the image plane.

When highly magnified, the pattern is observed to consist of a central spot or diffraction disk surrounded by a series of diffraction rings. In the nomenclature of diffraction, the bright central spot is called the 0th-order diffraction spot, and the rings are called the 1st-, 2nd-, 3rd-, etc.-order diffraction rings (see Fig. 5-4). When the objective is focused properly, the intensity of light at the minima between the rings is 0. As we will see, no lens-based imaging system can eliminate the rings or reduce the spot to a point. The central diffraction spot, which contains ~84% of the light from the point source, is also called the *Airy disk,* after Sir George Airy (1801–92), who described some of its important properties.

The Airy disk pattern is due to diffraction, whose effect may be described as a disturbance to the electric field of the wavefront in the aperture of the lens—the consequence of passing an extended electromagnetic wavefront through a small aperture. The disturbance continues to alter the amplitude profile of the wavefront as the front converges to a focus in the image plane. Diffraction (the disturbance) constantly perturbs the incident wavefront, and since no one has invented a lens design to remove it, we must accept its alteration of points comprising the image. The size of the central disk in the Airy pattern is related to the wavelength λ and the aperture angle of the lens. For a telescope or camera lens receiving an incident planar wavefront from an infinitely distant source such as a star, the aperture angle is given as the focal ratio f/D, where D is

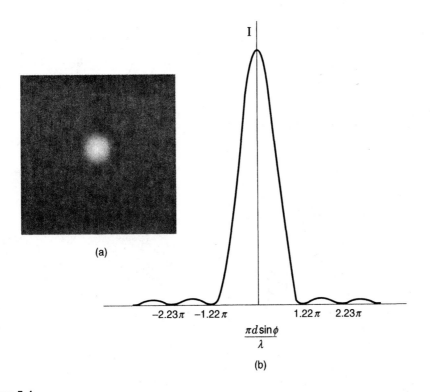

(a)

(b)

$$\frac{\pi d \sin \phi}{\lambda}$$

Figure 5-4

The diffraction pattern of a point source of light. (a) Diffraction pattern showing central diffraction disk surrounded by diffraction rings. (b) Intensity profile a diffraction spot. The central spot and surrounding rings are evident. The separation distance between the center of the spot and the first minimum depends on the angular aperture of the lens.

the lens diameter and f is the focal length. In this case, the *aperture angle* is the angular diameter of the lens as seen from a point in the image plane at focal length f. The size of the diffraction disk radius d is given by

$$d = 1.22\lambda\,(f/D).$$

In a microscope, the aperture angle is described by the numerical aperture NA, which includes the term sin θ, the half angle over which the objective can collect light coming from a nearby object. (NA is defined further in Chapter 6.) In the case of a microscope image, the radius d of the diffraction spot for a self-luminous point of light in the image plane is described by a related expression:

$$d = 1.22\,\lambda/2\text{NA}$$

In both situations, the size of the spot decreases with decreasing wavelength and increasing numerical aperture, but always remains a disk of finite diameter. The spot size produced by a 25× oil immersion objective with NA = 1 is about 30 μm. Obtaining an image whose resolution is limited by the unit diffraction spot, rather than by scattered light or lens aberrations, is what is meant by the term *diffraction limited*. We examine the relationship between diffraction and resolution in Chapter 6.

Demonstration: Viewing the Airy Disk with a Pinhole Aperture

It is easy to observe the Airy disk by examining a point source of light through a pinhole (Fig. 5-5). This is best done in a darkened room using a bright lamp or microscope illuminator whose opening is covered with a piece of aluminum foil containing a pinhole that will serve as a point source of light when viewed at a distance of several feet. The illuminator's opening should be completely covered so that the amount of stray light is minimal. A second piece of foil is prepared with a minute pinhole, 0.5 mm diameter or less, and is held up to the eye to examine the point source of light at the lamp. The pinhole–eye lens system (a pinhole camera) produces a diffraction pattern with a central Airy disk and surrounding diffraction rings. The same observation can be made outdoors at night examining distant point sources of light with just the pinhole held over the eye. This simple diffraction pattern is the basic unit of image formation. If the eye pinhole is made a little larger, the Airy disk becomes smaller, in accordance with the principle of angular aperture we have described. Now turn on the room lights and look through the pinhole at objects in the room, which look dim (not much light through the pinhole) and blurry (low resolution). The blurry image is caused by large overlapping diffraction spots of image components on the retina. When the pinhole is removed and the objects are viewed directly, the image is clearer because the larger aperture size afforded by the eye's iris and lens results in smaller diffraction spots and an increase in resolution and clarity. We begin to appreciate that an extended image can be viewed as the summation of a myriad of overlapping diffraction spots.

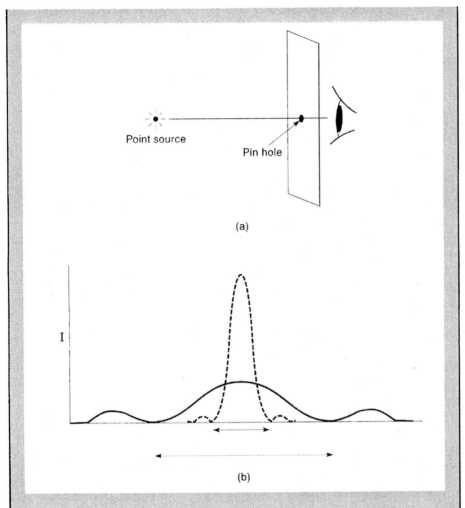

Figure 5-5
The image of a point source of light viewed through a pinhole is a perceptible disk. (a) Viewing a point source of light through a pinhole. (b) Relative sizes of the diffraction disk on the retina when a point source is viewed with normal vision (dotted line) and through a pinhole (solid line). With a pinhole, the diffraction disk expands about 10-fold to 30–40 μm, covering a patch of hundreds of cone cells on the retina, thereby allowing perception of a visible diffraction disk and surrounding rings.

In cameras and telescopes, the image of a star is likewise always a finite diffraction disk (not a point) with linear radius $r = 1.22\lambda \, (f/D)$, where f/D is called the *focal ratio* or *f*-number. Roughly, the diameter of the diffraction spot in μm is the focal ratio in millionths of a meter. Thus, the diameter of the diffraction spot in the primary focal plane of the 250 cm diameter, $f/5$ mirror of the Hubble space telescope is 5 μm. In the case of the human eye or photographic camera, the image of a point source on the retina or film has a diameter of about 3 μm. In all of these optical systems, the image of a point source corresponds to a diffraction disk, and the terms NA and f/D are measures of the effective aperture angle and light-gathering ability. As shown in this chapter, the spatial resolution of the microscope is limited by the smallest disk that it is possible to obtain by varying λ and NA. Only when the images of specimen details subtend diameters greater than this limit can you begin to obtain information regarding the size, shape, and periodicity of the object.

CONSTANCY OF OPTICAL PATH LENGTH BETWEEN THE OBJECT AND THE IMAGE

Before we examine the effect of lens diameter on diffraction spot size and spatial resolution, we need to consider the concept of optical path length for an imaging system containing a perfectly corrected lens. In practice, microscope objectives and other corrected lenses only approximate the condition of being perfectly corrected, and waves arrive at the conjugate image point somewhat out of place and out of phase. This is usually not a serious problem. Despite practical limitations in finishing lenses with spherical surfaces, most microscope lenses give diffraction-limited performance with an average error in phase displacement (wavefront error) of less than $\lambda/4$, and manufacturers strive to obtain corrections of $\lambda/10$ or better. As is known, light from a self-luminous point in an otherwise dark specimen plane radiates outward as an expanding spherical wavefront; waves collected by the objective are refracted toward the center line of the lens and progress as a converging spherical wavefront to a single point in the image plane. However, it is also true—and this point is critical for image formation—that *the number of vibrations as well as the transit time experienced by waves traveling between an object point and the conjugate image point are the same regardless of whether a given wave passes through the middle or the periphery of the lens.*

Ideally, all waves from an object point should arrive at the image point perfectly in phase with one another. Given the large differences in the spatial or geometric path lengths between central and peripheral waves, this seems improbable. The explanation is based on the concept of *optical path length,* a length distinct from the geometric path length. It can be used to calculate the number of vibrations experienced by a wave traveling between two points. As it turns out, variations in the thickness of the lens compensate for the differences in geometric paths, causing all waves to experience the same number of vibrations (Fig. 5-6). It can also be shown that the transit time required for light to travel between object and image points along different trajectories having the same optical path length is the same. These concepts become important when we discuss the spatial and temporal coherence of light later in the chapter.

In optics, the *optical path length (OPL)* through an object or space is the product of the refractive index n and thickness t of the object or intervening medium:

$$OPL = nt.$$

If the propagation medium is homogeneous, the number of vibrations of a wave of wavelength λ contained in the optical path is determined as

$$\text{Number of vibrations} = nt/\lambda.$$

Since the frequency of vibration remains constant and the velocity of light $= c/n,$ when a wave traverses a lens of higher refractive index than the surrounding medium,

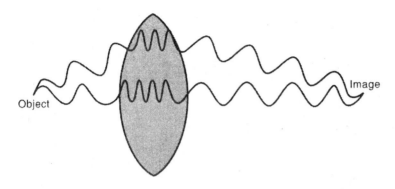

Figure 5-6
Preservation of the optical path length between the object and image. The optical path length may be regarded as the number of cycles of vibration experienced by a wave between two points. Two waves traveling in phase from a point in an object and entering the center and periphery of a lens cover different physical distances, but experience the same optical path length, and therefore arrive simultaneously and in phase at the conjugate point in the image plane. This occurs because the wave traveling the shorter geometric path through the middle of the lens is relatively more retarded in its overall velocity, owing to the longer distance of travel in a high-refractive-index medium (the glass lens). Note that the total number of cycles (the number of wavelengths) is the same for both waves.

the wavelength and velocity decrease during transit through the lens. Thus, the number of cycles of vibration per unit of geometrical distance in the lens is greater than the number of cycles generated over the equivalent distance in the surrounding medium. The overall optical path length expressed as the number of vibrations and including the portions in air and in glass is thus described as

$$\text{Number of vibrations} = n_1 t_1 / \lambda_1 + n_2 t_2 / \lambda_2,$$

where the subscripts 1 and 2 refer to parameters of the surrounding medium and the lens. As we will encounter later on, the *optical path length difference* Δ between two rays passing through a medium vs. through an object plus medium is given as

$$\Delta = (n_2 - n_1)t.$$

EFFECT OF APERTURE ANGLE ON DIFFRACTION SPOT SIZE

Now let us examine the effect of the aperture angle of a lens on the radius of a focused diffraction spot. We consider a self-luminous point P that creates a spherical wavefront that is collected by the objective and focused to a spot P' in the image plane (Fig. 5-7a). In agreement with the principle of the constancy of optical path length, waves passing through points A and B at the middle and edge of the lens interfere constructively at P'. (The same result is observed if Huygens' wavelets [discussed in the next section] are constructed from points A and B in the spherical wavefront at the back aperture of the lens.) If we now consider the phase relationship between the two waves arriving at another point P'' displaced laterally from P' by a certain distance in the image plane, we see that a certain distance is reached where the waves from A and B are now 180° out of

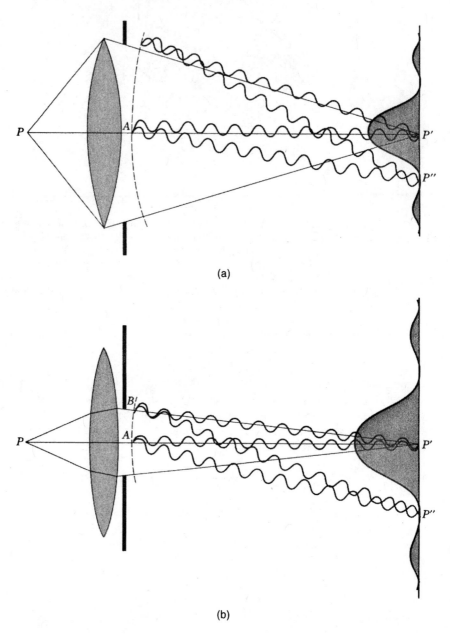

(a)

(b)

Figure 5-7

Aperture angle determines the size of the diffraction spot. Point source *P* and its conjugate image *P'* in the image plane. Point *P''* is moved laterally in the focal plane away from *P'* until destructive interference at a certain distance determines the location of the first diffraction minimum and thus the radius of the diffraction spot. (a) Points A and B in the wavefront with full lens aperture. (b) Points A and B in the wavefront with reduced aperture angle caused by partially stopping down the condenser iris diaphragm.

phase with each other and destructively interfere. The light intensity at this position (the first minimum in the diffraction pattern of the spot) is 0. Indeed, it can be shown that *the sum of contributions from all points in the aperture results in an amplitude of 0 at this location and nearly so at all other locations in the image plane other than in the central diffraction spot and surrounding diffraction rings.* A geometrical explanation of the phenomenon is given by Texereau (1963).

In Figure 5-7b we observe that the aperture (closing down the condenser diaphragm in a conjugate focal plane) reduces the angular aperture of the optical system, which increases the size of the diffraction spot and the distance $P'P''$ to the first diffraction minimum. Therefore, reducing the angular aperture decreases spatial resolution in the image. If we reexamine the optics of the pinhole camera, it now becomes clear why viewing a point source through a pinhole aperture held in front of the eye allows perception of an observable diffraction disk (Fig. 5-5). For the fully dilated eye the Airy disk of a point source covers ~ 2 cone cells on the retina, but with the reduced angular aperture using a pinhole, the disk diameter expands some 40-fold, stimulates dozens of receptor cells, and results in the perception of a disk.

DIFFRACTION BY A GRATING AND CALCULATION OF ITS LINE SPACING, *D*

We will now examine the diffraction of light at a specimen using a transparent diffraction grating as a model for demonstration and discussion. We should bear in mind that the principles of diffraction we observe at a grating on an optical bench resemble the phenomena that occur at specimens in the microscope. A *diffraction grating* is a planar substrate containing numerous parallel linear grooves or rulings, and like a biological specimen, light is strongly diffracted when the spacing between grooves is close to the wavelength of light (Fig. 5-8). If we illuminate the grating with a narrow beam from a monochromatic light source such as a laser pointer and project the diffracted light onto a screen 1–2 m distant, a bright, central 0th-order spot is observed, flanked by one or more higher-order diffraction spots, the 1st-, 2nd-, 3rd-, etc.-order diffraction maxima. The 0th-order spot is formed by waves that do not become diffracted during transmission through the grating. An imaginary line containing the diffraction spots on the screen is perpendicular to the orientation of rulings in the grating. The diffraction spots identify unique directions (diffraction angles) along which waves emitted from the grating are in the same phase and become reinforced as bright spots due to constructive interference. In the regions between the spots, the waves are out of phase and destructively interfere.

The *diffraction angle* θ of a grating is the angle subtended by the 0th- and 1st-order spots on the screen as seen from the grating (Fig. 5-9). The right triangle containing θ at the screen is congruent with another triangle at the grating defined by the wavelength of illumination, λ, and the spacing between rulings in the grating, d. Thus, $\sin \theta = \lambda/d$, and reinforcement of diffraction spots occurs at locations having an integral number of wavelengths—that is, 1λ, 2λ, 3λ, and so on—because diffracted rays arriving at these unique locations are in phase, have optical path lengths that differ by an integral number of wavelengths, and interfere constructively, giving bright diffraction spots. If $\sin \theta$ is calculated from the distance between diffraction spots on the screen and between the screen and the grating, the spacing d of the rulings in the grating can be determined using the *grating equation*

$$m\lambda = d \sin\theta,$$

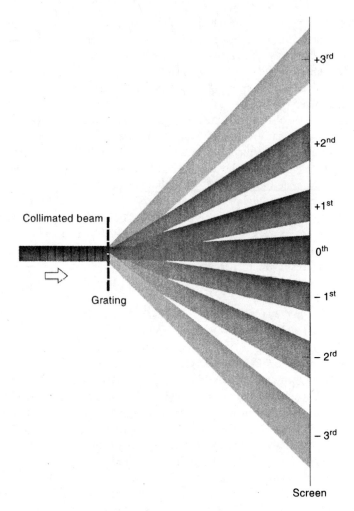

Figure 5-8
The action of a diffraction grating. Multiple orders of diffracted light are shown.

where λ is the wavelength and m is an integral number of diffraction spots. (For calculations based on the distance between the 1st- and 0th-order spots, $m = 1$; if the distance between the 2nd- and 0th-order spots is used, $m = 2$, etc.) Notice that the diffraction angle θ *increases* as the grating spacing d *decreases* and as the wavelength λ *increases*.

The effect of wavelength can be demonstrated by illuminating the grating with white light. Under this condition, the 0th-order spot appears white, while each higher-order diffraction spot appears as an elongated spectrum of colors. Thus, the diffraction angle depends directly on the wavelength. Blue light, being most energetic, is scattered the least, so the blue ends of the spectra are always located closest to the 0th-order central spot.

An alternate device, called *Hugyens' principle,* can also be used to determine the location of the diffraction spots and diffraction angle θ of a grating. Christiaan Huygens, the Dutch physicist (1629–92), used a geometrical construction for determining the location of a propagating wavefront, now known as the *construction of Huygens' wavelets* (Fig. 5-10). According to Huygens' principle, every point on a propagating

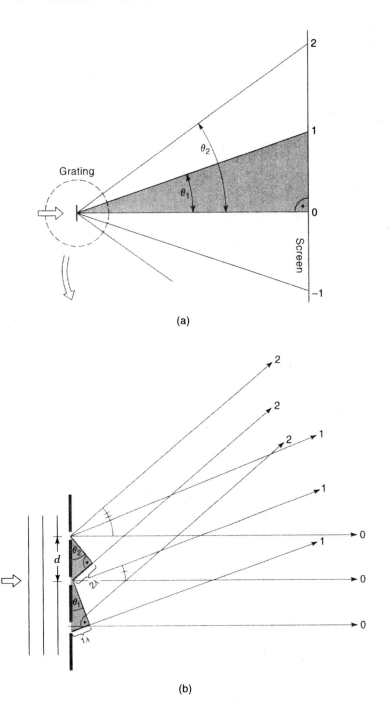

(a)

(b)

Figure 5-9

Dependence of scattering angle on grating spacing and wavelength. (a) Rays from a diffraction grating projected on a viewing screen. The angle of scattering of the 1st- and 2nd- order rays is shown as θ_1 and θ_2. The grating and 1st- and 2nd-order spots define a right triangle that includes θ_1 and is congruent with a triangle that can be delineated at the grating as shown in (b). (b) The diffracted rays at the grating define a right triangle that includes diffraction angle θ. For the 1st- and 2nd- etc.-order diffracted rays, the base of the triangle is an integral number of wavelengths, 1λ, 2λ etc. Thus, the angle of diffraction depends on two parameters: the grating spacing d and wavelength λ.

Wavefronts

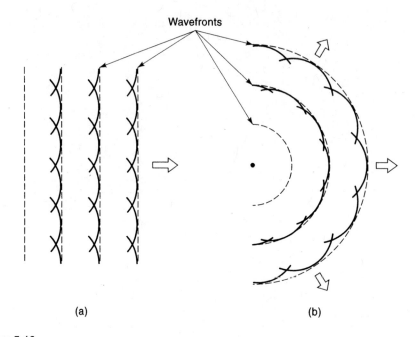

(a)   (b)

Figure 5-10

Huygens' wavelets are used to describe the propagation of (a) planar and (b) spherical wavefronts.

wavefront serves as the source of *spherical secondary wavelets,* such that the wavefront at some later time is defined by the envelope covering these wavelets. Further, if wave propagation occurs in an isotropic medium, the secondary wavelets have the same frequency and velocity as the original reference wavefront. The geometrical construction of Huygens' wavelets is a useful device for predicting the locations of wavefronts as modified by both refraction and diffraction; however, Huygens' theory does not account for many wave-optical phenomena of diffraction, which require the application of newer approaches in physical optics.

When applied to the diffraction grating, we can use the construction of Huygens' wavelets to determine the location of the diffraction spots and the diffraction angle. Take a moment to study the construction for a diffraction grating with spacing d in Figure 5-11, which emphasizes the concept that the diffraction spots occur at angles where the optical path lengths of constituent waves are an integral number of wavelengths λ. At locations between the diffraction spots, waves vary by a fraction of a wavelength and destructively interfere. Although Huygens' wavelet construction accounts for the locations of diffraction spots, it does not account for all aspects of the diffraction process. For example, the sum of all of the energy present in the luminous regions of the diffraction pattern is known to equal the energy incident on the grating. This is inconsistent with ray particle models and predictions from the geometrical construction of wavelets that photons are distributed uniformly on the diffraction screen, being annihilated where there is destructive interference. We enter here a murky area where the wave and particle natures of light are difficult to reconcile.

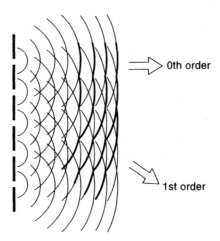

Figure 5-11

Geometrical determination of scattering angles in a diffraction grating using the construction of Huygens' wavelets.

Demonstration: The Diffraction Grating

It is possible to see the trajectories of diffracted rays by immersing a transparent diffraction grating replica in a dish of water. Fill a 9 × 12 inch glass baking dish with water, add 1–2 mL of milk, and place the dish on a dark background. Place a small laser or laser pointer up against the dish and direct the beam at a diffraction grating immersed in the water. The rulings on the grating should be oriented vertically. The 0th-order beam and higher order diffracted beams are made visible by the light-scattering properties of the miscelles of lipid and protein in the milk. Notice that the diffracted rays are linear and sharply defined. Each point along a ray represents a location where overlapping spherical wavefronts emergent from each illuminated diffracting ruling in the grating give constructive interference, meaning that the waves are in phase with one another, differing precisely by an integral number of wavelengths. The dark spaces in between the rays represent regions where the wavefronts are out of phase and give destructive interference. The geometry supporting destructive interference does not mean, however, that light self-annihilates in the dark zones. For any given point located in a dark zone, it can be seen that there is no visible light between the point and the illuminated spot on the grating. Instead, diffraction results in light energy being directed only along paths giving constructive interference. All of the light energy contained in the incident laser beam is contained in the 0th order and higher order diffracted rays. The diffraction angle θ that is subtended at the grating by the 0th-order ray and a higher order diffracted ray depends on the refractive index of the medium and on the wavelength, λ.

Illuminate a diffraction grating with a bright white light source and project the diffraction pattern on a wall or viewing screen (Fig. 5-12). It is convenient to use the I-beam optical bench so that the grating, filters, and masks can be stably mounted in holders clamped to the beam. If high-intensity xenon or mercury arc lamps are used, you should position a heat-absorbing filter (BG38 glass) or,

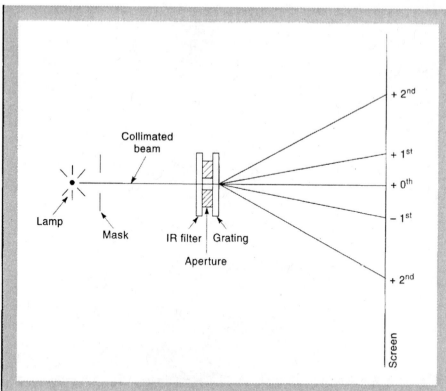

Figure 5-12

Demonstration of diffraction with a grating. An optical bench may be used to hold optical elements at fixed locations. A collimated bright white light source and a heat-cut filter are used to illuminate a diffraction grating covered by a closely apposed aperture mask containing a 2–3 mm hole or slit. If a slit is used instead of a hole, it should be perpendicular to the rulings in the grating.

better, a heat-reflecting mirror (hot mirror) between the lamp and the grating to keep the grating from melting. It is also useful to place an opaque metal mask (aluminum foil) containing a 3–4 mm diameter hole immediately in front of the grating in order to obtain a more sharply defined diffraction pattern.

On the viewing screen, notice the central 0th-order spot, white and very bright, flanked on two sides by 1st-, 2nd-, and higher-order diffraction spots, each of which appears as a bright spectrum of colors, with the blue ends of the spectra oriented toward the 0th-order spot. Higher-energy blue wavelengths are diffracted the least and are located closest to the 0th-order spot within each diffraction-order spectrum and in accordance with the relation already given, showing that the diffraction angle $\theta \propto \lambda/d$. It is easy to measure the angle θ by simple trigonometry, particularly when monochromatic light is used. When the grating is illuminated with monochromatic light from a laser pointer ($\lambda = 625-665$ nm), all of the diffraction spots appear as sharply defined points.

Using a white card as a screen, move the card closer to the grating. Observe that the diffraction angle defined by the grating and the 1st- and 0th-order diffraction spots remains constant and that the spots move closer together as the card approaches the grating. It is easy to see the linear diffracted rays by darkening the room and introducing a cloud of chalk or talcum dust between the grating and the screen.

Finally, examine the diffraction pattern of a sinusoidal (holographic) grating. The sinusoidal grating channels much of the incident light into the 1st-order diffraction spots, which are much brighter than the spots produced by a conventional ruled grating. Such gratings are very useful for examining the spectra of filters and illuminators.

ABBE'S THEORY FOR IMAGE FORMATION IN THE MICROSCOPE

Ernst Abbe (1840–1905) developed the theory for image formation in the light microscope while working in the Carl Zeiss optical workshop in Jena, Germany (Fig. 5-13). Abbe observed that diffracted light from a periodic specimen produces a diffraction pattern of the object in the back focal plane (the diffraction plane) of the objective lens. *According to Abbe's theory, interference between 0th- and higher-order diffracted rays in the image plane generates image contrast and determines the limit of spatial resolution that can be provided by an objective.* For a periodic object such as a diffraction grating, it is easy to demonstrate that light from at least two orders of diffraction must be captured by the objective in order to form an image (see Exercise, this Chapter). In the minimal case, this could include light coming from two adjacent diffraction spots, such as the 0th- and one 1st-order spot generated by the grating. If light from only a single diffraction order is collected by the lens (only the 0th order is collected), there is no interference, and no image is formed. Put another way, there is no image if light diffracted at the specimen is excluded from the lens. Extending the concept further, the larger the number of diffraction orders collected by the objective, the sharper and better resolved (the greater the information content) are the details in the image.

Objects that exhibit fine, periodic details (diffraction gratings, diatoms, striated muscle) provide important insights about the roles of diffraction and interference in image formation. Figure 5-14 shows a diffraction grating with periodic rulings illuminated by a collimated beam of light having planar wavefronts. Light diffracted by the rulings is collected by a lens, and an image of the rulings is created in the primary image plane.

Note the following:

- A certain amount of incident light does not interact with the specimen and is transmitted as undeviated (nondiffracted) rays. These rays form the central 0th-order diffraction spot in the diffraction plane and go on to evenly illuminate the entire image plane.

- Each ruling in the grating acts as an independent source of diffracted waves that radiate outward as a series of spherical wavefronts (Huygens' wavelets) toward the objective lens. Note that the aperture of the lens is large enough to capture some of the diffracted light. The effective NA of the objective is critically dependent on the setting of the condenser aperture. The diffracted light forms higher-order diffraction spots flanking the 0th-order spot in the diffraction plane.

- The 0th- and higher-order diffraction spots in the diffraction plane correspond to locations where there is constructive interference of waves that differ in their path

Figure 5-13

Ernst Abbe, 1840–1905. Principles of microscope and objective lens design, the theory of image formation in the microscope, and standardized lens manufacturing procedures all trace their beginnings to the work of Ernst Abbe and his collaborations with Carl Zeiss in Jena, Germany, in the 1860s. Until then, lens making was an art and a craft, but the new industrial philosophy demanded technical perfection, lens designs based on theory and research, and improvements in raw materials. At Abbe's initiative, lens curvatures were examined using an interference test with Newton's rings, and lens designs were based on Abbe's sine-squared condition to remove aberrations. He created the first planachromatic lens, and after much research, the apochromatic lens, which was commercially sold in 1886. After many false starts over a 20-year period, the research-theory-testing approach for manufacturing lenses proved to be successful. These improvements and new photographic lens designs required new types of glass with values of refractive index and color dispersion that were not then available. Abbe and Zeiss won grants and developed new glasses in collaborations with the industrialist, Otto Schott, owner of the Jena Glassworks. Other

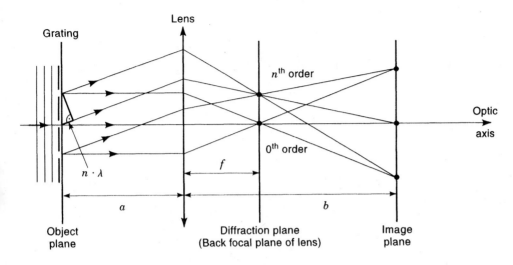

Figure 5-14

Abbe's theory for image formation in a light microscope. An objective lens focused on a grating ($2f > a > f$) in the object plane produces a magnified real image of the grating in the image plane. The diffraction plane is located at $1f$ in the back aperture of the lens. An incident planar wavefront is shown. Diffracted nth-order and nondiffracted 0th-order rays are separated in the diffraction plane, but are combined in the image plane.

lengths by exactly 1, 2, 3, . . . wavelengths, respectively. The waves at each spot are exactly in phase and are diffracted at the slits in the specimen at the same unique angle of diffraction (θ). Note that each point in the diffraction plane corresponds to a certain angle of light leaving the specimen. The absence of light between the spots is due to interference between waves that are in or out of phase with each other. The 0th- and higher-order diffraction spots are most clearly focused and separated from one another in the diffraction plane (the rear focal plane of the objective lens).

- Image formation in the image plane is due to the interference of undeviated and deviated components that are now rejoined and interfere in the image plane, causing the resultant waves in the image plane to vary in amplitude and generate contrast. Abbe demonstrated that at least two different orders of light must be captured by a lens for interference to occur in the image plane (Fig. 5-15).

---

inventions were the Abbe achromatic condenser, compensating eyepieces for removing residual color aberration, and many other significant items of optical testing equipment. Abbe is perhaps most famous for his extensive research on microscope image formation and his diffraction theory, which was published in 1873 and 1877. Using a diffraction grating, he demonstrated that image formation requires the collection of diffracted specimen rays by the objective lens and interference of these rays in the image plane. By manipulating the diffraction pattern in the back aperture, he could affect the appearance of the image. Abbe's theory has been summarized as follows: *The microscope image is the interference effect of a diffraction phenomenon.* Abbe also introduced the concept of numerical aperture ($n \sin\theta$) and demonstrated the importance of angular aperture on spatial resolution. It took 50 years for his theory to become universally accepted, and it has remained the foundation of microscope optical theory ever since. Ernst Abbe was also a quiet but active social reformer. At the Zeiss Optical Works, he introduced unheard-of reforms, including the 8-hour day, sick benefits, and paid vacations. Upon his death, the company was handed over to the Carl Zeiss Foundation, of which the workers were part owners.

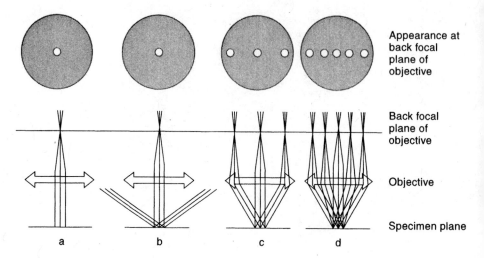

Figure 5-15

Generation of an image by interference requires collection of two adjacent orders of diffracted light by the objective lens. If diffraction at the specimen does not occur (a), or diffracted rays are not collected by the objective (b), no image is formed. In (c), a minimum of two adjacent diffraction orders (0th and 1st) are collected, and an image is formed. In (d), multiple diffracted orders are collected, leading to a high degree of definition in the image.

- A periodic specimen with an interperiod spacing d gives rise to a periodic pattern with spacing D in the diffraction plane, where $D \sim 1/d$. Therefore, the smaller the spacings in the object, the farther apart the spots are in the diffraction plane, and vice versa. The relationship is

$$d \approx f\lambda/D \cos\theta$$

where f is the focal length of the lens, λ is the wavelength, and θ is the acute angle at the principal plane in the objective lens from which the focal length is measured and which forms a right triangle together with the 0th- and 1st-order diffraction spots.

Abbe's theory of image formation explains the following important points: If a specimen does not diffract light or if the objective does not capture the diffracted light from an object, no image is formed in the image plane. If portions of two adjacent orders are captured, an image is formed, but the image may be barely resolved and indistinct. If multiple orders of diffracted light are captured, a sharply defined image is formed. The theory is also the basis for determining the spatial resolution of the light microscope, which is the subject of the next chapter.

DIFFRACTION PATTERN FORMATION IN THE BACK APERTURE OF THE OBJECTIVE LENS

Let us now consider the diffraction pattern in the microscope's diffraction plane in the back aperture (back focal plane) of the objective lens. Under conditions of Koehler illumination, a diffraction image of a specimen is formed just behind the objective in the objective's back (or rear) focal plane. You should take a moment to reinspect the figures in Chapter 1 and recall that this plane is conjugate with other aperture planes in the microscope—that is, the lamp filament, the front focal plane of the condenser, and the iris aperture of the eye. For specimens having periodic details (a diffraction grating) and under certain conditions of diaphragm adjustments, you can inspect the diffraction pattern of an object using a Bertrand lens. The diffraction pattern in the back aperture and the image in the image plane are called *inverse transforms of each other,* since distance relations among objects seen in one plane are *reciprocally* related to spacings present in the other plane, as will now be explained.

Demonstration: Observing the Diffraction Image in the Back Focal Plane of a Lens.

It is easy to confirm the presence of a diffraction image of an object in the back aperture of a lens using an I-beam optical bench, a 3 mW HeNe laser light source, and an electron microscope copper grid (Figure 5-16). A 50 mm or shorter focal length lens is placed just in front of the laser to spread the beam so as to illuminate the full diameter of the grid. The EM grid taped to a glass microscope slide and mounted on a lens holder serves as an object. Position the grid at a distance that allows full illumination of its surface by the laser. Confirm that the grid generates a diffraction pattern by inserting a piece of paper into the beam at various locations behind the grid. Now position an optical bench demonstration lens of 50–100 mm focal length just over 1 focal length from the grid to create a sharply focused image (real intermediate image) of the grid on a projection screen some 2–3 meters away from the lens. Adjust the lens position as required to obtain a focused image of the grid on the screen. The image reveals an orthogonal pattern of square empty holes delimited by metal bars of the grid, which is a stamped piece of copper foil. To see the focused diffraction pattern of the grid, insert a white sheet of paper in the optical path at one focal length away from the lens. The diffraction pattern consists of an orthogonal pattern of bright points of light (see Fig. 5-1). Confirm that the focused diffraction image is located precisely in the back focal plane at 1 focal length distance by moving the paper screen back and forth along the beam.

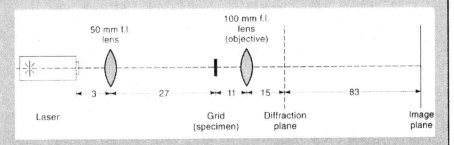

Figure 5-16

Demonstration of the diffraction image in the back focal plane of a lens.

Let us first consider the image of an object such as a diffraction grating with lines spaced just a few micrometers or less apart. Light that passes through the grating (we might imagine any microscope specimen) and does not interact with it is called direct light and forms a small central spot (the 0th-order diffraction spot) in the middle of the diffraction pattern. Light that becomes diffracted (diffracted waves) forms a pattern of widely separated 1st-, 2nd-, 3rd-, etc.-order spots flanking the 0th-order spot. The diffraction patterns of periodic objects such as diatoms or striated muscle can be observed by closing the condenser diaphragm down to a minimum and inspecting the diffraction plane with a Bertrand lens. It is understood that the light contained in the diffraction pattern goes on to form the image of the grating through interference in the real intermediate image plane in the eyepieces of the microscope. Looking at the image through the eyepieces, we do not see the diffraction pattern, but the inverse transform that is derived from it. Thus, we say that the *object image is the inverse transform of the diffraction pattern in the back aperture.* The reciprocal nature of the space relationships of the two images is described as follows: Fine periodic features separated by short distances in the object and image are seen as widely separated diffraction spots in the diffraction image; coarse features separated by relatively long distances in the object and real intermediate image take the form of closely separated diffraction spots close to the central 0th-order spot in the diffraction pattern.

Joseph Gall demonstrated this relationship by placing a transparent image (photographic negative) of a periodic polymer at the back aperture and by providing illumination from a pinhole placed at the lamp's field stop. There was no specimen on the stage of the microscope. Looking in the eyepieces, the diffraction pattern (the inverse transform) of the polymer is observed. Used this way, the microscope functions as an optical diffractometer (Gall, 1967). Thus, images located in the diffraction plane and in the image plane are inverse transforms of one another, and the two images exhibit space relationships that are related as the reciprocal of the space relations in the other image. The exercise at the end of this chapter reinforces these points. We will revisit this concept throughout the book, particularly in the chapters on phase contrast microscopy and in the section on fast Fourier transforms in Chapter 15.

PRESERVATION OF COHERENCE: AN ESSENTIAL REQUIREMENT FOR IMAGE FORMATION

Our discussion of the role of diffraction and interference in image formation would not be complete without considering the requirement for coherent light in the microscope. Object illumination with rays that are partially coherent is required for all forms of interference light microscopy (phase contrast, polarization, differential interference contrast) discussed in the following chapters.

Nearly all "incoherent" light sources, even incandescent filament lamps, are partially coherent—that is, the waves (wave bundle) comprising each minute ray emanating from a point on the filament vibrate in phase with each other. In reality, the degree of coherence within a given ray is only partial, since the photons vibrate slightly out of phase with each other. The distance over which the waves exhibit strong coherence is also limited—just a few dozen wavelengths or so—so that if you examined the amplitude of the ray along its length, you would see it alternate between high-amplitude states, where the waves vibrate coherently, and low-amplitude states, where waves are transiently out of phase with each other. In contrast, laser beams can be coherent over meters of distance, whereas waves emitted from a fluorescence source are completely incoherent.

The cause of coherence derives from the fact that atoms in a microscopic domain in the filament, excited to the point that they emit photons, mutually influence each other, which leads to synchronous photon emission. Thus, a tungsten filament or the ionized plasma in a mercury arc may each be considered as a large collection of minute atomic neighborhoods, each synchronously emitting photons. A discrete number of coherent waves following the same trajectory is thus called a ray or pencil of light. The action of the collector lens of the illuminator is to form an image of the filament in the front aperture of the condenser, which then becomes the source of partially coherent rays that illuminate the object in the specimen plane (Fig. 5-17).

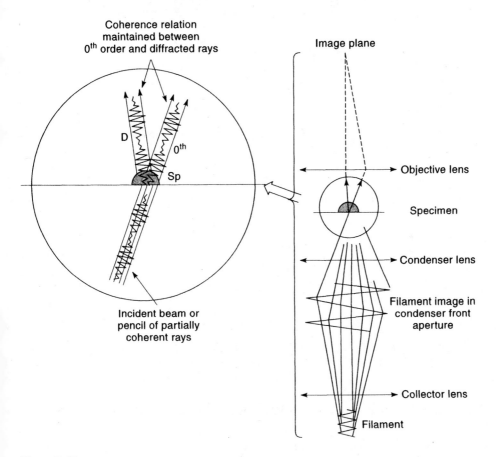

Figure 5-17

Partially coherent wave bundles in the light microscope. Small domains in the lamp filament emit partially coherent wave bundles that reform the image of the filament in the front aperture of the condenser. Rays (small beams or "pencils" of partially coherent photons) incident on a small particle in the specimen plane are broken up into diffracted and undeviated (0th-order) waves that maintain their coherence relationship. A myriad of ray pairs go on through the objective lens and combine incoherently with other rays in the image plane to form the image. The coherence relationship between undeviated and diffracted rays is essential for image formation in all forms of interference microscopy (phase contrast, polarization, differential interference) described in the following chapters.

For a given ray incident at an object, it is now possible to see that a coherence relationship exists between the diffracted and undiffracted (0th-order) beams. This coherence relationship is maintained between the object and the image plane, where waves recombine and interfere. At that site, the myriad coherent wave partners add incoherently with other waves to produce the final image. This important concept underlies many forms of light microscopy.

Exercise: Diffraction by Microscope Specimens

- Determine the periodicity of unknown diffraction gratings using the diffraction equation, $m\lambda = d\sin\theta$. Measure the distances in centimeters between the 1st- and 0th-order spots on a projection screen and the distance between one of the spots and the grating to calculate $\sin\theta$ and the grating spacing, d. It is easiest to perform this measurement using a narrow, monochromatic beam such as that provided by a laser pointer.

- Review Koehler illumination and check the focus of the illuminator of your microscope. Focus on a piece of diffraction grating mounted on a glass microscope slide with a 40× dry objective, and calculate the spacing using the eyepiece reticule and calibration factor determined in the exercise for Chapter 1. For accuracy, measure the number of eyepiece units covering 10 or more spacings on the stage micrometer, repeat the procedure 10 times, and determine the mean value. How well do the two numbers agree?

- Examine and compare the diffraction patterns of various other specimens on the projection screen and estimate their periods (suggestions: microscopic barbules on a bird feather, grooves on a semitransparent CD disk). If the CD is completely opaque, treat it as a reflection grating, tilting the CD at a 45° angle to form a right triangle between the 0th-order spot, the CD, and the light source using the equation for a 90° reflection: $d = \sqrt{2}\lambda/\sin\theta$. Use the laser pointer as a light source.

- Focus the grating in monochromatic green light using a 10× objective with your microscope, stop down (maximally constrict) the condenser aperture diaphragm, and look to see if you can still resolve the grating in normal viewing mode. Examine the diffraction pattern of the grating with an eyepiece telescope under conditions where the image is not resolved, barely resolved, and fully resolved. Do you agree with Abbe's conclusion that a minimum of two adjacent diffraction orders is required for resolution?

- Using the same condenser position, remove the green filter and examine the grating in red and blue light. For each color filter, examine the pattern of diffraction spots with the eyepiece telescope and note the corresponding changes in the spacing between the 1st-order diffraction spots. Do your observations agree with the relationship between spatial resolution and wavelength?

DIFFRACTION AND SPATIAL RESOLUTION

OVERVIEW

In this chapter we examine the role of diffraction in determining spatial resolution and image contrast in the light microscope. In the previous chapter we emphasized that Abbe's theory of image formation in the light microscope is based on three fundamental actions: diffraction of light by the specimen, collection of diffracted rays by the objective, and interference of diffracted and nondiffracted rays in the image plane. The key element in the microscope's imaging system is the objective lens, which determines the precision with which these actions are effected. As an example, examine the remarkable resolution and contrast in the image of the diatom, *Pleurosigma,* made with an apochromatic objective designed by Abbe and introduced by Carl Zeiss over 100 years ago (Fig. 6-1). To understand how such high-resolution images are obtained, we examine an important parameter of the objective, the numerical aperture, the angle over which the objective can collect diffracted rays from the specimen and the key parameter determining spatial resolution. We also investigate the effect of numerical aperture on image contrast. In examining the requirements for optimizing resolution and contrast, we make an unsettling discovery: It is impossible to obtain maximal spatial resolution and optimal contrast using a single microscope setting. A compromise is required that forces us to give up some spatial resolution in return for an acceptable level of contrast.

NUMERICAL APERTURE

Implicit in the Overview is an understanding that the objective aperture must capture some of the diffracted rays from the specimen in order to form an image, and that lenses that can capture light over a wide angle should give better resolution than an objective that collects light over a narrower angle. In the light microscope, the angular aperture is described in terms of the *numerical aperture* (NA) as

$$NA = n \sin\theta,$$

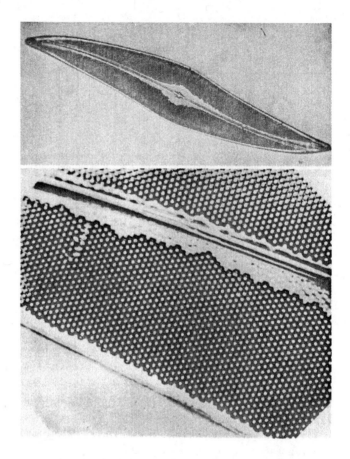

Figure 6-1

Resolution of the pores in a diatom shell with an apochromatic objective lens. Joseph Gall described these historic photographs of the diatom *Pleurosigma* prepared over 100 years ago using one of Abbe's lenses (*Molecular Biology of the Cell,* vol. 4, no. 10, 1993). "The photographs . . . are taken from a Zeiss catalog published in 1888 in which Abbe's apochromatic objectives were advertised. Both figures show the silica shell of the diatom *Pleurosigma angulatum.* Because of the regular patterns of minute holes in their shells, diatoms have long been favorite objects for testing the resolution of microscope objectives. The top figure shows an entire shell at 500×, a magnification beyond which many 19th-century objectives would show little additional detail. The bottom figure, reproduced here and in the original catalog at a remarkable 4900×, was made with an apochromatic oil immersion objective of 2.0 mm focal length and a numerical aperture of 1.3. The center-to-center spacing of the holes in the shell is 0.65 μm, and the diameter of the holes themselves is about 0.40 μm. Almost certainly this objective resolved down to its theoretical limit of 0.26 μm in green light."

where θ is the half angle of the cone of specimen light accepted by the objective lens and n is the refractive index of the medium between the lens and the specimen. For dry lenses used in air, $n = 1$; for oil immersion objectives, $n = 1.515$.

The diffraction angles capable of being accepted by dry and oil immersion objective lenses are compared in Figure 6-2. By increasing the refractive index of the medium between the lens and coverslip, the angle of diffracted rays collected by the objective is increased. Because immersion oil has the same refractive index as the glass coverslip

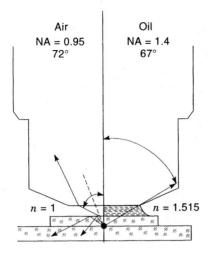

Air
NA = 0.95
72°

Oil
NA = 1.4
67°

$n = 1$ $n = 1.515$

Figure 6-2

Effect of immersion oil on increasing the angular extent over which diffracted rays can be accepted by an objective lens. Numerical aperture is directly dependent on the wavelength λ and the sine of the half angle of the cone of illumination θ accepted by the front lens of the objective. For dry lenses, NA is limited, because rays subtending angles of 41° or greater are lost by total internal reflection and never enter the lens (dotted line). The practical limit for a dry lens is ~39°, which corresponds to an acceptance angle of 72°, and an NA of 0.95. By adding high-refractive index immersion oil matching that of the glass coverslip ($n = 1.515$), an oil immersion objective can collect light diffracted up to 67°, which corresponds to NA = 1.4.

(1.515), refraction of specimen rays at the coverslip-air interface is eliminated, the effective half angle is increased, and resolution is improved. The reader can refer to Pluta (1988) for more details on this important phenomenon.

SPATIAL RESOLUTION

For point objects that are self-luminous (fluorescence microscopy, dark-field microscopy), or for nonluminous points that are examined by bright-field microscopy in transmitted light where the condenser NA is $\geq$ the objective NA, the *resolving power* of the microscope is defined as

$$d = 0.61\lambda/\text{NA},$$

where d is the minimum resolved distance in μm, λ is the wavelength in μm, and NA is the numerical aperture of the objective lens.

In the case of bright-field microscopy, where the condenser NA < objective NA (the condenser aperture is closed down and/or an oil immersion condenser is used in the absence of oil), the resolution is given as

$$d = \frac{1.22\lambda}{\text{condenser NA} + \text{objective NA}}.$$

These equations describe the *Rayleigh criterion* for the resolution of two closely spaced diffraction spots in the image plane. By this criterion, *two adjacent object points are defined as being resolved when the central diffraction spot (Airy disk) of one point coincides with the first diffraction minimum of the other point in the image plane* (Fig. 6-3). The condition of being resolved assumes that the image is projected on the retina or detector with adequate magnification. Recording the real intermediate image on film or viewing the image in the microscope with a typical 10× eyepiece is usually adequate, but detectors for electronic imaging require special consideration (discussed in Chapters 12, 13, and 14). The Rayleigh resolution limit pertains to two luminous points in a dark field or to objects illuminated by incoherent light. For a condenser and objective with NA = 1.3 and using monochromatic light at 546 nm under conditions of oil immersion, the limit of spatial resolution $d = 0.61\lambda/\text{NA} = 0.26$ μm. Numerical apertures are engraved on the lens cap of the condenser and the barrel of the objective lens.

An image of an extended object consists of a pattern of overlapping diffraction spots, the location of every point x,y in the object corresponding to the center of a diffraction spot x,y in the image. Imagine for a moment a specimen consisting of a crowded field of submicroscopic particles (point objects). For a given objective magni-

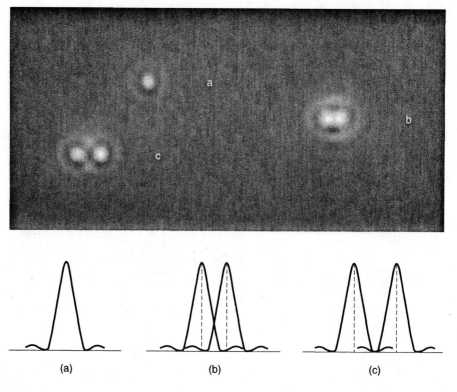

(a)   (b)   (c)

Figure 6-3

Rayleigh criterion for spatial resolution. (a) Profile of a single diffraction pattern: The bright Airy disk and 1st- and 2nd-order diffraction rings are visible. (b) Profile of two disks separated at the Rayleigh limit such that the maximum of a disk overlaps the first minimum of the other disk: The points are now just barely resolved. (c) Profile of two disks at a separation distance such that the maximum of each disk overlaps the second minimum of the other disk: The points are clearly resolved.

fication, if the angular aperture of a microscope is increased, as occurs when opening the condenser diaphragm or when changing the objective for one with the same magnification but a higher NA, the diffraction spots in the image grow smaller and the image is better resolved (Fig. 6-4). Thus, larger aperture angles allow diffracted rays to be included in the objective, permitting resolution of specimen detail that otherwise might not be resolved (Fig. 6-5).

The optical limit of spatial resolution is important for interpreting microscope images. Irregularities in the shapes of particles greater than the limiting size (0.52 µm diameter in the example cited previously) just begin to be resolved, but particles smaller than this limit appear as circular diffraction disks, and, regardless of their true sizes and shapes, always have the same apparent diameter of 0.52 µm. (The apparent variability in the sizes of subresolution particles is due to variations in their intensities, not to variability in the size of their diffraction spots.) Thus, whereas minute organelles and filaments such as microtubules can be *detected* in the light microscope, their apparent diameter (for the lens given previously) is always 0.52 µm, and their true diameters are *not resolved*. It should therefore be apparent that two minute objects whose center-to-center distance is less than 0.26 µm cannot be resolved, but that two objects with physical radii smaller than this size can easily be resolved from each other if they are farther apart than 0.26 µm.

It must be remembered that *adjusting the condenser aperture directly affects spatial resolution in the microscope.* Since a large aperture angle is required for maximum resolution, the front aperture of the condenser must be fully illuminated. Stopping down

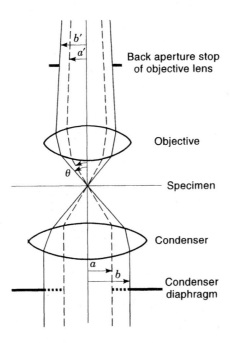

Figure 6-4

Role of the condenser diaphragm in determining the effective numerical aperture. Closing the front aperture diaphragm of the condenser from position *b* to *a* limits the angle θ of the illumination cone reaching the objective, and thus the effective numerical aperture. Notice that the back aperture of the objective is no longer filled at the reduced setting.

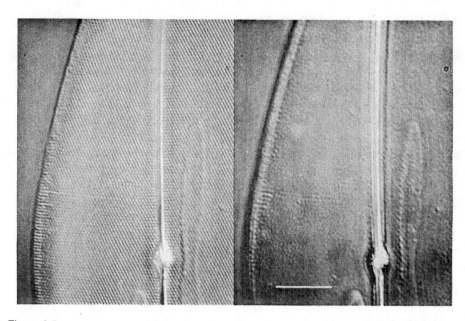

Figure 6-5

Effect of numerical aperture on spatial resolution. The diatom *Pleurosigma* photographed with a 25×, 0.8 NA oil immersion lens using DIC optics. (a) Condenser aperture open, showing the near hexagonal pattern of pores. (b) The same object with the condenser diaphragm closed. The 1st-order diffracted rays from the pores are not captured by the objective with a narrow cone of illumination. Spatial resolution is reduced, and the pores are not resolved. Bar = 10 μm.

the condenser diaphragm limits the number of higher-order diffracted rays that can be included in the objective and reduces resolution. In an extreme case, the condenser aperture may be nearly closed in a mistaken effort to reduce light intensity. Then the half angle of the light cone entering the lens is greatly restricted, and resolution in the image is reduced significantly. The proper way to reduce light intensity is to turn down the voltage supply of the lamp or insert neutral density filters to attenuate the illumination.

DEPTH OF FIELD AND DEPTH OF FOCUS

Just as diffraction and the wave nature of light determine that the image of a point object is a diffraction disk of finite diameter, so do the same laws determine that the disk has a measurable thickness along the z-axis. *Depth of field Z* in the object plane refers to the thickness of the optical section along the z-axis within which objects in the specimen are in focus; *depth of focus* is the thickness of the image plane itself. Our present comments are directed to the depth of field. For diffraction-limited optics, the wave-optical value of Z is given as

$$Z = n\lambda/NA^2,$$

where n is the refractive index of the medium between the lens and the object, λ is the wavelength of light in air, and NA is the numerical aperture of the objective lens. Thus, the larger the aperture angle (the higher the NA), the shallower will be the depth of field. The concept of depth of field is vivid in the minds of all of those who use cameras. Short-focal-length (fast) lenses with small focal ratios ($<f/4$) have a shallow depth of field, whereas the depth of field of long-focal-length (slow) lenses ($>f/16$) is relatively deep. At one extreme is the pinhole camera, which has an infinitely small NA and an infinite depth of field—all objects, both near and far, are simultaneously in focus in such a camera.

The depth of field along the z-axis is determined by several contributing factors, including principles of geometrical and physical optics, lens aberrations, the degree of physiological accommodation by the eye, and overall magnification. These variables and quantitative solutions for each are described in detail by Berek (1927), and are reviewed by Inoué (in Pawley, 1995) and Pluta (1988). Calculated values of the wave optical depth of field for a variety of objective lenses are given in Table 4-1.

The depth of field for a particular objective can be measured quickly and unambiguously using the microscope. A planar periodic specimen such as a diffraction grating is mounted obliquely on a specimen slide by propping up one end of the grating using the edge of a coverslip of known thickness. When the grating is examined in the microscope, it will be seen that only a narrow zone of grating will be in focus at any particular setting of the specimen focus dial. The depth of field z is then calculated from the width w of the focused zone (obtained photographically) and the angle of tilt α of the grating through the relationship

$$Z = nw \tan\alpha,$$

where n is the refractive index of the medium surrounding the grating.

OPTIMIZING THE MICROSCOPE IMAGE: A COMPROMISE BETWEEN SPATIAL RESOLUTION AND CONTRAST

Putting these principles into practice, let us examine two specimens using transmitted light illumination and bright-field microscope optics: a totally opaque object such as a copper mesh electron microscope grid and a stained histological specimen. These objects are called *amplitude objects,* because light obscuring regions in the object appear as low-intensity, high-contrast regions when compared to the bright background in the object image. In comparison, transparent colorless objects such as living cells are nearly invisible, because the amplitude differences in the image are generally too small to reach the critical level of contrast required for visual detection. We discuss methods for visualizing this important class of transparent objects in Chapter 7.

With the microscope adjusted for Koehler illumination using white light, begin by opening the condenser front aperture to match the diameter of the back aperture of the objective lens to obtain maximum aperture angle and therefore maximal spatial resolution. This operation is performed using an eyepiece telescope or Bertrand lens while inspecting the back focal plane of the objective lens. Since this plane is conjugate with the condenser aperture, the edges of the condenser diaphragm appear when the diaphragm is sufficiently stopped down. With the telescope lens focused on its edges, open the diaphragm until its margins include the full aperture of the objective to obtain the maximum possible aperture angle. Expecting maximum resolution in normal viewing mode, we are surprised to see that the image of the opaque grid bars is gray and not black, and that the overall contrast is reduced. Similarly, fine details in the histological specimen such as collagen bundles, cell organelles, and the edges of membranes and nuclei have low contrast and are difficult to distinguish. The overall impression is that the image looks milky and washed out—in short, unacceptable. The poor definition at this setting of the condenser diaphragm is due to polychromatic illumination, scattered light, and reduction in the degree of coherence of the contributing waves. However, considerable improvements in the quality of the image can be made by using a bandpass filter to restrict illumination to a limited range of wavelengths and by partially closing the condenser diaphragm.

Monochromatic light assures that chromatic aberration is eliminated and that unit diffraction spots in the image are all of uniform size. With white light, diffraction spot size varies by nearly a factor of 2, due to the presence of different color wavelengths, with the result that diffraction features in the image are somewhat blurred. Monochromatic illumination sharpens the image and increases contrast, particularly for objects with inherently low contrast. For color objects such as histological specimens that are to be examined visually or recorded with a gray-scale camera or black-and-white film, contrast can be improved dramatically by selecting filters with complementary colors: a green filter for pink eosin dye or a yellow filter for blue hematoxylin stain. A green filter, for example, removes all of the pink eosin signal, reducing the amplitude of eosin-stained structures and making them look dark against a bright background in a gray-scale image.

To our surprise, closing down the condenser diaphragm also has a pronounced effect: It increases the contrast and greatly improves the visibility of the scene (Fig. 6-6). The grid bars now look black and certain features are more readily apparent. There are several reasons why this happens: (1) Part of the improvement comes from *reducing the amount of stray light* that becomes reflected and scattered at the periphery of the lens. (2) Reducing the aperture *increases the coherence of light*; by selecting a smaller portion of the light source used for illumination, the phase relationships among diffracted rays are more defined, and interference in the image plane results in higher-amplitude differences, thus increasing contrast. (3) With reduced angular aperture, the unit diffraction spots comprising the image become larger, causing lines and edges to become thicker and cover a greater number of photoreceptor cells on the retina, thus making the demarcations appear darker and easier to see. Thus, the benefits of improved contrast and visibility might lead us to select a slightly stopped-down condenser aperture, even though spatial resolution has been compromised slightly in the process. If the condenser aperture is closed down too far, however, the image loses significant spatial resolution and the dark diffraction edges around objects become objectionable.

Thus, the principles of image formation must be understood if the microscope is to be used properly. A wide aperture allows maximal spatial resolution, but decreases contrast, while a smaller, constricted aperture improves visibility and contrast, but decreases spatial resolution. For all specimens, the ideal aperture location defines a balance between resolution and contrast. A useful guideline for beginners is to stop down the condenser aperture to about 70% of the maximum aperture diameter, but this is not a rigid rule. If we view specimens such as a diffraction grating, a diatom, or a section of striated muscle, stopping down the diaphragm to improve contrast might suddenly obliterate periodic specimen details, because the angular aperture is too small to allow diffracted light to be collected and focused in the image plane. In such a situation, the aperture should be reopened to whatever position gives adequate resolution of specimen detail and acceptable overall image visibility and contrast.

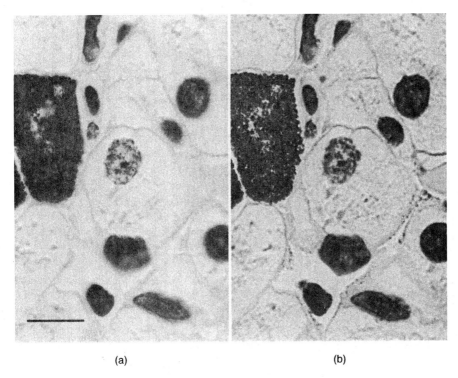

(a)   (b)

Figure 6-6

Effect of the condenser aperture on image contrast and resolution. (a) With unrestricted
aperture, resolution is maximal, but contrast suffers from stray light. (b) With the condenser
aperture stopped down, light fills ~70% of the diameter of the back aperture of the objective.
Contrast is improved, but resolution is reduced. Bright-field light micrograph of a
hematoxylin-stained section of *Amphiuma* liver containing hepatocytes and pigment cells.
Bar = 10 μm.

Exercise: Resolution of Striae in Diatoms

In this exercise we will use a diatom test plate to examine the effect of numerical
aperture and wavelength on resolution in the light microscope. Diatoms are unicellular algae that produce shells of silica containing arrays of closely spaced
pores. They have been used to determine the spatial resolution of objective lenses
in microscopes for well over a century.

Diatoms belong to the class Chrysophyta of yellow-green and golden-brown
algae. Their transparent quartz cell walls (valves or frustrules) are usually composed of two overlapping halves (valves) that contain semicrystalline and amorphous silica, which gives them a hard, brittle texture. The two overlapping valves
are ornamented with tiny dots or perforations (pores) that are organized into rows
(striae), both of which are commonly used to calibrate the magnification of microscopes. The ridge and pore spacings are constant for a given species. Figure 6-7
shows a diatom test plate obtained from Carolina Biological Supply Company,

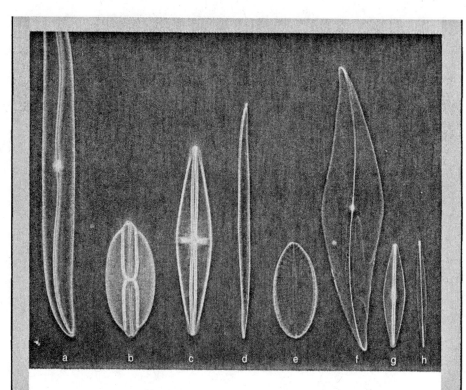

Specimen		Length (μm)	Period (μm/stria)
a	*Gyrosigma balticum*	280	0.67
b	*Navicula lyra*	160	1.25
c	*Stauroneis phenocenteron*	150	0.71
d	*Nitzschia sigma*	200	0.43
e	*Surirella gemma*	100	0.50
f	*Pleurosigma angulatum*	150	0.53
g	*Frustulia rhomboides*	50	0.29
h	*Amphipleura pellucida*	80-140	0.27

Figure 6-7

Diatom test plate and key.

Burlington, North Carolina. The accompanying table indicates the cell size and interstria spacing for the eight species of diatoms on the plate. While performing this exercise, compare the image of *Pleurosigma* produced with a modern lens with that produced by one of the first apochromatic lenses designed by Abbe (Fig. 6-1). If a test plate is not available, use a grain of dispersed diatomaceous earth (from a chemical supply house), or, failing that, some household scrubbing cleaner, which contains diatom shells as an abrasive.

1. Review the steps for adjusting the microscope for Koehler illumination. Identify and locate the positions of the four aperture planes and the four

field planes, and prepare a list of each set in order, beginning with the lamp.

2. Adjust the condenser for bright-field mode. Image the diatom test plate using green light (546 nm bandpass filter) and a 40× objective. Focusing is difficult due to the extremely small size and transparent nature of the specimen. First locate the diatoms under low power (10×). Carefully move the x and y coordinates of the stage until the specimen comes into the field of view. Then swing in the 40× lens and refocus on the diatoms. Compare the image of the diatoms with Figure 6-7 and note the indicated spacings. Using bright-field optics and green light, open the condenser aperture to the proper location using the telescope lens, and note the species with the smallest spacing that it is possible to resolve with your microscope. Indicate the species and note the distance between the striae from the figure.

3. Calculate the theoretical resolution of the microscope under these conditions. The NA is indicated on the barrel of the lens. Show your work. The apparent and calculated resolution limits should roughly agree.

4. Now examine the diatom *Pleurosigma* with a dry 100× objective and close down the field-stop diaphragm to illuminate just this one diatom and no other. Make an accurate sketch of the hexagonal arrangement of the pores. Close down the condenser diaphragm to its minimum size. Now examine its diffraction pattern in the diffraction plane of the microscope using the telescope lens. Make an accurate sketch of the diffraction pattern.

5. How does the diffraction pattern relate to the spacing pattern of the diatom pores in the image plane?

6. Examine the diatoms in red and blue light and with the condenser aperture open or maximally closed. Which pair of conditions gives the lowest and the best resolution of the striae?

7. Examine the diatom *Pleurosigma* or *Diatoma* with a 100× oil immersion lens. Can the striae and pores be resolved? Examine the diffraction plane under these conditions. Make sketches of both views. Now oil the condenser as well as the objective and repeat your observations. Make sketches. Do you notice a difference in resolution?

PHASE CONTRAST MICROSCOPY AND DARK-FIELD MICROSCOPY

OVERVIEW

Unstained objects such as cells present a unique problem for the light microscopist because their images generate very little contrast and are essentially invisible in ordinary bright-field microscopy. As we have seen, this is even true for transparent periodic specimens such as diffraction gratings and diatoms. Although transparent objects induce phase shifts to interacting beams of light due to scattering and diffraction, they remain nearly invisible, because the eye cannot detect differences in phase. In this chapter we examine two optical methods for viewing such objects: phase contrast microscopy, which transforms differences in the relative phase of object waves to amplitude differences in the image; and dark-field microscopy, where image formation is based solely on diffracted wave components. Phase contrast microscopy produces high-contrast images of transparent specimens such as cells and micro-organisms, tissue slices, lithographic patterns, and particles such as organelles. Living cells in tissue culture can also be examined directly, without fixation and staining (Fig. 7-1).

PHASE CONTRAST MICROSCOPY

In the case of stained, histological preparations or specimens with naturally occurring pigments, specific wavelengths are absorbed by dyes or pigments, allowing objects to appear in color when illuminated with white light. With monochromatic illumination using a color filter complementary to the color of the specimen—for example, a blue object examined through a yellow filter—object rays are significantly reduced in amplitude, resulting in a high-contrast image. Such objects are called *amplitude objects* because they directly produce amplitude differences in the image that are detected by the eye as differences in the intensity (Fig. 7-2). Although most transparent biological specimens do not absorb light, they do diffract light and cause a phase shift in the rays of light passing through them; thus, they are called *phase objects* (Fig. 7-2). The retardation imparted to a plane-wave front is shown in Figure 7-3. Phase contrast microscopes feature an optical design that transforms differences in the phase of object-diffracted waves to differences in the image, making objects appear as if they had been optically stained.

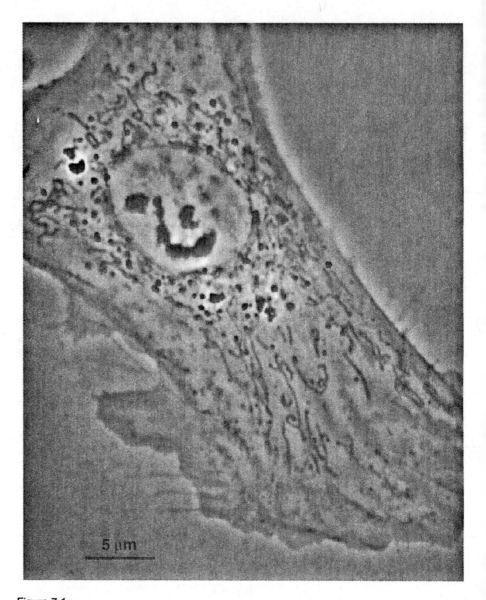

Figure 7-1

Phase contrast image of a living tissue culture cell. The cell nucleus, organelles, and membrane edges are clearly resolved in this normal rat kidney (NRK) cell. Phase-dense objects include mitochondria, lysosomes, and nucleoli, and domains of nuclear chromatin. Phase-light objects represent lipid droplets and small vesicles. Bar = 5 μm. (Specimen courtesy of Rodrigo Bustos, Johns Hopkins University.)

Because the method is dependent on diffraction and scattering, phase contrast optics also differentially enhance the visibility of the light scattering edges of extended objects and particles. The performance of modern phase contrast microscopes is remarkable. Under favorable conditions and with electronic enhancement and image processing, objects containing just a few protein molecules can be detected.

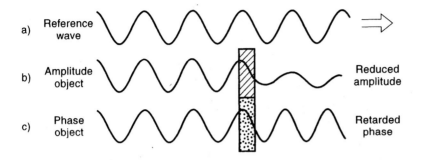

Figure 7-2

Effects of amplitude and phase objects on the waveform of light. (a) Reference ray with characteristic amplitude, wavelength, and phase. (b) A pure amplitude object absorbs energy and reduces the amplitude, but does not alter the phase, of an emergent ray. (c) A pure phase object alters velocity and shifts the phase, but not the amplitude, of an emergent ray.

In the 1930s, Frits Zernike, a Dutch physicist at the University of Groningen, created an optical design that could *transform differences in phase to differences in amplitude.* The development of phase contrast optics is a brilliant example of how basic research in theoretical optics led to a practical solution for viewing unstained transparent objects in the light microscope. The Zeiss optical works in Jena introduced phase contrast objectives and accessories in 1942, which transformed research in biology and medicine. For his invention and theory of image formation, Zernike won the Nobel prize in physics in 1953 (Fig. 7-4). Excellent descriptions of the technique are found in Bennett et al. (1951), Françon (1961), Slayter (1976), Slayter and Slayter (1992), and Pluta (1989).

In this chapter we first examine the process of image formation using the terminology and diagrams that are commonly employed to explain phase contrast optics. Then we examine the relative contributions of diffraction and differences in optical path length in generating the phase contrast image.

The Behavior of Waves from Phase Objects in Bright-Field Microscopy

WAVE TERMINOLOGY AND THE IMPORTANCE OF COHERENCE
Upon transit through a phase object, an incident wavefront of an illuminating beam becomes divided into two components: (1) an undeviated (0th-order) wave or *surround wave* (S wave) that passes through the specimen, but does not interact with it, and (2) a

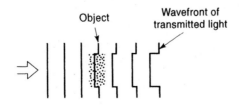

Figure 7-3

Disturbance by a phase object to an incident planar wavefront.

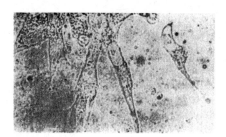 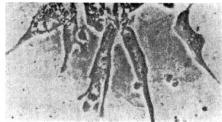

Figure 7-4

"How I Discovered Phase Contrast." Modified excerpts from Dr. Zernike's Nobel prize address delivered in 1953 in Stockholm, Sweden, and published in the March 11, 1955, issue of *Science* (Zernike, 1955). Top: Dr. Zernike in his laboratory, November 1953. Bottom: Living tissue culture cells as seen with bright field (left) and phase contrast (right).

Phase contrast was not discovered while I was working with a microscope, but originated in my interest in diffraction gratings. About 1930 our laboratory obtained a large concave grating ruled by Robert Wood at Johns Hopkins University in Baltimore. Periodic errors in the grating lines made by ruling machines at that time caused the grating to exhibit a strongly striped surface, but when the grating was examined with a telescope at some 6 m distance and was exactly focused on the surface of the grating the stripes disappeared! By a succession of experiments and calculations I soon succeeded in explaining this. In a simpler case, a telescope was used to examine the phases of lines in a diffraction pattern of a vertical line-source of light after placing a 2 mm wide slit close behind the objective of the telescope. The diffraction maxima were observed but their phases could not be distinguished. However, the phases could be observed by throwing the diffraction image on a *coherent background* that served as a reference surface. Now I happened to know of a simple method Lord Rayleigh described in 1900 for making what I called *phase strips*—glass plates with a straight groove 1 mm wide and etched with acid to a uniform depth of half a wavelength. When a phase plate was placed in the spectrum of the faulty grating and examined with the telescope, the strips on the grating surface now stood out clearly.

deviated or *diffracted wave* (D wave) that becomes scattered in many directions. Typically, only a minority of incident waves are diffracted by cellular objects. Both S and D waves are collected by the objective lens and focused in the image plane at the site corresponding to the image of the particle, where they undergo interference and generate a resultant *particle wave* (P wave). The relationship among waves is thus described as P = S + D. Detection of the object image depends on the intensities, and hence on the amplitudes, of the P and S waves. *Only when the amplitudes of the P and S waves are significantly different in the image plane can we see the object in the microscope.* Before beginning our explanation of the interference mechanism, we should note our earlier discussion of the coherence of light waves in the myriad small beams (wave bundles) illuminating the specimen (Chapter 5). This condition is of great practical importance for phase contrast microscopy, because image formation through constructive and destructive interference requires coherent illumination such that: (1) A definite phase relationship exists between the S and D waves, and (2) the phase relationship must be preserved between the object and the image.

DEPICTION OF WAVE INTERACTIONS WITH SINE WAVE AND VECTOR DIAGRAMS

With this general scheme in mind, let us now examine Figure 7-5a, which shows the S, D, and P waves as sine waves of a given wavelength in the region of the object in the image plane. The S and P waves, whose relative intensities determine the visual contrast, are shown as solid lines, whereas the D wave (never directly observed) is shown as a dashed line. The amplitude of each wave represents the sum of the E vectors of the component waves. The D wave is lower in amplitude than the S wave, because there are fewer D-wave photons than there are S-wave photons at the image point. Notice that the *D wave is retarded in phase by λ/4 relative to the S wave* due to its interaction with the object particle. The P wave resulting from interference between the D and S waves is retarded relative to the S wave by only a small amount (~λ/20) and has an amplitude similar to that of the S wave. *Since the S and P waves have close to the same amplitude*

For a physicist interested in optics it was not a great step to change over from this subject to the microscope. Remember that in Ernst Abbe's remarkable theory of the microscope image the transparent object under the microscope is compared with a grating. Abbe examined gratings made of alternating opaque and transparent strips (amplitude gratings). Gratings made of transparent alternating thick and thin strips (phase gratings) produce diffraction spots that show a phase difference of 90°. For a phase object, my phase strip in the focal plane of the microscope objective brought the direct image of the light source into phase with the diffracted images, making the whole comparable to the images caused by an amplitude object. Therefore the image in the eyepiece appears as that of an absorbing object—that is, with black and white contrast, just as if the object has been stained.

On looking back on these events I am impressed by the great limitations of the human mind. How quick we are to learn—that is, to imitate what others have done or thought before—and how slow to understand—that is, to see the deeper connections. Slowest of all, however, are we in inventing new connections or even in applying old ideas in a new field. In my case, the really new point was the diffraction pattern of lines of the grating artifacts, the fact that they differed in phase from the principal line, and that visualization of phases required projection of the diffraction image on a coherent background. The full name of the new method of microscopy might be something like "phase-strip method for observing phase objects in good contrast." I shortened this to "phase contrast."

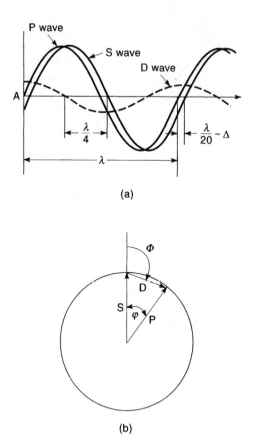

Figure 7-5

Phase relations between S, D, and P waves in bright-field microscopy. S and D waves, generated at the object, recombine through interference to generate the resultant particle image wave (P) in the image plane of the microscope (P = S + D). Relative to S, the D wave has a lower amplitude and is retarded in phase by ~λ/4. The slight phase shift of λ/20 in the resultant P wave is related to the optical path length difference and is typical for small object details in a cell. Since the amplitudes of the S and P waves are the same, the contrast is 0, and the object remains invisible against the background. (a) Wave components shown as sine waves. In this chapter, sine waves represent the amplitude and phase of a whole population of waves, not single photons. Thus, the lower amplitude of the D wave does not mean energy was absorbed, but rather that there are fewer D waves than there are S waves. (b) Vector diagram of S, P, and D waves. For explanation, see text.

in simple light microscopy, there is no contrast in the image and the object remains invisible.

The same situation can be described vectorially using a system of polar coordinates, where the length of a vector represents the amplitude of a wave, and the angle of rotation of the vector relative to a fixed reference (angular phase shift ϕ) represents the amount of phase displacement (Fig. 7-5b). Phase retardations are shown as clockwise rotations, whereas phase advancements are depicted as counterclockwise rotations. In this plot (technically a phasor diagram), the reconstructed P wave is shown as the vector sum of the S and D waves. This form of presentation is convenient because you can see more clearly how different degrees of phase shift in the D wave affect the phase of the resultant P wave, and vice versa. A complete description of this form of wave analysis is given by Hecht (1998), Pluta (1989), and Slayter (1976), but for our purposes, a brief definition will be sufficient to explain the diagram. The phase shift of D relative to S on the graph is shown as Φ, where $\Phi = \pm 90° + \phi/2$, and ϕ is the relative phase shift (related to the optical path difference) between the S and P vectors. For objects with negligible optical path differences (phase shifts ϕ), Φ is $\pm 90°$. As shown in the figure, a D wave of low amplitude and small phase shift results in a P wave with an amplitude that is nearly equal to that of the S wave. With similar amplitudes for S and P, there is no contrast, and the object remains invisible.

The Role of Differences in Optical Path Lengths

We encountered the concept of optical path length previously when we discussed the action of a lens in preserving the constancy of optical path length between object and image for coherent waves emerging from an object and passing through different regions of the lens (Fig. 5-6). For phase contrast microscopy, we are concerned with the role of the object in altering the optical path length (relative phase shift ϕ) of waves passing through a phase object.

Since the velocity of light in a medium is $v = c/n$, where c is the speed of light in a vacuum, rays of light passing through a phase object with thickness t and refractive index n greater than the surrounding medium travel slower through the object and emerge from it retarded in phase relative to the background rays. The difference in the location of an emergent wavefront between object and background is called the *phase shift* δ (same as ϕ above), where δ in radians is

$$\delta = 2\pi\Delta/\lambda,$$

and Δ is the *optical path difference*, which was defined in Chapter 5 as

$$\Delta = (n_2 - n_1)t.$$

The Optical Design of the Phase Contrast Microscope

The key element of the optical design is to (1) isolate the surround and diffracted rays emerging from the specimen so that they occupy different locations in the diffraction plane at the back aperture of the objective lens, and (2) advance the phase and reduce the amplitude of the surround light, in order to maximize differences in amplitude between the object and background in the image plane. As we will see, the mechanism for generating relative phase retardation is a two-step process: D waves are retarded in phase by $\sim\lambda/4$ at the object, while S waves are advanced in phase by a phase plate positioned in or near the diffraction plane in the back aperture of the objective lens. Two special pieces of equipment are required: a condenser annulus and an objective lens bearing a phase plate for phase contrast optics.

The *condenser annulus*, an opaque black plate with a transparent annulus, is positioned in the front aperture of the condenser so that the specimen is illuminated by beams of light emanating from a ring (Fig. 7-6). (In some texts, the illuminating beam emergent from the condenser is described as a hollow cone of light with a dark center—

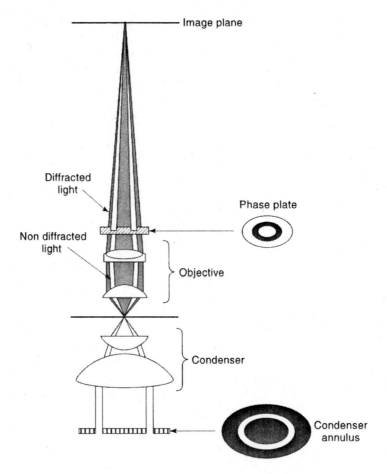

Figure 7-6

Path of nondiffracted and diffracted beams in a phase contrast microscope. An annular aperture in the front focal plane of the condenser generates a hollow cone of light that illuminates the specimen and continues (approximately) as an inverted cone that is intercepted by a phase plate at the back aperture of the objective lens. The image of the annulus is in sharp focus in this plane because it is conjugate to the front aperture plane of the condenser. Diffracted specimen rays fill the shaded region of the illumination path.

a concept that is useful but not strictly true.) The condenser annulus replaces the variable diaphragm in the front aperture of the condenser. Under conditions of Koehler illumination, S waves that do not interact with the specimen are focused as a bright ring in the back focal plane of the objective (the diffraction plane). Remember that under these conditions the objective's back focal plane is conjugate to the condenser's front aperture plane, so nondiffracted (0th-order) waves form a bright image of the condenser annulus at the back aperture of the objective. Light that is diffracted by the specimen (D waves) traverses the diffraction plane at various locations across the entire back aperture, the amount and location depending on the number, size, and refractive index differential of light-scattering objects in the specimen. Since the direct (0th-order light) and diffracted light become spatially separated in the diffraction plane, you can selectively manipulate the phase of either the S- or D-wave components.

To differentially alter the phase and amplitude of the direct (undeviated) light, a *phase plate* is mounted in or near the back focal plane of the objective (Figs. 7-6 and 7-7). In some phase contrast objectives, the phase plate is a plate of glass with an etched ring of reduced thickness to selectively advance the phase of the S wave by $\lambda/4$. The same ring is coated with a partially absorbing metal film to reduce the amplitude of the light by 70–75%. In other lenses the same effect is accomplished by acid etching a lens surface that is in or near the back focal plane of the objective lens. Regardless of the method, it is important to remember that phase contrast objectives are always modified in this way and thus are different from other microscope objectives.

The optical scheme for producing positive and negative phase contrast images is given in Figure 7-8. As discussed in the preceding section, the D wave emergent from the object plane is retarded by $\lambda/4$ relative to the phase of the S wave. In positive phase contrast optics (left side of the diagram), the S wave is advanced in phase by $\lambda/4$ at the phase plate, giving a net phase shift of $\lambda/2$, which now allows destructive interference with D waves in the image plane. Generally, the manipulation of relative phase advancement, while essential to phase contrast optics, is still unable to generate a high-contrast image, because the amplitude of the S wave is too high to allow sufficient contrast. For this reason, the ring in the phase plate is darkened with a semitransparent metallic coating to reduce the amplitude of the S wave by about 70%. Since $P = S + D$, interference in the image plane generates a P wave with an amplitude that is now considerably less than that of S. Thus, the difference in phase induced by the specimen is transformed into a difference in amplitude (intensity). Since the eye interprets differences in intensity as contrast ($C = \Delta I/I_b$), we now see the object in the microscope. (See Chapter 2 for discussion of formula.) *Positive phase contrast* systems like the one just described *differentially advance the phase of the S wave relative to that of the D wave.* Cellular objects having a higher refractive index than the surrounding medium are dark in appearance, whereas objects having a lower refractive index than the surrounding medium appear bright.

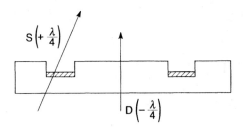

Figure 7-7

The action of a phase plate at the rear surface of the objective lens. Surround or background rays (S) are advanced in phase relative to the D wave by $\lambda/4$ at the phase plate. Relative phase advancement is created by etching a ring in the plate that reduces the physical path taken by the S waves through the high-refractive-index plate. Since diffracted object rays (D) are retarded by $\lambda/4$ at the specimen, the optical path difference between D and S waves upon emergence from the phase plate is $\lambda/2$, allowing destructive interference in the image plane. The recessed ring in the phase plate is made semitransparent so that the amplitude of the S wave is reduced by 70–75% to optimize contrast in the image plane.

It is also possible to produce optics giving *negative phase contrast,* where the S wave is retarded relative to the D wave, causing high-refractive-index objects to appear bright against a gray background. In this case, the phase plate contains an elevated ring that retards the phase of the 0th-order S wave relative to the phase of the D wave. The effect of this action in generating negative phase contrast is shown on the right-hand side of Figure 7-8.

Alignment

To form a phase contrast image, the rings of the annulus and phase plate must have matching diameters and be perfectly aligned. A multiple-position condenser with a rotating turret may contain two or three annuli intended for use with different phase contrast objectives. Small annuli are used for low-power dry objectives, whereas large annuli are employed with high-power, oil immersion lenses. The nomenclature used by different microscope companies varies, but the proper selection can always be made by matching the designation on the edge of the turret with the corresponding designation on the barrel of the objective lens. Whenever a lens is changed and a new annulus is brought into position, it is important to inspect the objective back aperture to make sure the annulus and the phase plate match and that they are aligned. Since the objective lens is usually fixed, alignment is performed by moving the condenser annulus with special annulus positioning screws on the condenser. The annulus adjustment screws, not to be confused with the condenser centration screws, are either permanently mounted on the condenser turret or come as separate tools that must be inserted into the condenser for this purpose. After bringing the rings into sharp focus with the telescope focus, move the bright image of the annulus to exactly coincide with the dark ring on the phase plate (Fig. 7-9). Improper alignment gives a bright, low-contrast image, because the bright background rays are not properly attenuated or advanced in phase as required by phase contrast theory.

Interpreting the Phase Contrast Image

Phase contrast images are easiest to interpret when the cells are thin and spread out on the substrate. When such specimens are examined in positive contrast mode, the conventional mode of viewing, objects with a higher refractive index than the surrounding medium appear dark. Most notably, phase contrast optics differentially enhance the contrast of the edges of extended objects such as cell margins. Generally, positive phase contrast optics give high-contrast images that we interpret as density maps. As an approximation, this interpretation is usually correct, because the amplitude and intensity in an object image are related to refractive index, and optical path length. Thus, a series of objects of increasing density (such as cytoplasm, nucleus, and nucleolus) are typically seen as progressively darker objects. However, the size and orientation of asymmetric objects also affect intensity and contrast. Further, there are optical artifacts we need to recognize that are present in every phase contrast image.

Interpreting phase contrast images requires care. In positive phase contrast optics, cell organelles having a lower refractive index than the surrounding cytoplasm generally appear bright against a gray background. Examples include pinocytotic vesicles, lipid droplets, and vacuoles in plant cells and protozoa. For objects that introduce relatively

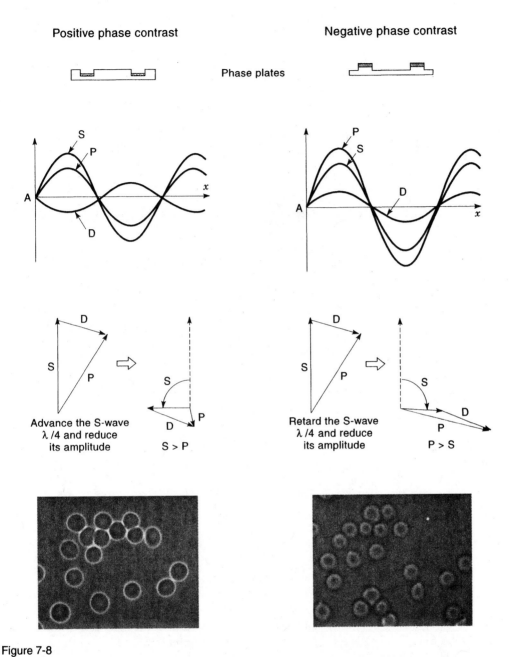

Figure 7-8
Comparison of positive and negative phase contrast systems. Shown in pairs, from the top down: phase plates for advancing (positive contrast) or retarding (negative contrast) the surround wave; amplitude profiles of waves showing destructive interference (positive phase contrast) and constructive interference (negative phase contrast) for a high-refractive-index object. Notice that the phase plate advances or retards the S wave relative to the D wave. The amplitude of the resultant P wave is lower or higher than the S wave, causing the object to look relatively darker or brighter than the background. Vector diagrams showing advancement of the S wave by λ/4, which is shown as a 90° counterclockwise rotation in positive phase contrast, and retardation of the S wave by λ/4, which is shown as a 90° clockwise rotation in negative phase contrast. Addition of the S and D wave vectors gives P waves whose amplitudes vary relative to the S waves. Images of erythrocytes in positive and negative phase contrast optics.

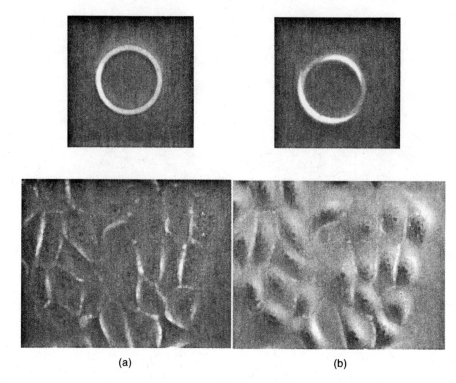

(a)   (b)

Figure 7-9

Alignment of condenser and objective annuli. An eyepiece telescope or Bertrand lens is used to examine the back aperture of the objective lens. (a) The dark ring of the phase plate must be perfectly centered with the bright ring of light from the condenser annulus. The adjustment is made using two condenser plate-centering screws. These screws are distinct from the condenser centration screws, which are used to center the condenser lens with respect to the optic axis of the microscope. (b) Notice the low-contrast shaded image resulting from a misaligned annulus.

large phase retardations (phase shift Φ of the diffracted wave $\sim\lambda/2$), interference becomes constructive, making the objects appear brighter than the background.

To avoid confusion regarding bright and dark contrast in phase contrast images, it is useful to reconsider the term *optical path difference,* which is the product of refractive index and object thickness, and is related to the relative phase shift between object and background waves. It is common to hear microscopists refer to high- and low-refractive-index objects in a phase contrast image, but this is technically incorrect unless they know that the objects being compared have the same thickness. Thus, a small object with a high refractive index and a large object with a lower refractive index can show the same optical path difference and yet appear to the eye to have the same intensity (Fig. 7-10). In particular, conditions that cause shrinking or swelling of cells or organelles can result in major differences in contrast. Likewise, replacement of the external medium with one having a different refractive index can result in changes in image contrast.

Finally, phase contrast images show characteristic patterns of contrast—halos and shade-off—in which the observed intensity does not correspond directly to the optical path difference of the object. These patterns are sometimes referred to as phase artifacts or distortions, but should be recognized as a natural result of the optical system. *Phase halos* always surround phase objects and may be dark or light depending on whether the optical

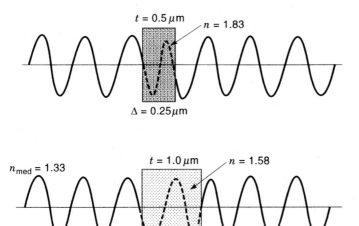

Figure 7-10

Effects of refractive index and specimen thickness on the optical path length. The phase contrast image reveals differences in optical path length as differences in light intensity, thus providing contrast. Since optical path length difference Δ is defined as the product of thickness t and refractive index n difference such that $\Delta = (n_1 - n_2)t$, two objects that vary both in size and refractive index can have the same optical path length and the same intensity in the phase contrast microscope.

path through an object is greater or less than that of the medium. (For examples, see Figures 7-1 and 7-8.) For objects that appear dark, the phase halo is light, and vice versa. Halos occur because the ring in the phase plate in the objective back aperture also receives some diffracted light from the specimen—a problem accentuated by the fact that the width of the annulus generated by the 0th-order surround waves is smaller than the actual width of the annulus of the phase plate. Due to requirements of optical design, the difference in width is usually about 25%. Since diffracted rays corresponding to low spatial frequencies pass through the annulus on the plate, they remain 90° out of phase relative to the 0th-order light. The absence of destructive interference by these low-spatial-frequency diffracted waves causes a localized contrast reversal—that is, a halo—around the object. Halos are especially prominent around large, low-spatial-frequency objects such as nuclei and cells. Another contributing factor is the redistribution of light energy that occurs during interference in the image plane. As in the case of the diffraction grating and interference filter, interference results in a redistribution of light energy, from regions where it is destructive to regions where it is constructive. High contrast halos can be objectionable for objects generating large optical path differences such as erythrocytes, yeast cells, and bacteria. In many cases it is possible to reduce the amount of phase shift and diffraction and therefore the amount of halo, by increasing the refractive index of the medium using supplements such as glycerol, mannitol, dextran, or serum albumin. As will be seen in the next exercise, changing the refractive index of the medium can even reverse image contrast, turning phase-dark objects into phase-bright ones.

Shade-off is another optical effect that is particularly apparent in images of extended phase objects. In a phase contrast mechanism based solely on optical path differences (one that does not consider the effects of diffraction), you might expect that the image of a large phase object of constant optical path length across its diameter would appear uniformly dark or light, but this is not the case. As shown schematically in Figure 7-11, the intensity profile of a phase-dark object gradually increases toward the center of the object. If the object is large enough, the light intensity in the central regions approaches that of the surrounding background. Shade-off is frequently observed on large, extended objects such as extended or flattened cells (Fig. 7-1), flattened nuclei, and planar slabs of materials, for example, mica or glass. The phenomenon is also known as the *zone-of-action effect,* because central uniform zones and refractile edge zones of an object diffract light differently. In central regions of an object, the amount of diffraction and the angle of scattering are greatly reduced. Object rays, although retarded in phase, deviate only slightly from the 0th-order component, and fall within the annulus of the phase plate. As a result, the amplitude and intensity of the central region are essentially the same as the background. The presence of shade-off in extended objects and the high image contrast at edges remind us that the phase contrast mechanism is principally one of diffraction and scattering.

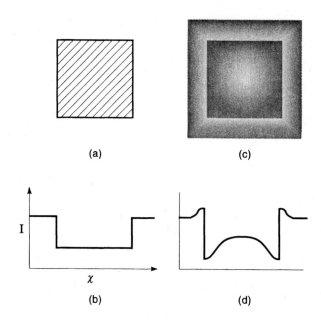

(a)   (c)

(b)   (d)

Figure 7-11
The effects of halo and shade-off in the phase contrast image. (a) Phase object. (b) Intensity profile of object shown in (a). (c) Phase object showing phase halo and shade-off.
(d) Intensity profile of (c).

Exercise: Determination of the Intracellular Concentration of Hemoglobin in Erythrocytes by Phase Immersion Refractometry

In this exercise you will practice the proper alignment of phase contrast optics, examine blood cells in positive and negative contrast, and, using a series of solutions of varying refractive index, calculate the concentration of hemoglobin in your erythrocytes.

The phase shift of light, δ, which occurs when light passes through a cell with refractive index n_o in a medium with refractive index n_m, is given by $\delta = (n_o - n_m)\ t/\lambda$, where t is the thickness of the object and λ is the wavelength. The phase contrast microscope converts the phase shift into an observable change in amplitude. When $n_o > n_m$ (the case for most cellular structures), objects appear dark against a gray background (positive contrast). When $n_m > n_o$, objects look bright against a gray background (negative contrast). When $n_o = n_m$, no relative retardation occurs and the object is invisible. If blood cells are placed in a series of albumin solutions of increasing concentration (made isotonic by adjusting the concentration of NaCl to prevent shrinking and swelling due to osmotic effects), positive and negative cells can be counted under the phase contrast microscope and a curve can be constructed, and the isotonic point can be determined from the point at which 50% dark and bright cells are observed. This is a sensitive null method that can be used to obtain the concentration of solids in cells. In this case, we will calculate the intracellular molarity of hemoglobin in erythrocytes. *Note:* For erythrocytes, the intracellular concentration of hemoglobin is so high that cells look bright against a gray background when examined in normal isotonic saline. In the following exercise on erythrocytes, positive cells appear bright, and negative cells look dark.

1. Swab the tip of your finger with 70% ethanol and prick it with a sterile disposable lancet. Place a small drop of *fresh* blood (5 µL) on a microscope slide. Immediately place one drop (~60 µL) of albumin test solution on the droplet of blood, cover with a coverslip, and count the number of positive and negative cells—100 cells total—for each sample. *Count only single cells that are seen face on. Do not count cell aggregates.* If necessary, you can check for "invisible" cells by turning away the condenser annulus, but perfect refractive index matching occurs only rarely, because the biconcave shape of the cells results in optical path differences through a single cell. Midpoint cells will appear both black and white simultaneously. It is recommended that you prepare and score one slide at a time. To recognize typical positive and negative cells, you should test the extreme albumin concentrations first. You should place your slides in a moist chamber (a sealed container with a moist paper towel) so that they do not dry out while you are busy counting other slides. Calculate the number of negative cells for each solution. Show your work.

2. Plot the % negative cells vs. mg/mL albumin on a piece of graph paper, and determine the concentration of albumin giving 50% positive and negative cells.

3. Given that hemoglobin has about the same specific refractive index increment as albumin and comprises 97% of the cell solids, having determined the concentration of albumin that is isotonic to the cells, calculate the molar concentration of hemoglobin in the erythrocytes. The molecular weight of native hemoglobin tetramer is 64,000 daltons. Treat the concentration (mg/mL) of bovine serum albumin (molecular weight, 67,000 daltons) as if it were hemoglobin, and use the molecular weight (of either) to calculate the molarity of hemoglobin in erythrocytes. How many times

more concentrated is hemoglobin in an erythrocyte than tubulin (2 μM) or aldolase (20 nM) in a fibroblast?

4. To what do you attribute the range of phase densities observed in the same erythrocyte sample? What are the estimated extremes of molarity of hemoglobin in your cells? How would you test if this is an artifact or represents real variation?

DARK-FIELD MICROSCOPY

In most forms of transmitted light microscopy, both the diffracted rays (rays that interact with the specimen) and nondiffracted rays (rays that pass undeviated through the specimen) are collected by the objective lens and contribute to image formation. For unstained transparent specimens, we have seen that the component of nondiffracted background light is very large, resulting in bright, low-contrast images in which details are poorly visible. Another solution for viewing such objects is *dark-field microscopy,* in which the nondiffracted rays are removed altogether so that the image is composed solely of diffracted wave components. This technique is very sensitive because images based on small amounts of diffracted light from minute phase objects are seen clearly against a black or very dark background. Dark-field microscopy is most commonly used for minute light-diffracting specimens such as diatoms, bacteria and bacterial flagella, isolated organelles and polymers such as cilia, flagella, microtubules, and actin filaments, and silver grains and gold particles in histochemically labeled cells and tissues. An example of a dark-field image of labeled neurons is shown in Figure 7-12. The number of scattering objects in the specimen is an important factor, because the scattering of light from too many objects may brighten the background and obscure fine details.

Theory and Optics

Dark-field conditions are obtained by illuminating the specimen at an oblique angle such that direct, nondiffracted rays are not collected by the objective lens. The effect of dark-field optics can be obtained quickly with bright-field optics by rotating the condenser turret so that rays illuminate the specimen obliquely. Only diffracted light from the specimen is captured by the objective, and the direct waves pass uncollected off to one side of the lens. The disadvantage of this technique is that unidirectional illumination of highly refractile objects can introduce large amounts of flare. Much better images are obtained with a special dark-field condenser annulus, which is mounted in the condenser turret. Special oil immersion dark-field condensers must be used for oil immersion objectives. Dark-field microscopy resembles phase contrast microscopy in that the specimen is illuminated by rays originating at a transparent annulus in the condenser. However, in dark-field optics only diffracted rays are collected by the objective and contribute to the image; nondiffracted rays are pitched too steeply and do not enter the lens (Fig. 7-13). Since nondiffracted background light is absent from the image, light-diffracting objects look bright against a dark field.

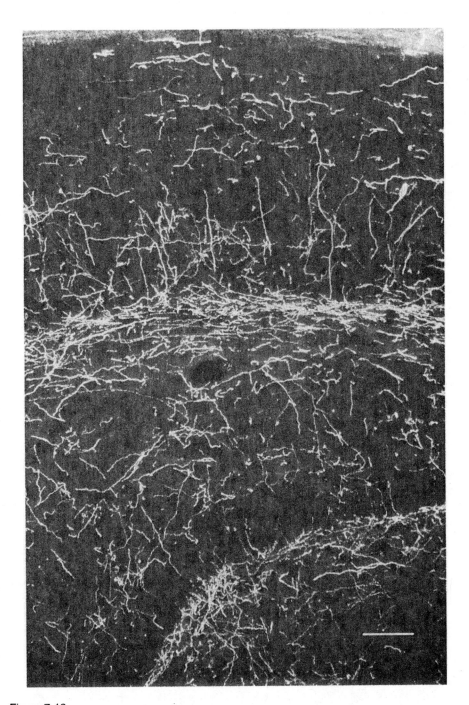

Figure 7-12

Dark-field image of neurons in a section of rat brain. Neurons were labeled with an axon-specific antibody conjugated to horseradish peroxidase. The section was developed with diaminobenzidine. The brown reaction product on the axons appears bright white in this dark-field light micrograph, which only reveals diffracted light components. Bar = 100 μm. (Image courtesy of Mark Molliver, Johns Hopkins University.)

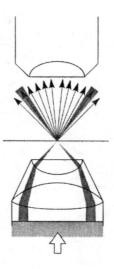

Figure 7-13

Optical scheme for dark-field microscopy. The geometry allows only diffracted light to be collected by the objective lens. Direct, nondiffracted rays are inclined at a steep angle and miss the objective entirely.

There are several ways to create a dark-field image:

- Use the dark-field condenser stop, frequently labeled D on the condenser aperture turret, in combination with a medium power lens with NA < 0.8. If the NA of the objective is lower than the NA of the illuminating beam generated by the condenser and dark-field annulus, nondiffracted waves are excluded from the objective. If the objective lens contains an adjustable diaphragm, this can be stopped down slightly to help block any scattered direct rays that enter the lens.

- An effective and economical approach is to use a phase contrast annulus that is intentionally oversized so that nondiffracted illuminating rays do not enter the objective lens—that is, a high NA condenser annulus with a low NA objective.

- For high magnification work requiring oil immersion objectives, you can employ special oil immersion dark-field condensers with parabolic or cardioid reflective surfaces (Fig. 7-14). These condensers reflect beams onto the specimen at a steeply pitched angle, giving a condenser NA of 1.2–1.4. A *paraboloid condenser* receives a planar wavefront and reflects it off a polished paraboloidal surface at the periphery of the condenser to a point in the specimen plane. The front aperture of the condenser contains an opaque glass with transparent annulus similar to that used in phase contrast microscopy. A *cardioid condenser* receives a collimated beam that is reflected at two surfaces to generate a steeply pitched beam for specimen illumination: A central convex spherical mirror reflects rays to a peripheral concave cardioidal mirror that defines the shape of the beam. Illumination by this condenser is aplanatic and thus free of both spherical aberration and coma. Since both condensers work by reflection, there is no chromatic aberration. The oil immersion objective lens used with these condensers should contain a built-in iris

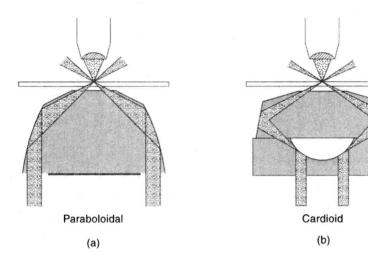

Paraboloidal   Cardioid

(a)   (b)

Figure 7-14

Two oil immersion dark-field condensers. (a) A paraboloid condenser receives a planar wavefront that is reflected off a polished paraboloidal surface to a spot in the specimen plane. The front aperture of the condenser contains an opaque glass with transparent annulus similar to that used in phase contrast microscopy. (b) A cardioid condenser receives a hollow cylinder of light, which is reflected by two spherical surfaces to generate a steeply pitched cone of light for specimen illumination: A central convex mirror reflects rays to a peripheral circumferential mirror with the figure of a cardioid, which reflects a steeply pitched cone of light onto the object.

diaphragm so that the numerical aperture can be reduced to 0.9–1.0 to exclude direct rays.

Image Interpretation

The appearance of a dark-field image is similar to one of self-luminous or fluorescent objects on a dark background, but with the difference that edges of extended, highly refractile objects diffract the greatest amount of light and dominate the image, sometimes obscuring the visibility of fainter, smaller objects. In addition, details in dark-field images are broader and less distinct compared to other imaging modes such as phase contrast, because removal of one entire order of light information from the diffraction plane makes edge definition less distinct in the image. Further, if the NA of the objective selected is too restricted, many diffracted waves are also eliminated, resulting in a loss of definition of fine details in the specimen.

In summary, dark-field optics are advantageous because they allow detection of weak diffracted light signals, and may be the method of choice for viewing fine structural details. Specimens as small as lysosomes, bacterial flagella, diatom striae, and microtubules are all easily seen in well-adjusted dark-field optics, even though these structures have dimensions that are ~20-times less than the resolution limit of the light microscope. Dark-field optics are also inexpensive, simple to employ, and generally do not require special equipment such as DIC prisms, strain-free lenses, or phase contrast objectives.

Exercise: Dark-Field Microscopy

1. Adjust the microscope for dark-field mode using a low-NA objective (10× or 20×) and a high-NA phase contrast or dark-field annulus. These annuli produce steeply pitched cones of light that are not accepted by low-NA, 10× or 20× objectives. Only the scattered light is accepted, which is what you want.

2. With the microscope adjusted for Koehler illumination, focus on a few isolated buccal epithelial cells (obtained by scraping of the underside surface of the tongue) and compare with the image obtained by phase contrast microscopy. Notice that object edges are enhanced by both of these modes of microscopy. In dark-field, each speck of dirt is imaged as a beautiful Airy disk surrounded by higher-order diffraction rings. This is an excellent chance to see the Airy disk, the diffraction pattern of a point source of light.

3. Using an oil immersion dark-field condenser (NA 1.4) and a 60× or 100× oil immersion objective, focus on the cilia and flagella of unicellular protozoa and algae. Cultures of protozoa can be obtained commercially, but a drop of pond water will do as well. *Note:* If the objective lens contains an aperture diaphragm, try stopping it down to improve contrast and visibility, as minute structures are difficult to detect.

The following specimens are suggested for dark-field examination:

- Buccal epithelial cells
- Culture of protozoa
- Axons and glial cells in sections of rat brain labeled with antibodies adsorbed on gold colloid particles
- Blood cells in a 60 µL drop of phosphate-buffered saline
- Taxol-stabilized microtubules in 15 mM imidazole, 1 mM Mg-GTP, 5 µM taxol. The microtubule stock is ~5–10 mg/mL. Dilute to 0.1 mg/mL for observation. Difficult specimen!
- Flagella of *Escherichia coli* bacteria. Difficult specimen!

PROPERTIES OF POLARIZED LIGHT

OVERVIEW

In this chapter we turn our attention to polarization microscopy and a unique class of molecularly ordered objects that become visible upon illumination with polarized light. Figure 8-1 demonstrates the unique ability of a polarizing microscope to reveal molecular order in crystals and starch grains found in plant cell cytoplasm. Polarized light is also used in interference microscopy, including differential interference contrast (DIC) microscopy. Although we can observe high-contrast images of ordered objects using a polarizing microscope, it is remarkable that the eye has no ability in the usual sense to distinguish polarized light from random light. For this we require special filters called polarizers, retarders, and compensators. The relationships between the physics of polarized light and images of molecularly ordered specimens are remarkable in their economy and precision and are well worth mastering. Since the topic of polarized light is technically demanding, we use this chapter to describe its generation, properties, and interaction with different objects and optical devices. Our goal is to understand the working principles of the polarizing microscope, which is described in Chapter 9. Our reward will be in appreciating how the polarizing microscope reveals patterns of molecular order that otherwise can only be studied using more expensive, technically difficult methods such as electron microscopy or X-ray diffraction that operate at the resolution limit of molecules and atoms.

THE GENERATION OF POLARIZED LIGHT

The bulk light from most illuminators used in light microscopy is nonpolarized, the E vectors of different rays vibrating at all possible angles with respect to the axis of propagation (Fig. 8-2a). In a ray or beam of *linearly polarized light,* the E vectors of all waves vibrate in the same plane; the E vectors of beams of polarized light covering an extended area are *plane parallel.* Since the plane of vibration of the E vector can occur at any angle, to describe the orientation of the plane in a beam cross section we describe the angle of tilt relative to a fixed reference plane designated 0° (Fig. 8-2b). A device

Figure 8-1
Crystals and starch grains in *Amaryllis* cytoplasm as seen in a polarizing microscope. The high contrast indicates a high degree of molecular order in these structures.

that produces polarized light is called a *polarizer;* when used to determine the plane of vibration, the same filter is called an *analyzer.* The most efficient polarizers are made of transparent crystals such as calcite, but polarized light can also be generated simply and economically using a partially light-absorbing sheet of linear polarizing material of the type originally introduced by the Polaroid Corporation. Linear polarizers have a unique transmission axis (usually marked on the plate) that defines the plane of vibration of the transmitted rays. Polarizers also play an important role as an analytic tool for determin-

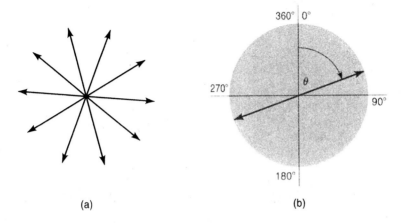

(a)   (b)

Figure 8-2

Random vs. linearly polarized light. The drawings show two beams of light, each containing several photons, as they would appear in cross section, looking down the axis of the beam. Refer to Figure 2-2 for orientation. (a) Random light: The E vectors of the waves are randomly oriented and vibrate in all possible planes. (b) Linearly polarized light: The E vectors of all waves comprising the beam vibrate in a single plane. The angle of the plane of vibration of the E vector relative to the vertical reference line is designated θ. The angle of tilt is called the azimuthal angle.

ing the orientation of polarized light whose plane of vibration is not known. Two linear polarizers—a polarizer and an analyzer—are incorporated in the optics of a polarizing microscope.

Demonstration: Producing Polarized Light with a Polaroid Filter

A *Polaroid sheet*, or *polar,* is a polarizing device that can be used to demonstrate linearly polarized light. The sheet has a transmission axis, such that incident waves whose E vectors vibrate in a plane parallel to the axis pass through the filter, while other rays are absorbed and blocked (Fig. 8-3). Because of its unique action, the Polaroid sheet can be used to produce linearly polarized light or to determine the plane of vibration of a polarized beam whose orientation is not known. Used in these ways, the sheet is then called, respectively, a polarizer or an analyzer. To become familiar with polarized light, perform the following operations using a pair of Polaroid sheets:

• Place two polars on top of each other on a light box and rotate one of the polars through 360°. At two azimuths separated by 180°, light transmission reaches a maximum, while at two azimuths separated from the first two by 90°, transmission is substantially blocked, and the field looks black. In the first case, light is transmitted because the transmission axes of the two polars are parallel, and all of the light transmitted by the first filter passes through

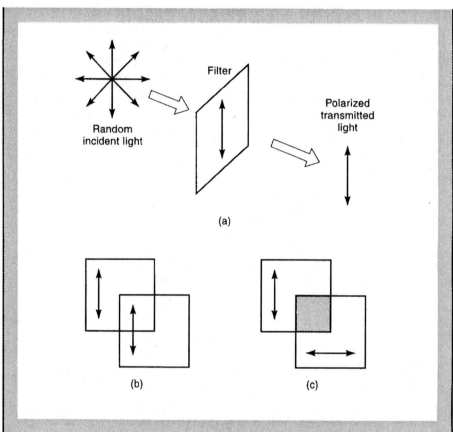

Figure 8-3
A Polaroid sheet generates linearly polarized light. (a) Only rays whose E vectors vibrate in a plane parallel with the transmission axis of the sheet are transmitted as a linearly polarized beam; other rays are partially transmitted or blocked. (b) A second overlapping polar transmits light of the first polar if its transmission axis is parallel to that of the first polar. (c) Transmission is completely blocked if the transmission axes of the two polars are crossed.

the second. The removal of light at the unique crossed position is called *extinction*. Here, polarized light from the first filter vibrates in a plane that is perpendicular to the transmission axis of the second filter and is substantially blocked by the second filter.

- Examine a beam of light reflected off a flat nonconducting surface (glass or plastic) with a polar while rotating it through 360°. The reflected beam is minimized at two azimuths, indicating that it is partially linearly polarized. Used this way, the Polaroid sheet is called an analyzer. If the transmission axis of the polar is known beforehand, you should be able to observe that the plane of vibration of the reflected rays is parallel to the plane of the reflecting surface. If the transmission axis of a polar is not known, inspection of a reflected beam can be used to determine the approximate orientation of its transmission axis.

- Using two crossed polars positioned on a light box, examine a piece of glass such as a microscope slide and a sheet of cellophane, while rotating them between the polars. The glass does not interact with polarized light and remains essentially invisible. Such materials are said to be *optically isotropic.* However, rotation of the cellophane sheet through 360° reveals four azimuths separated by 90° at which the entire sheet looks very bright. The unique ability of cellophane to interact with polarized light is due to the presence of aligned parallel arrays of cellulose molecules and birefringence, which is described later in the chapter. Materials of this type are said to be *optically anisotropic.* All objects suitable for polarization microscopy exhibit some degree of molecular orientation and optical anisotropy.

POLARIZATION BY REFLECTION AND SCATTERING

Polarized light is also produced by a variety of physical processes that deflect light, including refraction, reflection, and scattering. Light reflected from the surfaces of dielectric materials is partially linearly polarized, with the E vectors of the reflected waves vibrating parallel to the reflecting surface and the extent of polarization increasing with decreasing angles of incidence. For light incident on a transparent material such as water or glass, there is a unique angle known as *Brewster's angle,* at which the reflected waves are completely plane polarized (Fig. 8-4). For the simple case of a beam of incident light travelling through air ($n = 1$), the critical angle is given as

$$\tan\theta = n.$$

For water ($n = 1.33$) and glass ($n = 1.515$) the critical angles are 53° and 57°, respectively. As an interesting note on reflection polarization, manufacturers of Polaroid sunglasses mount sheets of polarizing material in the frames with the transmission axis of the Polaroids oriented perpendicularly in the field of view. Bright reflections off horizontal surfaces, such as the roofs of cars or water on a lake, are very efficiently blocked, while the random light is partially blocked, since only vertically polarized rays can reach the eye.

Linearly polarized light is also produced by light scattering. A common example is the polarization of light in the northern sky caused by the scattering of sunlight by air molecules. The extent of polarization ($\sim$50%) is readily appreciated on a clear day by rotating a polarizer held close to the eye. In accordance with principles of scattering, polarization is maximal at an angle 90° from the sun. For additional information on polarization by refraction, reflection, and scattering, see the interesting discussions by Minnaert (1954) and Hecht (1998).

VECTORIAL ANALYSIS OF POLARIZED LIGHT USING A DICHROIC FILTER

The Polaroid sheet just described commonly consists of a film of parallel arrays of linear polyvinyl alcohol molecules with embedded polyiodide microcrystals (H-ink)

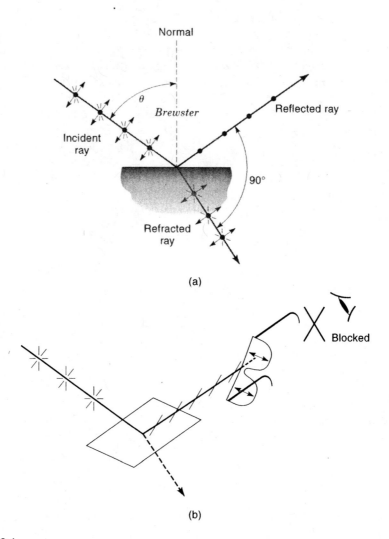

(a)

(b)

Figure 8-4

Reflection polarization and Brewster's critical angle. (a) The drawing shows an incident ray of random light (E vectors at random orientations) being reflected off a horizontal surface (glass or other dielectric material) as a linearly polarized beam to the eye. At a certain critical angle (Brewster's angle θ), the degree of polarization of the reflected ray is 100% and the E vectors of reflected rays vibrate in a plane that is perpendicular to the plane of incidence (the plane defined by the incident, reflected and refracted rays). The refracted ray, oriented at a 90° angle with respect to the reflected ray, is only partially polarized. (*Note:* Brewster's angle is defined by the condition that exists when the reflected wave is at 90° to the refracted wave; the phenomenon occurs because in the second medium the electric vector of the component in the plane of incidence is pointing in the direction of propagation of the reflected wave and consequently has no resultant wave in that direction.) (b) A Polaroid sheet can be used to demonstrate the polarized nature of the reflected ray. If the transmission axis of the Polaroid is oriented at 90° with respect to the vibrational plane of the reflected ray, transmission is blocked. This principle is used to advantage in the design of Polaroid sunglasses to reduce or eliminate reflective glare.

aligned in the same direction as the organic matrix. The transmission axis of the filter is perpendicular to the orientation of the crystals and linear polymers in the filter. Thus, rays whose E vectors vibrate parallel to the crystal axis are absorbed. Waves with oblique orientations are partially absorbed, depending on the azimuths of their vibrational planes. Filters such as the Polaroid sheet that differentially transmit rays vibrating in one plane while absorbing those in other planes are said to exhibit *dichroism;* hence the common name, *dichroic filter.* As we will see in later sections, the Polaroid sheet is a special kind of beam splitter designed to transmit rays vibrating at a certain azimuthal angle as linearly polarized light. For H-series filters that are commonly employed in microscopy, only about 25% of incident random light is transmitted, but the degree of polarization of the transmitted rays is > 99%.

If two polarizers are oriented so that their transmission axes are perpendicular to each other, they are said to be crossed, and all of the light transmitted by the polarizer (now linearly polarized) is extinguished by the analyzer (Fig. 8-3). The extent to which incident random light is extinguished by two crossed polars is called the *extinction factor* and is defined as the ratio of the intensity of transmitted light observed for two polars when positioned in parallel and in crossed orientations ($I_\parallel/I_\times$). Extinction factors of 10^3–10^5 or greater are required for polarization microscopy and can be obtained using two dichroic filters.

The role of the analyzer in controlling the transmission of polarized light can be understood from vector diagrams showing the angular orientation (azimuthal angle) and magnitude of the E vectors of rays drawn from the perspective of viewing a ray end on, down its axis of propagation. Figure 8-5 shows a vertically oriented analyzer, four incident waves of linearly polarized light that are equal in amplitude but vibrating in different planes, and the amplitudes of those waves after transmission through the analyzer. If each incident wave is resolved into its horizontal and vertical components, it can be seen that the entire vertical component of each ray passes through the analyzer. Further, the amplitudes of the vertical component and the transmitted ray rapidly decrease as the plane of vibration of the incident ray approaches an azimuth perpendicular to the transmission axis of the analyzer.

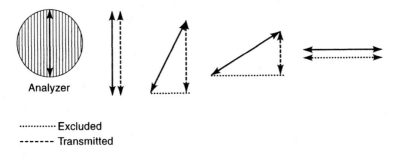

Analyzer

·············· Excluded

------- Transmitted

Figure 8-5

Transmission through an analyzer of linearly polarized rays vibrating in different planes. A linearly polarized ray vibrating at an azimuthal angle different from the transmission axis of an analyzer can be resolved into its horizontal and vertical components using principles of vector math. The amplitude of the transmitted ray is equal to the vertical vector component.

The amount of light transmitted through two polars crossed at various angles can be calculated using *Malus' law,* which states that

$$I = I_o \cos^2\theta,$$

where I is the intensity of light passed by an analyzer, I_o is the intensity of an incident beam of linearly polarized light, and θ is the angle between the azimuth of polarization of the incident light and the transmission axis of the analyzer. Thus, for two polars crossed at a 90° angle, $\cos^2\theta = 0$, and $I = 0$, so polarized light produced by the first filter (the polarizer) is completely excluded by the second filter (the analyzer). For polarizers partially crossed at 10° and 45°, the light transmitted by the analyzer is reduced by 97% and 50%, respectively.

DOUBLE REFRACTION IN CRYSTALS

Many transparent crystals and minerals such as quartz, calcite, rutile, tourmaline, and others are optically anisotropic and exhibit a property known as *double refraction.* When letters on a printed page are viewed through a calcite crystal, remarkably each letter appears double (Figure 8-6). Calcite is therefore said to be *doubly refracting,* or *birefringent.* Birefringent materials split an incident ray into two components that traverse different paths through the crystal and emerge as two separate rays. This occurs because atoms in crystals are ordered in a precise geometric arrangement causing direction-dependent differences in the refractive index. In contrast, a sheet of glass, such as a microscope slide, which is an amorphous mixture of silicates, is usually optically isotropic and does not exhibit double refraction.

When a ray of light is incident on a birefringent crystal, it usually becomes split into two rays that follow separate paths. One ray, the *ordinary ray* or *O ray,* observes the laws of normal refraction, while the other ray, the *extraordinary ray* or *E ray,* travels along a different path. Thus, for every ray entering the crystal there is a pair of O and E rays that

Figure 8-6

Double refraction in a calcite crystal. A letter viewed through a plane surface of the crystal appears double, the two images corresponding to the ordinary and extraordinary rays. As the crystal is rotated, the E ray rotates around the O ray. The O ray obeys the normal laws of refraction and does not rotate.

emerges, each of which is linearly polarized. *The electric field vectors of these two rays vibrate in mutually perpendicular planes.* A sketch depicting this phenomenon is shown in Figure 8-7a. These features are easily observed by placing a crystal of calcite on a printed page and looking down on the crystal while rotating a dichroic filter held in front of the eye (see Demonstration); the double letters become alternately visible and invisible as the filter is rotated through an angle of 90°. There are two unique angles of incidence on the crystal for which the behavior is different (1) The calcite crystal and others of its class contain a single unique axis known as the *optic axis.* Incident beams that are perpendicular to the optic axis are split into O and E rays, but the trajectories of these rays are coincident (Fig. 8-7b). At this unique angle of incidence, the O and E rays emerge at the same location on the crystal surface, but have different optical path lengths and are therefore shifted in phase. *This geometry pertains to most biological specimens that are examined in a polarizing microscope.* (2) Incident rays that follow trajectories parallel to this axis behave as ordinary rays and are not split into O and E rays (Fig. 8-7c). Under these conditions of illumination, calcite behaves as if it were optically isotropic, like glass. (It is difficult to demonstrate the optic axis of calcite because it runs obliquely across the diameter of the crystal, and it is necessary to look down crystal edges. One solution is to examine a specially prepared slab cut perpendicularly to the optic axis of the crystal.) These principles are displayed clearly in Hecht (1998), Pluta (1988), and Wood (1964).

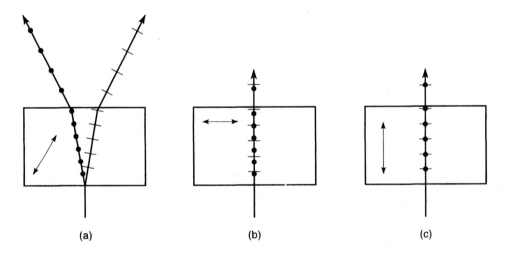

(a)	(b)	(c)

Figure 8-7

Splitting of an incident ray into O- and E-ray components by a birefringent crystal. The rectangular slabs shown in a, b, and c have been cut from parent crystals in such a way that the optic axes are oriented differently. Incident light is linearly polarized. Dots and dashes indicate the planes of vibration of linearly polarized O and E rays. Dots indicate the vibrations of E vectors that are perpendicular to the plane of the page, while the space between the dots represents one wavelength; dashes indicate vibrations parallel to the plane of the page. (a) A ray incident on a crystal at an angle oblique to the optic axis of the crystal is split into O and E rays that traverse different physical paths through the crystal. The emergent O and E rays are linearly polarized, vibrate in mutually perpendicular planes, and exhibit an optical path difference. (b) An incident ray whose propagation axis is perpendicular to the optic axis is split into O and E rays, but the two rays follow the same trajectory through the crystal and do not diverge. Emergent rays can exhibit an optical path difference. This is the usual case for birefringent biological specimens. (c) An incident ray whose propagation axis is parallel to the optic axis is not split and behaves as an ordinary ray. The optical path lengths of the emergent rays are the same.

Demonstration: Double Refraction by a Calcite Crystal

- Place a crystal on a page of printed text and rotate the crystal while looking down on it from above. One ray does not move as the crystal is rotated (the O ray), whereas the other ray (the E ray) rotates around the fixed ray according to the angle of rotation of the crystal (Fig. 8-6). The O ray does not move because it obeys the laws of normal refraction and passes through the crystal on a straight, undeviated trajectory. The E ray, in contrast, travels along a tilted or oblique path, and its point of emergence at the surface of the crystal changes depending on the orientation of atoms in the crystal and therefore on the orientation of the crystal itself. Because the angle of divergence is so great, wedges of calcite crystal can be cut and glued together in such a way that one of the two components is reflected and removed while the other is transmitted, making calcite an ideal linear polarizer.

- Examine the crystal resting on a black spot on a blank page through a dichroic polarizing filter held close to the eye, while rotating the crystal through 360°. First one ray and then the other ray alternately comes into view and becomes extinguished, and the black spot appears to jump back and forth as the filter is rotated through increments of 90°. The double images correspond to the O and E rays, each of which is linearly polarized and vibrates in a plane perpendicular to the other. For most crystals the two rays are close together and must be magnified to be observed, but in calcite the rays are so widely separated that no magnification is necessary. As we will see, this is due to large differences in the refractive index along different paths through the crystal. Notice too that polarized light cannot be distinguished from ordinary random light and that an analyzer is required to distinguish the different planes of polarization.

Materials such as calcite, quartz, and most molecularly ordered biological structures that contain a single optic axis are called *uniaxial*. Another class of *biaxial* crystals with two optic axes also exists, but is rarely encountered in biological systems. Biological examples of ordered macromolecular assemblies that can be seen in the polarizing microscope include such objects as lipid bilayers, bundles of microtubules and actin filaments, plant cell walls, crystalline granules of starch, lignin, and other materials, chromosomes from certain organisms, DNA kinetoplasts in trypanosomes, chloroplasts, and many other structures.

We have used the terms *double refraction* and *birefringence* to refer to the ability of molecularly ordered objects to split an incident ray of light into two components, the O and E rays, but the two terms refer to different aspects of the same process. Double refraction refers to the visible phenomenon: the splitting of a single incident ray into two resultant rays as exemplified by a crystal of calcite. Birefringence refers to the cause of the splitting: the existence of direction-dependent variation in the refractive index in a molecularly ordered material. Birefringence B also refers to a measurable quantity, the difference in the refractive index $(n_e - n_o)$ experienced by the O and E rays during transit through an ordered object such that

$$B = (n_e - n_o).$$

Depending on the values of n_e and n_o, the *sign of birefringence* may be positive or negative, and specimens are therefore said to be either positively or negatively birefringent. Note also that birefringence is not a fixed value, but varies, depending on the orientation of the birefringent object relative to the illuminating beam of polarized light. We return to this important point in later sections.

Birefringence is related to another term, the *optical path difference* Δ (or in the field of polarized light, the *relative retardation* Γ), and both are defined as the relative phase shift expressed in nanometers between the O and E waves emergent from a birefringent object. As described in Chapters 5 and 7, the optical path length δ is given as $\delta = nt$, where n is the refractive index of a homogeneous medium between points A and B and t is the object thickness. Notice that the optical path length is a distance given in units of parameter t and that this term is equal to the geometric distance only when $n = 1$. The optical path difference Δ for two objects spanning the same distance is

$$\Delta = (n_1 - n_2)t.$$

Relative retardation and birefringence are related by the analogous expression

$$\Gamma = (n_e - n_o)t,$$

where t is the thickness, the physical distance traversed through the specimen. Accordingly, small and large objects may give the same retardation depending on the magnitude of their birefringence and physical size. Retardation can also be expressed as the mutual *phase shift* δ between the two wavelengths, and is given (in radians) by

$$\delta = 2\pi\Delta/\lambda.$$

Double refraction or birefringence is a property of polarizers used in a polarizing microscope and of microscope specimens that are active in polarized light. Its presence in a specimen allows us to measure the pattern and extent of molecular alignments, refractive index differences, and specimen thickness.

KINDS OF BIREFRINGENCE

Birefringence can be an inherent physical property of specimens such as the calcite crystal, or can be generated through cell biosynthetic activities (cellulose fibrils and starch granules in plant cells), or can arise from outside physical forces (cytoplasmic flow, cell movements) acting on components of an otherwise disorganized system. Various kinds of birefringence have been defined:

- *Intrinsic birefringence* is based on the asymmetry of polarizability of chemical bonds within naturally occurring materials such as crystals. Examples: crystals of calcite and quartz.
- *Form birefringence* or *structural birefringence* is associated with extensive ordered arrangements of macromolecular assemblies having dimensions and spacings comparable to the wavelength of light. Form birefringence is sensitive to the refractive index of the surrounding medium. Examples: parallel arrangements of actin and myosin filaments in muscle myofibrils, microtubule bundles in mitotic spindle fibers of dividing cells.

- *Flow birefringence* refers to the induced alignment of asymmetric plate- or rod-shaped molecules in the presence of fluid flow or agitation. Examples: stirred solutions of detergents (shampoo), DNA, or flagella.

- *Strain birefringence* describes the forced alignment of molecules in a transparent solid deformed by an externally applied force. Example: stretched films of polyvinyl alcohol.

- *Stress birefringence* is descriptive when electronic deformation occurs in response to an external mechanical force without there being significant gross deformation. Example: a stressed lens caused by compression during mounting in a retaining ring or lens barrel.

PROPAGATION OF O AND E WAVEFRONTS IN A BIREFRINGENT CRYSTAL

In Chapter 3 we employed the concept of Huygens' wavelets to describe the location of a secondary wavefront generated by spherical wavelets and originating from a point source of light in a homogeneous medium. In a transparent birefringent material, the ordinary or O ray behaves in the same way, generating a spherical wavefront. However, the extraordinary or E waves behave differently. As described by Huygens in 1690, the expanding wavefront of the E ray at a time t can be described as the surface of an ellipsoid (Fig. 8-8). An *ellipsoid* is the figure generated by rotating an ellipse about its major or minor axis. The ellipsoidal form indicates the presence of different velocities for the E ray along different trajectories in the crystal, where the upper- and lower-limit velocities define the long and short axes of the wavefront ellipsoid. The long axis corresponds to the direction along which the wavefront reaches its greatest possible velocity through the crystal, and is termed the *fast axis,* while the short axis corresponds to the direction giving the smallest velocity, and is called the *slow axis.* The velocities of waves traveling in all other directions have intermediate values. Since the velocity of light in a medium is described as $v = c/n$, where c is the speed of light in a vacuum and n is the refractive index, we may infer that n is not constant in a birefringent crystal, but varies, depending on the path taken by a ray through the crystal. Several additional points about the propagation of wavefronts in a crystal are worth noting:

- For uniaxial crystals, the O and E wavefronts coincide at either the slow or the fast axis of the ellipsoid, and the difference in surface wavefronts along the propagation axis is termed the optical path difference or relative retardation.

- If the O and E wavefronts coincide at the major axis of the ellipsoid, $n_e > n_o$ in directions other than along the optic axis, and the specimen is said to be positively birefringent (Fig. 8-9). This is the case for crystals such as quartz and most ordered biological materials. For materials such as calcite, whose O and E wavefronts meet at the minor axis of the ellipsoid, $n_o > n_e$ in directions other than along the optic axis. Such materials are said to exhibit negative birefringence.

- For the unique case that the incident illuminating beam is parallel or perpendicular to the optic axis of the crystal, the paths of O and E rays follow the same trajectory and exit the crystal at the same location. If the incident ray is parallel to the optic axis, the E ray behaves as an O ray; if the incident ray is perpendicular to the optic axis, the O and E ray components experience different optical paths.

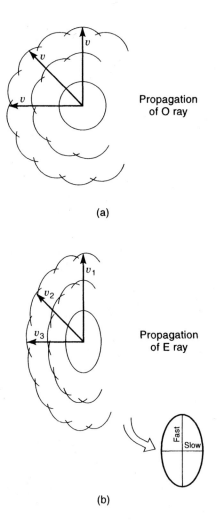

(a)

Propagation
of O ray

(b)

Propagation
of E ray

Figure 8-8

O and E rays emanating from a point in a birefringent material define spherical and
ellipsoidal wavefronts. (a) The wavefront defined by O rays is spherical because the
refractive index is uniform in all directions and waves propagate at a velocity given by the
expression $v = c/n$. Circles are used to draw Huygens' wavelets to depict the spherical
wavefront. (b) The wavefront defined by the E rays is ellipsoidal because the refractive index
n varies in different directions depending on the three-dimensional distribution of atoms and
molecules. Since $v = c/n$, the velocity of the E rays is direction dependent (shown as v_1, v_2,
v_3), resulting in a surface wavefront with the shape of an ellipsoid. Huygens' wavelets are
drawn using ellipses instead of circles to depict the advancing wavefront.

Double refraction is based on Maxwell's laws of electromagnetism. An explanation
requires vector calculus and is beyond the scope of this text, but we can make a brief
qualitative explanation of the principles involved. Since light contains both electric and
magnetic components, the velocity of light in a medium depends in part on the electrical conductivity of the material and the interaction of light with electric fields in that
medium, a property called the *dielectric constant* ϵ. For most dielectric substances and

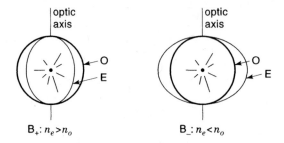

Figure 8-9
The relation of O and E wavefronts in specimens showing positive and negative birefringence.

biological materials, magnetic permeability is close to 1 and can be discounted; however, for materials with reduced magnetic permeability (metallic films), this would not be the case. The dielectric constant is related to the refractive index n by the simple relationship

$$\epsilon = n^2.$$

Therefore, in crystals having a particular lattice structure of atoms and molecules, the value of n is different in different directions, depending on the orientations of chemical bonds and electron clouds around atoms and molecules. The short axis (slow axis) of the wavefront ellipsoid corresponds to the axis defining the highest refractive index value; the long axis (fast axis) corresponds to the axis having the lowest refractive index value. The ellipsoid describing the orientation and relative magnitude of the refractive index in a crystal is called the *refractive index ellipsoid* or simply *index ellipsoid.*

BIREFRINGENCE IN BIOLOGICAL SPECIMENS

We have seen that the velocity of light during transit through a specimen is retarded by perturbations and interactions with electrons in the transmitting medium. The susceptibility of an electronic configuration to distortion is termed its *polarizability.* The more polarizable the electronic structure of a molecule, the more extensively light can interact with it, and, thus, the more slowly light is transmitted. In the extreme case, the interaction may be so strong that the specimen absorbs quanta of light.

Most chemical groups are asymmetric with respect to polarizability. For example, the electrons of carbon-carbon bonds in molecules with long carbon chains are most easily displaced along the direction of the bond (Fig. 8-10). A structure containing parallel chains of carbon-carbon bonds such as the cellulose polymers in cellophane is most polarizable in a direction parallel to the carbon chains. Since $\epsilon = n^2$, the velocity of light is lowest and the refractive index is highest in this direction. Therefore, when long carbon chain molecules are regularly oriented in biological structures as they are for example in cellulose fibers comprising a plant cell wall, the polarizability of the structure as a whole varies with its orientation in the illuminating beam of polarized light. As shown in Chapter 9, the orientation of the wavefront and index ellipsoids and the sign of birefringence of the specimen can be determined with the help of a compensator such as a full waveplate.

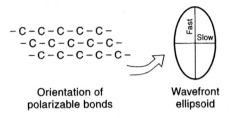

Orientation of   Wavefront
polarizable bonds   ellipsoid

Figure 8-10
The axis of polarizability corresponds to the slow axis of the *wavefront ellipse.* In biological materials composed of linear arrays of macromolecules, the axis of strongest polarizability is usually determined by the direction of carbon-carbon bond alignments. Because polarizability and refractive index are highest in the direction parallel to the axis of bond alignment, this direction corresponds to the slow axis of the wavefront ellipsoid for this material.

GENERATION OF ELLIPTICALLY POLARIZED LIGHT BY BIREFRINGENT SPECIMENS

Birefringent objects split rays of incident light into separate O- and E-ray components whose E vectors vibrate in mutually perpendicular planes. The O and E waves comprising each ray or wave bundle also become mutually shifted in phase owing to differences in refractive index experienced by each wave during transit through the specimen. With the calcite crystal, we observed that the O and E rays follow different trajectories and emerge at widely separated locations on the surface of the crystal, making it relatively easy to examine each linearly polarized component separately with the aid of a Polaroid sheet. Let us now consider the circumstance where the optic axis of the specimen is perpendicular to the incident beam, which is the usual case for retardation plates and most biological specimens that are mounted on a microscope slide and examined in the object plane of the microscope. Because the incident ray is perpendicular to the optic axis of the specimen, the O and E rays follow the same trajectory as the incident ray, with one ray lagging behind the other and exhibiting a phase difference according to the amount of birefringence. Since the O and E waves vibrate in mutually perpendicular planes, they cannot interfere to produce a resultant wave with an altered amplitude. This point becomes important in polarization microscopy, because interference is required to generate a contrast image. As a convenience for describing the behavior of the superimposed wave pair, we may add the two waves together to form a single resultant ray. The mutually perpendicular orientation of the two E vectors and phase difference between the two rays result in a three-dimensional waveform called *elliptically polarized light.*

As seen in Figure 8-11, the E vector of the recombined wave does not vibrate in a plane over the course of its trajectory, but progressively rotates about its propagation axis. When viewed perpendicular to its propagation axis, the E vector appears to follow the course of an elliptical spiral, and when viewed end-on, the E vector sweeps out the shape of an ellipse. The resultant elliptical wave is reconstructed using simple vector addition in three-dimensional space. For O and E waves of equal amplitude, the amount of ellipticity depends on the amount of phase shift (Fig. 8-12). For phase shifts of exactly $\lambda/4$ and $3/4\lambda$, the shape of the resultant wave can be described as a circular spiral; for phase shifts of λ or $\lambda/2$, linearly polarized light results; for all other phase differences (the vast

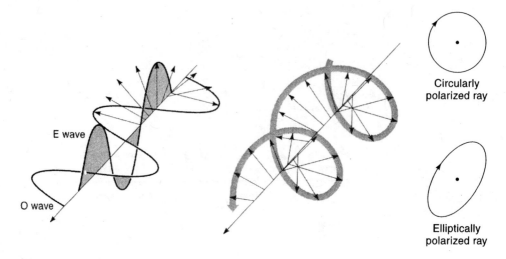

Figure 8-11

The waveforms of elliptically and circularly polarized light. O and E rays following the same propagation axis but vibrating in mutually perpendicular planes cannot interfere, but can be combined by vector addition. Depending on the relative phase difference between the two rays, the resultant wave may be linear or take on the form of a spiraling ellipse or circle. With a phase displacement of $\lambda/4$, the waveform is a circle.

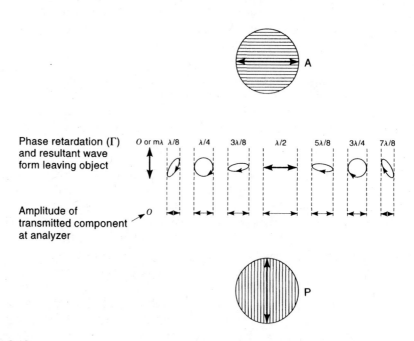

Figure 8-12

Effect of relative phase shift between O and E rays on the waveform of polarized light. Waves resulting from the combination of superimposed O and E rays have elliptical, spherical, or planar waveforms, depending on the amount of relative phase shift between the two rays. The orientations of the transmission axes of the polarizer and analyzer are indicated. The amplitudes of the components of vibration passing the analyzer are also shown.

majority), the shape is that of an elliptical spiral of varying degree of ellipticity. The component of elliptically polarized light that is able to pass through an analyzer varies depending on the amount of phase shift and is shown in Figure 8-12. The description of circular or elliptical waveforms is simply a convenient device for visualizing how O- and E-wave pairs interact with analyzers and optical elements called retardation plates.

Interference between two intersecting waves of light occurs only when their E vectors vibrate in the same plane at their point of intersection. Only when interference causes a change in the amplitude in the resultant wave can an object be perceived due to differences in intensity and contrast. The observed intensity from the O and E waves vibrating in mutually perpendicular planes emergent from a birefringent object is simply the sum of their individual intensities; no variations in intensity are observed because interference cannot occur and the object remains invisible. A sheet of cellophane held against a single polarizer on a light box is an example of this behavior. Cellophane is a birefringent sheet made up of parallel bundles of cellulose. The optic axis is parallel to the axis of the bundles and is contained in the plane of the sheet. When examined in polarized light without an analyzer, elliptically polarized light emerges from the cellophane, but since there is no interference or change in amplitude, the sheet remains invisible against the polarizer. However, if the cellophane is examined between two crossed polars, components of elliptical waves that are parallel to the analyzer are transmitted and emerge as linearly polarized light. Background rays from the polarizer are blocked by the analyzer, so the cellophane stands out as a bright object against a dark background. The sheet appears brightest when its optic axis is oriented at 45° with respect to the transmission axes of the two crossed polars.

POLARIZATION MICROSCOPY

OVERVIEW

Image formation in the polarizing microscope is based on the unique ability of polarized light to interact with polarizable bonds of ordered molecules in a direction-sensitive manner. Perturbations to waves of polarized light from aligned molecules in an object result in phase retardations between sampling beams, which in turn allow interference-dependent changes in amplitude in the image plane. Thus, image formation is based not only on principles of diffraction and interference, but also on the existence of ordered molecular arrangements. The degree of order encountered in objects ranges from near-perfect crystals to loosely ordered associations of asymmetric molecules or molecular assemblies. In the polarizing microscope, such structures generally appear bright against a dark background (Fig. 9-1). Polarization microscopy has been used to study the form and dynamics of many ordered cellular structures, including:

- Mitotic spindle fibers in dividing cells
- Actin filament bundles in a variety of cell types
- Actin and myosin filaments in the myofibrils of striated muscle cells
- Condensed DNA in certain sperm nuclei
- Kinetoplast DNA in the mitochondria of trypanosomes
- Helical strands of cellulose fibers in plant cell walls
- Condensates of starch and lignin in plant cells
- Virus crystalloids and crystals of organic compounds in the cytoplasm of plant cells
- Lipid bilayers of the cell plasma membrane and mitochondria

In many cases, polarization microscopy is the only available method for studying the structure, formation, and dynamics of labile macromolecular assemblies or examining the effects of chemicals, drugs, or environmental conditions on cellular structures in vivo. For additional examples of the application of polarized light in studies of mitosis, see Inoué and Oldenbourg (1998) and Oldenbourg (1996, 1999).

Figure 9-1

Polarized light image of DNA nucleoids in the trypanosomatid, *Crithidia fasciculata*. The discrete white and black bodies are kinetoplasts, highly condensed assemblies of mitochondrial DNA. The high degree of molecular order causes the bright and dark patterns of contrast in a polarizing microscope. (Image courtesy of Mark Drew, Johns Hopkins University.)

The polarizing microscope is also a remarkable analytical instrument, capable of providing quantitative measurements of differences in optical path length (retardation), which in turn can be used to calculate refractive index differences and the thicknesses of ordered specimens. Geologists use these parameters together with a reference chart to determine the identities of unknown crystalline minerals. These capabilities distinguish polarization microscopy from other forms of light microscopy and account for its popularity in biology, chemistry, geology, and materials science. Polarized light is also used for many forms of interference microscopy, including differential interference microscopy. In this chapter we discuss the function and alignment of polarizing optics and the action and method of deployment of several compensators.

OPTICS OF THE POLARIZING MICROSCOPE

A polarizing microscope is a compound light microscope fitted with a polarizer, an analyzer, and, if quantitative measurements of birefringence are to be made, a compensator (see Fig. 9-2). A *compensator* (also called a *retarder*) is a birefringent plate that is used to measure optical path differences and improve visibility. The polarizer is placed between the light source and the specimen, commonly near the front aperture of the condenser; the analyzer is placed between the specimen and the eye, usually some distance behind the back aperture of the objective lens. The polarizer is mounted with its transmission axis fixed in a horizontal (east-west or right-left) orientation as seen facing the

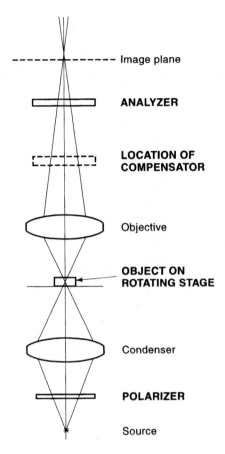

Image plane

ANALYZER

LOCATION OF COMPENSATOR

Objective

OBJECT ON ROTATING STAGE

Condenser

POLARIZER

Source

Figure 9-2

Optical components of a polarizing microscope. Notice the presence of a polarizer and analyzer, a rotatable stage, and a slot for accommodating a compensator. Polarization microscopy requires an intense light source.

microscope and looking down the optic axis of the microscope; the analyzer is then rotated at 90° to the polarizer in a north-south or up-down orientation to obtain extinction. It is desirable that the polarizer or analyzer be rotatable. The rotatable holder is often marked in degrees to facilitate adjustment to extinction. Having one of the polarizers in a rotatable holder assures proper orientation of the crossed polars along east-west and north-south axes and allows precise control for obtaining extinction. The use of a compensator also requires that the polarizers be adjusted precisely, because it is inserted into the optical path at a fixed orientation. It is also highly desirable that the specimen stage be rotatable, with its center of rotation adjusted to coincide with the optic axis of the microscope so that the fixed polarized light beam can illuminate a specimen at various angles simply by rotating the stage. Finally, there is an insertion slot for a compensator, which is located near and in front of the analyzer. For applications where birefringence is small and object contrast is very low, it is important to use strain-free, polarization-grade condenser and objective lenses. Strain birefringence exists in most lenses as a result of manufacture and mounting in a lens barrel. When severe, birefringence from a lens can obscure faint signals in images of weakly birefringent objects. For this reason, manufacturers select out low strain objectives for use in polarization microscopy.

ADJUSTING THE POLARIZING MICROSCOPE

- Focus a specimen and adjust for Koehler illumination with both polarizing filters and compensator removed from the optical path.

- Insert the fixed polarizer (depending on the microscope, either the polarizer or analyzer might be fixed—i.e., glued into position in the holder so that it cannot rotate). If the polarizer is fixed, its transmission axis should be oriented horizontally east-west as seen looking in the microscope. If the polarizer is rotatable, check that the filter is positioned correctly by looking for the mark indicating the orientation of the transmission axis or checking the degrees of rotation on the holder.

- Insert the rotatable analyzer (or polarizer), and rotate it until the two polars are crossed and maximum extinction is obtained. Do this by examining a blank region on a specimen slide and making fine adjustments to the analyzer until the image is maximally dark. The transmission axis of the analyzer should now be oriented vertically or north-south. The critical adjustment for extinction can only be made while looking in the microscope. Extinction indicates that the polars are crossed, but does not guarantee that their azimuths are perfectly oriented east-west and north-south.

- Insert a telescope eyepiece or Bertrand lens and focus on the back aperture of the objective lens. A dark *polarization cross* is observed at extinction (Fig. 9-3), with brighter regions of intensity between the horizontal and vertical arms of the cross. Commonly, the horizontal and vertical components of the cross are broader in the center of the field and narrower at the periphery. This shape is caused by the depolarizing effects of spherical lens surfaces in the condenser and the objective lenses, and is normal for most microscope systems. The cross should be centered in the field of view and should be perfectly upright. If not, the polarizer and analyzer may need to be rotated a bit to achieve proper orientation. Perfect alignment with respect to an azimuth at $0°$ is essential if quantitative measurements of azimuthal angles or birefringence are to be made with a compensator. Reference to the polarization

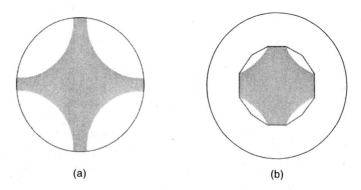

(a)   (b)

Figure 9-3

View of the polarization cross in the back aperture of the objective lens. Views before (a) and after (b) proper adjustment of the condenser aperture diaphragm.

cross is also useful in case a polar becomes misaligned in its holder, requiring the operator to readjust the polar's orientation.

- Using the telescope eyepiece or Bertrand lens, partially close the condenser aperture diaphragm so that bright outer regions of depolarized light visible at the edge of the aperture field are blocked. This greatly improves the extinction factor of the microscope optics. Switching back to viewing mode, the field of view at extinction should now be very dark. Normally it should only be necessary to rotate the specimen slide to examine the object at different azimuthal angles, leaving the positions of the polarizer and analyzer fixed and unchanged. Because it is often necessary to rotate specimens during examination, it is convenient to use a specially designed rotating stage that is marked in degrees at its periphery. Rotating stages must be adjusted so that their center of rotation is superimposed on the optic axis of the microscope.

APPEARANCE OF BIREFRINGENT OBJECTS IN POLARIZED LIGHT

Birefringent specimens exhibit characteristic patterns and orientations of light and dark contrast features that vary, depending on the shape and geometry of the object (linear or elongate vs. spherical) and the molecular orientation. In the absence of a compensator, spherical objects with radially symmetric molecular structure exhibit a dark upright polarization cross superimposed on a disk composed of four bright quadrants. Thus, there are eight alternating bright and dark contrast regions distributed around the circumference of the sphere. If a compensator such as a λ/4 (quarter-wave) plate is inserted into the beam so that its slow axis is oriented at 45° with respect to the transmission axes of the polarizer and analyzer, a pattern of four quadrants is observed, with one pair of opposite quadrants showing bright contrast and the other pair dark contrast. Instructions for performing this operation are given at the end of the chapter.

Linear objects such as elongate striated muscle cells with coaxial alignments of linear filaments have a different appearance. In the absence of a compensator, rotation of the specimen stage through 360° reveals eight angular azimuths at which the muscle cells alternately appear bright (45°, 135°, 225°, and 315°) or dark (0°, 90°, 180°, 270°); with a compensator present, there are four azimuths at 45°, 135°, 225°, and 315° at which the object alternately appears light or dark with respect to the background.

PRINCIPLES OF ACTION OF RETARDATION PLATES AND THREE POPULAR COMPENSATORS

With the addition of a retardation plate or compensator, the polarizing microscope becomes an analytical instrument that can be used to determine the *relative retardation* Γ between the O and E waves introduced by a birefringent specimen. Since $\Gamma = t(n_e - n_o)$, either the birefringence or the thickness of a specimen can be determined if the other parameter is known (see Chapter 8). An excellent description of the action of compensators is given by Pluta (1993).

Transparent plates of birefringent materials such as quartz, mica, or plastic that introduce a fixed amount of retardation between the O- and E-ray pairs are called *retardation plates* or *retarders*. Retarders are prepared at a certain thickness and with the optic axis contained in the plane of the plate. Since incident rays of linearly polarized light are perpendicular to the optic axis of the plate, the O- and E-ray components follow the same trajectory through the plate, but become retarded in phase relative to one another; waves emerge as linearly, circularly, or elliptically polarized beams, depending on the amount of relative phase retardation. Refer to Figure 8-12 to review how changes in the amount of phase retardation affect the waveform and plane of vibration of resultant waves emergent from the plate. The orientation of the optic axis of the plate relative to the plane of vibration of the incident polarized light beam is important. The most common retarders introduce phase retardations of 1λ, λ/2, and λ/4 (2π, π, or π/2 radians) for light of a specific wavelength and are called, respectively, full-wave, half-wave, and quarter-wave plates. As shown in Figure 9-4, a ray of the designated wavelength passing through a full-wave or λ plate remains linearly polarized and retains its original plane of vibration. The λ/2 plate rotates the plane of linearly polarized light by 90°, while the λ/4 plate converts linearly polarized light into light that is circularly polarized, and vice versa. Retarders producing less than λ/4 phase retardation produce elliptically polarized light.

When a retarder is used as a nulling device to determine the relative retardation Γ in a specimen, it is known as a *compensator*. Commonly, the compensation plate is mounted in a device that permits rotation or tilting of the plate by a variable number of degrees. Another method, known as de Sénarmont compensation, uses a fixed λ/4-plate compensator and requires the operator to rotate the analyzer. This action allows the

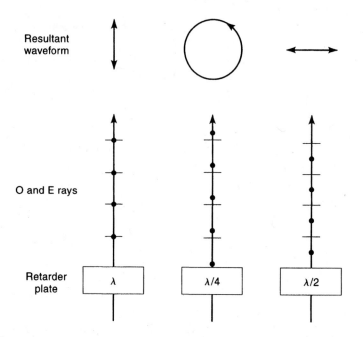

Figure 9-4

The action of three retarder plates. Retarders are special birefringent plates that introduce a fixed amount of relative retardation between O and E rays, whose wavelength spacings are shown here as dots and dashes, respectively. The incident rays are linearly polarized. Since the optic axis of the retarder is in the object plane and perpendicular to the incident ray, the O and E rays follow trajectories that are superimposable, but the waves are relatively retarded in phase.

compensator to be used as a nulling device to undo the phase shift imparted by the specimen through the introduction of an equal but opposite phase shift. The number of degrees of rotation on the compensator required to null the birefringence of an object and bring it to extinction is used to calculate Γ. If other variables are known, the retardation value Γ can be used to determine if birefringence is positive or negative, measure the amount of birefringence—the difference in refractive index ($n_e - n_o$) as experienced by the O and E rays—or determine the specimen thickness. It is even possible to deduce the geometry and patterns of molecular organization within an object (e.g., tangential vs. radial orientation of polarizable bonds of molecules in a spherical body). It is the compensator that makes the polarizing microscope a quantitative instrument.

Compensators are also used for qualitative purposes to control background illumination and improve the visibility of birefringent objects. In a properly adjusted polarizing microscope, the image background looks very dark, approaching black. Inoué has shown that the visibility improves if 5–10 nm retardation is introduced with a compensator, which increases the intensity of the background by a small amount (Inoué and Spring, 1997). The compensator can also be used to increase or decrease the amount of phase displacement between the O and E rays to improve the visibility of details in the object image. Thus, birefringent objects are sometimes seen with greater contrast using a compensator than they are using the polarizer and analyzer alone.

λ-Plate Compensator

Retardations of a fraction of a wave to up to several waves can be estimated quickly and economically using a λ-plate retarder as a compensator (Fig. 9-5). The λ plate is also known as a full-wave plate, first-order red plate, red-I plate (read "red-one plate"), or color tint plate. The plate is composed of a film of highly aligned linear organic polymers or of a sheet of mica, and thus is birefringent. The axis of polarizable bonds in the material defining the slow axis (higher refractive index) of the wavefront ellipsoid is usually marked on the plate holder. When placed between two crossed Polaroid sheets at a 45° angle and back-illuminated with white light, the plate exhibits a bright 1st-order red interference color, hence its designation as a first-order red plate. Full-wave plates introduce vivid interference colors to the image of a birefringent object and are useful for making rapid quantitative assessments of relative retardation, as well as for determining the orientation of index ellipsoids. In geology and materials science, full-wave plates are commonly used to identify birefringent crystalline minerals and determine specimen thickness (see Color Plate 9-1). For retardations of ~λ/3 or less, phase retardations can be measured with an accuracy of ±2 nm.

When placed in front of the analyzer so that its slow axis is oriented 45° with respect to the crossed polars, a red-I plate introduces a relative retardation between O and E rays of exactly one wavelength for green wavelengths of 551 nm (Pluta, 1988). Green wavelengths therefore emerge from the retardation plate linearly polarized in the same orientation as the polarizer and are blocked at the analyzer. O and E waves of all other wavelengths experience relative phase retardations of less than 1λ; they emerge from the plate as elliptically polarized waves and are only partially blocked by the analyzer. The color of white light minus green is bright magenta red, thus accounting for the color of the red-I plate. (To review the principles governing the formation of interference colors, see Chapter 2.) You can confirm that the green wavelength has been removed by examining the red interference color with a handheld spectroscope. All of

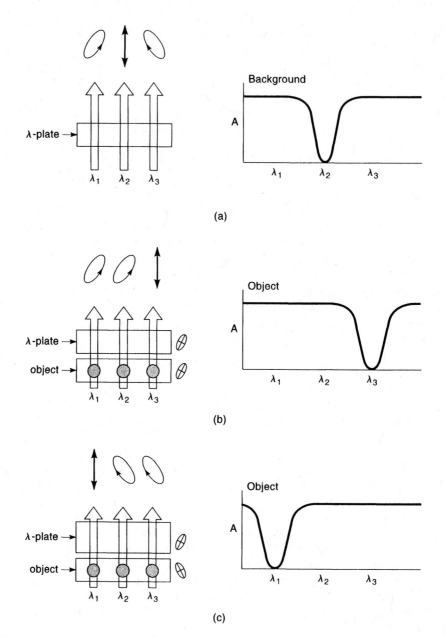

Figure 9-5

Action of a 1st-order full-wave plate in white light. (a) A red-I plate introduces a relative retardation for green light (λ_2), resulting in linearly polarized waves that are eliminated at the analyzer. Shorter (λ_1) and longer (λ_3) wavelengths produce elliptically polarized waves that are partially passed by the analyzer. The resulting spectrum shows extinction of green wavelengths, resulting in the perception of a magenta-red interference color (background color of the field of view). (b) The same plate with an object inserted whose wavefront ellipsoid has its slow axis parallel to that of the red plate. The relative retardation between O and E rays is increased so that now the wavelength giving linearly polarized light is shifted to a longer wavelength (λ_3) such as yellow, and the object exhibits a blue interference color, called an addition color. (c) The same plate with an object whose wavefront ellipsoid has its slow axis crossed with that of the red plate. In this case, the relative retardation between O and E rays is decreased, the wavelength giving linearly polarized light is shifted to a shorter wavelength (λ_1) such as blue, and the object exhibits a yellow interference color, called a subtraction color.

the spectral colors are seen except for a band of wavelengths in the green portion of the spectrum. If now a birefringent object such as a sheet of cellophane is inserted together with the red-I plate between two crossed polars and rotated, the color shifts to blue and yellow. This is because the relative retardation for the combination of red plate plus object is now greater or less than 555 nm, causing the wavelength of linearly polarized light to be shifted and resulting in a different interference color. By comparing object colors with the background color of the plate and referring to a Michel Lèvy color chart (Color Plate 9-1), it is possible to determine the relative retardation introduced by the object.

For birefringent biological materials such as small spherical inclusion bodies, full-wave plates can be used to determine patterns of molecular arrangement (see exercise on plant cell inclusion bodies). The shifted blue and yellow colors are called, respectively, addition and subtraction colors, and immediately tell about the orientation and magnitude of the fast and slow axes of the specimen's index ellipsoid. Thus, the λ plate is a useful semiquantitative device for estimating minute retardations and the orientations of index ellipsoids. As a survey device used to examine complex specimens, it may be more convenient than other more precise compensators.

Demonstration: Making a λ Plate from a Piece of Cellophane

Crumple up a piece of cellophane and place it between two crossed polars on a light box. It is immediately obvious that cellophane is strongly birefringent. In regions where the sheet is a single thickness, the color of the birefringence is white, but in regions where the sheet is folded over on itself and overlapping, a surprising display of interference colors is observed. The colors indicate regions where the phase retardation between O and E waves subtends a whole wavelength of visible light, resulting in the removal of a particular wavelength and the generation of a bright interference color. In these colorful regions, cellophane is acting as a full-wave plate. The different colors represent areas where the amount of relative retardation varies between the O and E waves. The optical path corresponding to any of these colors can be determined using a color reference chart (Color Plate 9-1). If we now insert an unknown birefringent object in the path, the color will change, and using the chart and the cellophane wave plate as a retarder, we can examine the color change and make a fairly accurate estimate of the path length of the unknown specimen. Used this way, the cellophane wave plate acts like a compensator. The following demonstration shows how to make a cellophane wave-plate retarder, understand its action, and use it as a compensator:

- Place a sheet of clear unmarked cellophane between two crossed polars on a light box and rotate the cellophane through 360°. Use a thickness of the type used for wrapping CD cases, boxes of microscope slides, pastry cartons, and so forth. Cellophane dialysis membranes are also very good. Do not use the plastic wrap made for food products or thick, stiff sheets. There are four azimuthal angles (located at 45° with respect to the crossed polars) at which the cellophane appears bright against the dark background. Mark the locations of these axes by drawing a right-angled X on the sheet with a marking pen. Cellulose is clearly birefringent, and the pattern of its action in polarized light is consistent with cellophane containing arrays of long-chain cellulose molecules arranged in parallel across the diameter of the sheet. Notice that the color of the birefringence of a single thickness of cellophane is white.

- Cut the sheet into the shape of a square or rectangle with a scissors or utility knife so that the arms of the X are perpendicular to the edges of the rectangle. At one corner of the sheet place another mark—this time, a straight line 2 cm long parallel to and 1 cm in from the edge of the sheet. Fold the corner of the sheet at a 45° angle, folding over the black line, and reexamine the folded cellophane between the crossed polars. The area of overlap now appears dark like the background and remains dark as the sheet is rotated. If not completely dark, slide the folded corner slightly while holding it between the crossed polars. Extinction occurs because the phase displacement of O and E rays induced by the first layer is exactly reversed and undone by the action of the second folded layer, whose orientation, relative to the main sheet, has been changed by 90°. This is confirmed by looking at the folded black line, which makes a 90° angle. This is the action of compensation, and the folded corner is acting as a compensator to nullify its own birefringence.

- Now fold the sheet in half so that opposite edges remain parallel to each other. Placed between crossed polars at a 45° angle, a brilliant color is seen, which in most cases will be bright yellow. If the polars are parallel, the interference color will be blue. The folded sheet now has the properties of a full-wave plate and is capable of generating interference colors. If the sheet is rotated by incremental amounts between crossed polars while the analyzer is rotated back and forth, you will see the various colors of the 1st-order interference spectrum. Since the retardation introduced by a single thickness of cellophane is ~230 nm, folding the sheet in this way doubles the retardation to 460 nm, a deep blue color. Remember that for a full-wave plate, rays of a particular wavelength (here 460 nm) retarded by exactly 1λ emerge from the plate linearly polarized and vibrating in the same plane as the original incident rays. This component is removed completely at the analyzer. The O- and E-ray pairs of other wavelengths, being retarded by more or less than 1 wavelength, emerge from the plate as elliptically polarized light and are partially passed by the analyzer. Thus, all visible wavelengths excepting the band of wavelengths near 460 nm are transmitted and are collectively perceived as the complementary color to blue, which is a yellow interference color. Inspection of a Michel Lèvy color chart indicates that the yellow color corresponds to removal of wavelengths near 460 nm, and one-half this amount (230 nm) is the amount of retardation for a single thickness of cellophane. The reason that the color of birefringence of a single sheet of cellophane looks white is also apparent: The relative phase retardation is too small to allow removal of a visible wavelength by the analyzer.

- The orientation of the index ellipsoid of the cellophane λ plate—a yellow-I plate—must still be determined. For this we require a birefringent reference object whose index ellipsoid is known. We will use a strip of cellophane tape, since this is manufactured with the strands of cellulose (and direction of the slow axis of the wavefront ellipsoid) parallel to the length of the tape. Remove two 1 cm length pieces of tape and draw a wavefront ellipse on each piece. The slow (short) axis of the ellipse is parallel to the long axis of the strip of tape; the fast (long) axis is parallel to the cut edge of the tape. Place the two pieces of tape at 90° to each other on the folded piece of cellophane, with each piece of tape being parallel to the edge of the cellophane sheet. Place the folded sheet between the crossed polars and orient the sheet at a 45° angle. The regions covered with tape show two new interference colors—pale yellow and sky blue—superimposed on a bright yellow background. Reference to the polarization color chart shows that the blue color corresponds to the removal of a wavelength of 690 nm (460 + 230 nm). This can only occur if the retardation caused by the folded cellophane is further retarded by the tape. The slow axes of the tape and cellophane must be parallel to each other. Conversely, the pale yellow interference color corresponds to removal of a much shorter wavelength (460 − 230 = 230 nm). The slow axis of the tape must be perpendicular to the slow axis of the folded cellophane sheet. This causes the net retardation to be reduced, thus producing a pale yellow interference color. The ellipse drawn for the tape that appears blue can now be marked, retaining the same orientation, on the original cellulose sheet. We now have a calibrated cellulose yellow-I retarder that can be used to determine the amount of retardation and the orientation of the slow and fast axes of other birefringent materials.

- Further practice with the yellow-I plate will reinforce the concept of the compensator as a device for increasing or diminishing the relative phase retardation between the O and E rays and measuring the relative retardation Γ of unknown birefringent objects.

Sénarmont Compensator

Compensation by the method of de Sénarmont requires a fixed quarter-wave plate ($\lambda/4$ plate) and a rotatable analyzer. Since $\lambda/4$ plates are designed for use at a specific wavelength, microscope illumination must be monochromatic, typically at the 546 nm green line of a mercury lamp, although $\lambda/4$ plates intended for use with other wavelengths, such as the 589 nm yellow line of the mercury lamp, are also available. Sénarmont compensation is commonly used to measure the amount of retardation in biological specimens, such as cell organelles, plant cell walls, and muscle fibers, that induce retardations between $\lambda/20$ and 1λ. As explained, this compensator is also used to improve the visibility of birefringent specimens, because birefringent objects can be made to produce bright and dark contrast patterns against a medium gray background (Fig. 9-1).

The retarder ($\lambda/4$ plate) is mounted in a *fixed* orientation between two crossed polars, usually between the back aperture of the objective lens and the analyzer (Fig. 9-6). The analyzer is *rotatable* and is marked in degrees around its circumference so that the angle of rotation can be precisely determined. The $\lambda/4$ plate is made of a birefringent material such as quartz and is prepared so that the optic axis of the plate is parallel to the

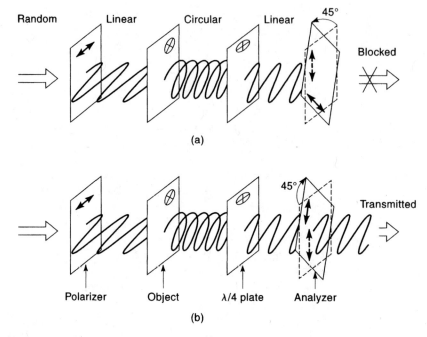

Figure 9-6

Compensation by the method of de Sénarmont. The equipment includes a fixed polarizer and λ/4-plate and a rotatable analyzer. The fixed λ/4-plate is inserted in the microscope so that the slow axis of its wavefront ellipsoid is oriented parallel to the transmission axis of the analyzer when the polarizer and analyzer are crossed. A λ/4-compensator produces circularly polarized light from incident linearly polarized waves and vice versa. To measure the amount of retardation from a specimen, the slow axis of the specimen is oriented at 45° with respect to the crossed polars. For convenience in the drawing, an object giving a relative retardation of λ/4 (the same as the λ/4-plate) is shown; hence, emergent waves are circularly polarized. The λ/4-plate converts these waves into linearly polarized light whose plane of vibration is now tipped 45° relative to the plane of the initial wave produced by the polarizer. When examined with the analyzer in crossed position, the object appears medium gray. Rotating the analyzer from its 0 position counterclockwise 45° blocks light transmission and gives extinction (a), while a clockwise 45° rotation gives maximum transmission (b). Since the relative retardation Γ by this method $= 2\theta$, we calculate that the retardation by the object is $2 \times 45° = 90°$, or λ/4. In a more typical case, a birefringent cellular object giving λ/20 retardation would require a rotation of the analyzer of 9° to give extinction.

plane of the plate. When inserted into the optical pathway, the slow axis of the wavefront ellipsoid of the plate is oriented parallel to the transmission axis of the analyzer (the orientation is north-south when the analyzer is in the crossed position). The plate is fixed and does not rotate. As described in Chapter 8, a plane-polarized beam incident on the birefringent plate is split into separate O- and E-ray components. With the plane of vibration of incident polarized waves parallel to the slow axis of the λ/4 plate, O and E waves from background regions in the specimen emerge linearly polarized, are blocked by the analyzer, and cause the field to look maximally dark. As the analyzer is rotated, the field brightens until it becomes maximally bright at an analyzer setting 90° away from that giving extinction. At 45° rotation, the field looks medium gray. Colors are not observed with incident white light because the amount of retardation introduced is considerably less than the one wavelength retardation required to generate an interference color. Thus, biological specimens such as living cells are frequently examined with the green line of a mercury arc lamp with a λ/4 plate designated for this wavelength. The use of monochromatic green light increases image contrast, allows more precise measurements of retardation, and helps protect living cells.

If a birefringent specimen is positioned diagonally between two crossed polars, the combined action of the specimen plus the λ/4 plate generates linearly polarized waves whose E vectors are tilted at some azimuth θ depending on the amount of retardation at the specimen. This action occurs because the λ/4 plate produces linearly polarized light from incident, elliptically polarized waves. A rotation of the analyzer through 180° is equivalent to a relative retardation by the specimen of one wavelength. With every 90° of rotation of the analyzer, a birefringent object such as a fiber looks alternately dark or bright. Since retardations of most cells and organelles are usually less than 1λ, and because the compensation method is so sensitive, Sénarmont compensation is commonly used to analyze cellular specimens. Compensation with a λ/4 plate allows for measurement of relative retardations of up to 1λ with an accuracy of ±0.15 nm.

To measure Γ, the λ/4 plate is inserted into the optical path between two crossed polars as described. An elongated birefringent object (e.g., a muscle fiber containing aligned myofibrils) is first oriented at a 45° angle with respect to the transmission axes of the crossed polars with the rotating stage of the microscope. At this orientation, the fiber generates elliptically polarized light, which is converted into linearly polarized light by the λ/4 plate. These waves are partially transmitted by the compensator, causing the fiber to look bright against a maximally dark background. The analyzer is then rotated from its crossed position through an angle θ until the fiber's intensity appears maximally dark (extinction) against a gray background. Notice that as the analyzer is rotated to obtain extinction of the specimen, the background is brightening. Since the angle of rotation of the analyzer θ at extinction is equal to one-half of the full phase shift between the O and E rays, the relative retardation Γ is given as

$$\Gamma_{obj} = 2\theta.$$

Γ can be used to calculate the values of the refractive index ellipsoid or the thickness of the object if one of these two parameters is known, since $\Gamma = t\,(n_e - n_o)$.

Brace-Koehler Compensator

For measuring very small phase retardations that occur in fine ultrastructural features such as mitotic spindles, cleavage furrows, and stress fibers in living cells, a Brace-Koehler compensator can be employed. This rotating compensator usually contains a thin birefringent plate made of mica, whose optic axis is contained within the plane of the plate. The compensator is used with monochromatic illumination—generally the 546 nm green line of the mercury arc lamp—and is capable of measuring retardations up to the limit of the compensator (a fixed value, but ranging from λ/10 to λ/30 depending on the particular compensator) with an accuracy of ±0.3 nm. These features make the Brace-Koehler compensator a highly sensitive measuring device.

In this method of compensation the analyzer and polarizer remain fixed in position, because the compensator itself is rotatable. The slow axis of the plate (marked γ on the compensator mounting) is oriented parallel to the north-south transmission axis of the analyzer and corresponds to the 0° position on the compensator dial. A linear birefringent specimen, such as a bundle of actin filaments or microtubules, is first oriented at a 45° angle with respect to the transmission axis of the analyzer by rotating the stage of the microscope to give maximum brightness. Because the maximum optical path difference of the compensator is small ($\sim\lambda/20$), the background appears dark and remains relatively dark through different angles of rotation of the compensator. The compensator is then rotated counterclockwise from its 0° position until light from the specimen is extinguished, matching the dark background. The angle of rotation from the zero position on the compensator is used to calculate the relative retardation between the O and E rays using the equation

$$\Gamma_{obj} = \Gamma_{comp} \sin 2\theta,$$

where θ is the angle of rotation of the compensator and Γ_{comp} is the maximum optical path difference of the compensator ($\sim\lambda/20$). The precise value of Γ_{comp} must be determined by calibration and is a constant in the equation. Depending on the particular compensator, retardations of $\sim\lambda/2000$ can be measured under optimal conditions.

Exercise: Determination of Molecular Organization in Biological Structures Using a Full Wave Plate Compensator

First prepare the microscope for polarization microscopy. The analyzer and polarizer might already be installed in the microscope, in which case it is only necessary to bring them into position as described in the text. If not, strips of dichroic polarizing film can be used in the vicinity of the specimen slide, although this is not standard practice for a research grade instrument. The polarizer is mounted between the specimen slide and the condenser lens with its transmission axis oriented east-west. The analyzer is placed between the specimen slide and the objective lens with its transmission axis oriented north-south. The analyzer is rotated slightly until the background is maximally dark (extinction). If a red-I plate is used, it is inserted between the crossed polars with its slow and fast axes oriented at a 45° angle with respect to the transmission axes of the polars. (See Appendices II and III for sources and instructions on how to prepare red-I plates for this exercise.) The blackened edge of the red-I plate marks the direction of the slow axis of the wavefront ellipsoid.

For orientation, examine several birefringent materials between crossed polars without the red plate, including grains of corn starch, plant cell crystalloids, insect leg and flight muscle, and prepared slides of striated muscle, buttercup root, and pine wood. Notice that linear birefringent structures such as myofibrils in striated muscle and plant cell walls are brightest when their long axes are oriented at 45° with respect to the transmission axes of the crossed polars. Spherical particles like starch grains are bright and exhibit a dark upright extinction cross through their centers. Re-examine the specimens after inserting the red plate. If the waveplate is a red-I plate, the background has a bright magenta-red (first order red) interference color. (*Note:* Other interference colors are possible if the thickness of the plate is not carefully controlled; in this case the interference colors of objects will vary somewhat from the colors described in the exercise.) Muscle fibers, collagen bundles and elongate plant crystals look purple (an addition color) if their long axes are parallel to the slow axis of the wavefront ellipsoid of the red-I plate, and golden yellow (a subtraction color) if their long axes are perpendicular to the slow axis of the plate. Looking at the Michel Lèvy chart of polarization colors, it can be seen that the relative retardation between the purple and gold colors is about 100 nm. Further explanation of the interference colors is given below and in the text.

We will now use the red-I plate to determine the molecular orientation in two plant cell inclusion bodies. Amylose-containing starch storage granules and lignin-containing wood pits are birefringent spherical objects in plant cells consisting of parallel bundles of long-chain molecules and polymers. The two likely patterns for molecular order in these structures might be compared, respectively, to the needles projecting radially from a pincushion (radial pattern) or to surface fibers in the layers of an onion (tangential pattern). The axis of carbon-carbon bonds in these models differs by 90° and is easily observed in a polarizing microscope equipped with a λ plate even though the inclusion bodies are extremely minute. The microscope is set up with the slow and fast axes of the wavefront ellipse of the λ plate oriented at 45° with respect to the transmission axes of the crossed polars. The specimen background exhibits a bright magenta-red color, whereas the granular inclusion bodies appear as spheroids with opposite pairs of yellow-orange and purple-blue quadrants. It is remarkable that the yellow and blue color patterns are reversed for the two types of bodies, indicating differences in the pattern of molecular alignment! Each pair of yellow quadrants and blue quadrants depends on the orientation of the slow and fast axes of the wavefront ellipsoid of the λ plate. The blue quadrants (the addition color) indicate the azimuth along which the slow axes of the specimen and plate are parallel to one another; the yellow quadrants (the subtraction color) indicate the azimuth along which the slow axes of the plate and object are perpendicular. By constructing a diagram where molecular alignment in each quadrant is shown as a series of parallel lines, you can deduce whether the molecules project radially like a pincushion or are ordered tangentially like the layers of an onion.

Roots of herbaceous plants contain an epidermis, a cortex, a pericycle (proliferative tissue), and a vascular cylinder or stele that runs along the axis of the root (Fig. 9-7). Inside the vascular cylinder, identify the xylem—long, longitudinally oriented elements for conducting water and dissolved nutrients and minerals principally upward to the leaves and branches. The phloem transports macromolecules and metabolites (principally downward toward the roots). These are surrounded by a sheath of pericycle and endodermis cells. Outside the pericycle is an extensive layer of cortex containing starch storage granules. Notice the specific stains for the xylem and phloem cells. The walls of plant cells contain ordered filaments of cellulose and lignin and thus are highly birefringent in a polarizing microscope.

The section of pine wood contains mostly xylem (the water transport tissue), plus some vascular rays and pitch canals where pitch accumulates (Fig. 9-7). The

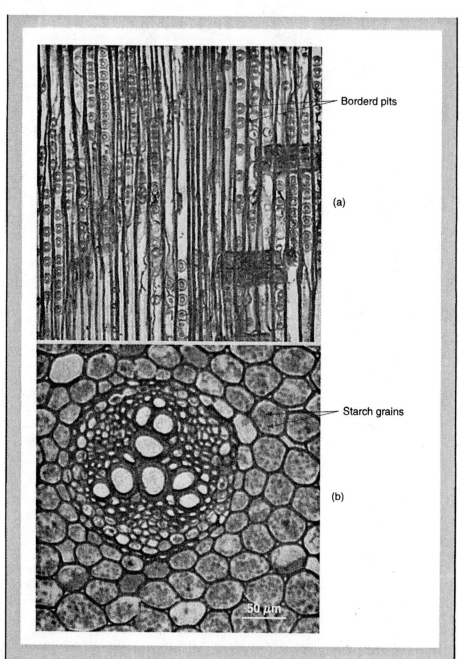

Borderd pits

(a)

Starch grains

(b)

50 μm

Figure 9-7

Histological sections of plant tissues with birefringent objects. (a) Radial section of pine wood containing cellulose walls and bordered pits. (b) Cross section of a buttercup root, an herbaceous plant. The cortical cells are filled with starch gains. (Specimen slides are from Carolina Biological Supply Company, Inc.)

xylem cells are called tracheids and each is interconnected with neighboring tracheids by valvelike structures called bordered pits that regulate hydrostatic pressure in the xylem during fluid transport. The pits are made mostly of cell wall materials, mainly lignin, a very dense phenolic polymer that gives wood its strength. Pine wood is about one-third lignin, the rest cellulose.

Examine the test slides of two plant tissues at 20–40×: a radial section of pine wood (*Pinus strobus*) containing bordered pits, and a transverse section of a buttercup root (*Ranunculus*) containing starch grains.

1. Examine each tissue by bright-field microscopy, carefully adjusting the microscope for Koehler illumination and positioning the condenser diaphragm. Identify bright spherical bodies in the cortical cells of the root and in the tracheid elements in pine wood.

2. Now examine the sections between two crossed polars by polarization microscopy. Insert the polarizer with its transmission axis east-west and the analyzer with its transmission axis north-south. Rotate the analyzer until the field of view is maximally dark (extinction). Now insert the focusing telescope eyepiece to examine the polarization cross in the back aperture of the objective lens and close down the condenser diaphragm to isolate the dark central portion of the cross, masking out the bright quadrants of light at the periphery; then replace the eyepiece to return to visual mode. What structures stand out? Make a sketch of the polarization cross in a bordered pit and explain the pattern of light and dark contrasts in the polarization image.

3. Place a red-I plate on top of the specimen slide with its slow axis at a 45° orientation to the polarizer and analyzer. Make sketches of a starch grain and a bordered pit indicating the colors seen in each of the polarization quadrants using the red-I plate.

4. Make drawings that show your interpretation of the molecular organization in starch grains and bordered pits as revealed by polarized light. Starch and lignin are crystals of long carbon chain macromolecules. Assume that both structures exhibit positive birefringence.

5. Make sketches showing your interpretation for the orientation of cellulose in the xylem tracheids in pine wood. What evidence supports your conclusion?

DIFFERENTIAL INTERFERENCE CONTRAST (DIC) MICROSCOPY AND MODULATION CONTRAST MICROSCOPY

OVERVIEW

In Chapter 7 we examined phase contrast microscopy, an optical system that converts optical path differences in a specimen to contrast differences in the object image. In this chapter we examine two optical systems for viewing gradients in optical path lengths: differential interference contrast (DIC) microscopy and modulation contrast microscopy (MCM). The images produced by these systems have a distinctive, relieflike, shadow-cast appearance—a property that makes the image appear deceptively three-dimensional and real (Fig. 10-1). The techniques were introduced within the past few decades and have gained in popularity to the point that they are now widely used for applications demanding high resolution and contrast and for routine inspections of cultured cells. Although the ability to detect minute details is similar to that of a phase contrast image, there are no phase halos and there is the added benefit of being able to perform optical sectioning. However, there is no similarity in how the images are made. The DIC microscope uses dual-beam interference optics based on polarized light and two crystalline beam-splitting devices called Wollaston prisms. The generation of contrast in modulation contrast microscopy is based on oblique illumination and the placement of light-obscuring stops partway across the aperture planes.

THE DIC OPTICAL SYSTEM

The DIC microscope employs a mode of dual-beam interference optics that transforms local *gradients in optical path length* in an object into regions of contrast in the object image. As will be recalled from Chapter 5 on diffraction, optical path length is the product of the refractive index n and thickness t between two points on an optical path, and is directly related to the transit time and the number of cycles of vibration exhibited by a photon traveling between the two points.

In DIC microscopy the specimen is sampled by pairs of closely spaced rays (coherent wave bundles) that are generated by a beam splitter. If the members of a ray pair traverse a phase object in a region where there is a gradient in the refractive index or thickness, or both, there will be an optical path difference between the two rays upon **153**

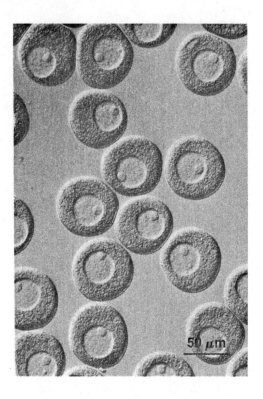

Figure 10-1
Primary oocytes of the surf clam, *Spissula solidissima,* in DIC optics. The gradients of
shading in the image indicate regions of rapidly changing optical path length in the cell. The
large specialized nucleus and prominent dense nucleolus comprise the germinal vesicle in
these meiotic cells.

emergence from the object, and that optical path difference is translated into a change in
amplitude in the image. Since an optical path difference corresponds to a relative phase
difference between the two rays, the presence of phase gradients is acknowledged in a
DIC specimen. Since optical path length is the product of refractive index and thickness,
we cannot tell from the image alone whether a phase gradient in the object is due to differences in refractive index or thickness, or both. Strictly speaking, amplitude differences in the image should be referred to as representing optical path differences, not
refractive index differences or differences in physical thickness, unless other information about the object is known. Refer back to Figure 7-9 to distinguish among these
points.

In the DIC microscope, two optically distinct planar wavefronts traversing a phase
object become deformed and vary in their optical path length in the region of the object;
differential interference between the two wavefronts produces high-contrast patterns of
the phase gradient (e.g., between the edge of the object and the surrounding medium).
The method uses polarized light and special beam-splitting prisms called Wollaston
prisms to generate and recombine the two wavefronts. Although both DIC and phase
contrast optics depend on relative phase differences between sampling beams, there are
fundamental differences. In phase contrast, the amplitude corresponds directly to the
optical path difference between a specimen ray and a general reference ray; in DIC

microscopy, amplitudes correspond to the derivative of the optical path difference profile and not to the optical path difference directly. Thus, if we could make a curve showing optical path length vs. distance across the diameter of a specimen and determine the first derivative of that curve, we would obtain the amplitude profile of the specimen as seen by DIC microscopy; hence the name *differential interference contrast microscopy* (Fig 10-2). The method was described in 1952 and 1955 by Georges Nomarski, a French optics theoretician, and was later developed as an optical system in the mid 1960s by Carl Zeiss, Inc., of Oberkochen, Germany. Excellent descriptions of the method are provided by Allen et al. (1969), Galbraith and David (1976), Lang (1970, 1975) and Padawer (1968).

DIC Equipment and Optics

Differential interference contrast optics are available for most research microscopes. The arrangement of four essential optical components is shown in Figure 10-3. (Instructions for aligning optics and evaluating imaging performance are given later in the chapter.) In order of their location on the optical pathway from the illuminator to the image plane, these include:

- A *polarizer* in front of the condenser to produce plane polarized light. The plane of vibration of the E vector of the waves is oriented horizontally on an east-west or right-left line when looking into the microscope just as in polarization microscopy. (The properties of polarized light and its application in polarization microscopy are discussed in Chapters 8 and 9.)
- A *condenser DIC prism* mounted close to the front aperture of the condenser to act as a beam splitter. The design and action of this prism, technically known as a Wollaston prism, are described in this chapter. Every incident ray (wave bundle) of polarized light entering the prism is split into two rays—O and E rays—that function as the dual beams of the interference system.

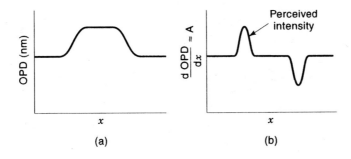

Figure 10-2

Gradients in optical path length yield differences in amplitude. (a) Plot of optical path length across the diameter of a phase-dense object. (b) Derivative of the optical path length curve shown in (a) added to a constant gives the amplitude profile perceived using DIC optics. Positive and negative slopes in (a) correspond to regions of higher and lower amplitude. Regions of the object exhibiting no change in slope have the same amplitude as the background.

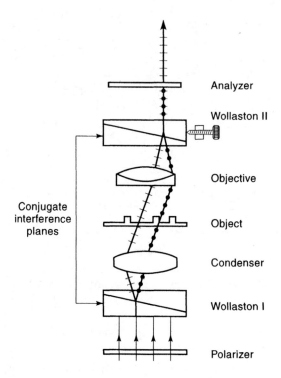

Figure 10-3

Optical components of a DIC microscope. Two polarizers (polarizer and analyzer) and two modified Wollaston prisms (DIC or Nomarski prisms) are required. The condenser DIC prism acts as a beam splitter, producing two closely spaced parallel beams that traverse the object and are recombined by the objective DIC prism. The dots and dashes indicate the mutually perpendicular vibrations of the two components of the split ray as defined in Figure 8-7.

- An *objective DIC prism* mounted close to the back aperture of the objective lens to recombine the two beams in the objective back aperture. The action of this prism is essential for interference and image formation.

- An *analyzer* to "analyze" rays of plane and elliptically polarized light coming from the objective and to transmit plane polarized light that is able to interfere and generate an image in the image plane. It is located near the objective back aperture with its vibrational plane oriented vertically in a north-south or top-bottom orientation when facing and looking in the microscope.

Alternative configurations based on the incorporation of a compensator provide greater control for adjusting image contrast. Since the system uses polarized light, special strain-free pol lenses are highly desirable, because ordinary objectives contain stress signatures in the glass from pressure points in the lens mounting and inhomogeneities in the glass that are birefringent and decrease contrast. Since the physical distance of separation of the wave pairs is as small as 0.18 μm for certain high-power oil immersion objectives (somewhat less than the diffraction-limited resolution of the lens itself), the specifications for lens performance are critical and must be met.

The DIC Prism

The DIC prism, known technically as a *Wollaston prism,* is a beam splitter made of two wedge-shaped slabs of quartz (Fig. 10-4). Since quartz is birefringent, incident rays of linearly polarized light are split or sheared into two separate O- and E-ray components. The shear axis (direction of shear) and the separation distance between the resultant O and E rays are the same for all O- and E-ray pairs across the face of the prism. The E vectors of the resultant O and E rays vibrate in mutually perpendicular planes as they do for other birefringent materials.

In a standard Wollaston prism the optic axes of two cemented wedges of calcite or quartz are oriented parallel to the outer surfaces of the prism and perpendicular to each other. If such a prism is placed between two crossed polars and examined face-on, a pattern of parallel interference fringes is observed due to interference between obliquely pitched wavefronts of the O- and E-ray components. The interference fringes appear to lie inside the prism at a location termed the *interference plane.* This makes it difficult to use a conventional Wollaston prism for certain objective lenses, where the interference plane of the prism must lie within the back focal plane (the diffraction plane) of the lens. This is especially problematic when the diffraction plane lies within the lens itself. Note, however, that the Smith T system of Leica incorporates Wollaston prisms into specially modified objectives. Therefore, *modified Wollaston prisms* are generally used for the objective lens, since the interference plane is displaced from the center of the prism to a distance several millimeters away from the prism itself (refer to the next demonstration). Such a prism does not have to be physically located in the aperture plane of the objective and is much easier to employ. Interference fringes in the field of view are also avoided, and uniform contrasting is achieved.

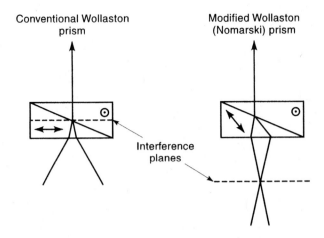

Figure 10-4

Design and action of a Wollaston prism. Wollaston and modified Wollaston (or Nomarski) prisms are used in DIC microscopy to generate and recombine pairs of O and E rays. The optic axes of the two quartz wedges comprising an objective prism are indicated (↔, ⊙). The oblique orientation of the optic axis of one of the wedges in a modified prism displaces the interference plane to a site outside the prism. Interference and visualization of the interference fringes require the action of the analyzer.

Nomarski cemented a conventional wedge with its optic axis parallel to the surface of the prism to another specially cut wedge whose optic axis was oriented obliquely with respect to the outside surface of the prism. For the condenser lens, there are fewer spatial constraints, and a conventional Wollaston prism can sometimes be used. In many cases the condenser prism is a modified Wollaston prism as well, although its interference plane is usually closer to the prism. Thus, the two Wollaston prisms of the DIC microscope are cut differently and are not interchangeable. The condenser prism acts as a beam splitter, while the objective prism recombines the beams and regulates the amount of retardation between O and E wavefronts. *Note the requirement for Koehler illumination to correctly position the interference planes of the DIC prisms in the conjugate aperture planes of the condenser and objective lenses.* The central or 0th-order interference fringe seen in the interference pattern just described is used to determine the proper orientation of the prisms during microscope alignment and is discussed later in this chapter.

Demonstration: The Action of a Wollaston Prism in Polarized Light

- Place a DIC prism between two crossed polars on a light box (white light) and rotate the prism until a dark interference fringe is observed. The intense dark band seen in the middle of the prism is the central 0th-order fringe that results from destructive interference between the O and E wavefronts that have equal path lengths in the middle of the prism. Prisms intended for low-magnification, low-NA optics reveal several parallel interference fringes. Prisms intended for high-magnification, high-NA work reveal a single dark interference fringe.

- Observe that the higher-order fringes appear in the colors of the interference spectrum if white light is used. Higher-order fringes appear dark gray instead of black when illuminated with monochromatic light.

- Also notice that the interference fringes appear to float in space some millimeters above or below the prism. This is because modified Wollaston (Nomarski) prisms have their interference planes located outside the prism itself.

- Rotate the prism between the crossed polars again and notice that there is a unique position giving extinction. At this orientation the phase retardation introduced between O and E rays by one slab of the prism is exactly reversed by phase displacements in the other slab, so light emerges plane parallel and vibrating in the same plane as the incident polarized light. The resultant rays are blocked by the analyzer, giving extinction.

- In the microscope examine the action of a DIC prism between crossed polars (the other prism has been removed) in the aperture plane using a telescope eyepiece or Bertrand lens. A bright field is seen with a dark fringe running across it. When both prisms are in position and the back aperture is observed, the field looks dark (extinction) because light emerging from each position in the condenser prism is now exactly compensated for by the objective prism. All beams emerging from the objective prism are again linearly polarized in the direction of the original polarizer, and thus are blocked by the analyzer (extinction).

Formation of the DIC Image

Both ray tracing and wave optics are useful for explaining image formation in the DIC microscope. By tracing the trajectories of rays from polarizer to image plane, we observe the actions of optical components and understand the DIC microscope as a double-beam interference device. By examining the form and behavior of wavefronts, we also come to appreciate that optimal image definition and contrast are affected by the amount of phase displacement (bias retardation) between the two wavefronts introduced by the operator. (Note, however, that properly adjusted DIC equipment can only ensure good visibility of details that are within the resolution limit of the optics of the bright-field microscope.) Although complex, these details will help you to understand where essential actions occur along the microscope's optical path, allow you to align and troubleshoot the optics, and help you to use the microscope effectively. Before proceeding, keep in mind that the potential spatial resolution is limited, but not guaranteed, by the NA of the objective and condenser lenses.

Ray tracing of the optical pathway shows that an incident ray of linearly polarized light is split by the condenser prism into a pair of O and E rays that are separated by a small distance (Figs. 10-3, 10-4, and 10-5). The E vectors of the two rays vibrate in mutually perpendicular planes. Between the condenser and objective lenses, the trajectories of the ray pair remain parallel to one another and are separated by 0.2–2 µm—the *shear distance*—which is as small or smaller than the spatial resolution of the microscope objective being employed. In fact, as the shear distance is reduced, resolution improves, although at some expense to contrast, until the shear distance is about one-half the objective's maximum resolution. Thus, every point in the specimen is sampled by pairs of beams that provide dual-beam interference in the image plane. Notice that there is no universal reference wave generated by an annulus and manipulated by a phase plate as in phase microscopy, where the distance separating the object and background rays can be on the order of millimeters in the objective back aperture.

In the absence of a specimen, the coherent O and E waves of each ray pair subtend the same optical path length between the object and the image; the objective prism recombines the two waves, generating waves of linearly polarized light whose electric field vectors vibrate in the same plane as the transmission axis of the polarizer; the resultant rays are therefore blocked by the analyzer and the image background looks black, a condition called *extinction* (Fig. 10-5a, b). Thus, the beam-splitting activity of the condenser prism is exactly matched and undone by the beam-recombing action of the objective prism. Note that the axes of beam splitting and beam recombination of both DIC prisms are parallel to each other and fixed at a 45° angle with respect to the transmission axes of the crossed polarizer and analyzer. This axis is called the *shear axis* because it defines the axis of lateral displacement of the O and E wavefronts at the specimen and at all locations between the specimen and the image.

If, however, the O- and E-ray pair encounters a phase gradient in an object, the two beams will have different optical paths and become differentially shifted in phase. We treat the situation the same as we do in standard light microscopy: Waves emanating from the same object particle in the specimen meet at their conjugate location in the image plane, with the difference that the waves must first pass through the objective DIC prism and analyzer. These waves emerge from the prism as *elliptically polarized light* (Fig. 10-5c). The E vector of the resultant ray is not planar, but sweeps out an elliptical pathway in three-dimensional space. These rays partially pass through the analyzer, resulting in a linearly polarized component with a finite amplitude. This information

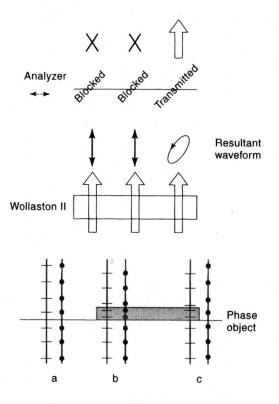

Figure 10-5

Progression of rays through the DIC microscope. An incident beam of linearly polarized light is split by the condenser DIC prism into O- and E-ray components that are focused by the condenser lens onto the specimen. The two rays follow separate parallel trajectories between the condenser and objective lenses. (a, b) In the absence of an optical path difference, the O and E rays are combined by the objective prism, giving linearly polarized light that vibrates in the same plane as the polarizer and is completely blocked by the analyzer. (c) If an optical path difference (phase shift) exists, the prism recombines the beams, giving elliptically polarized light that is partially transmitted by the analyzer.

plus our knowledge of diffraction and interference tells us that image formation will occur, but still does not provide a complete explanation for the unique shadow-cast appearance of the DIC image. For this we need to examine the formation and behavior of wavefronts.

Interference Between O and E Wavefronts and the Application of Bias Retardation

As just described, incident rays of linearly polarized light are split by the condenser DIC prism into O- and E-ray pairs, traverse the specimen, and are recombined by the objective DIC prism, generating linearly and elliptically polarized waves that are differentially transmitted by the analyzer according to the azimuths of their vibrational planes. Since transmitted rays are linearly polarized and are plane parallel, they interfere in the image plane and generate an amplitude image of the object. Another useful way of viewing the situation is to decompose the transmitted rays into their corresponding O- and E-wave components so that we can appreciate the importance of phase displacements between the waves and the role of the objective DIC prism as a contrasting device. Knowledge of the action of the objective DIC prism is important, because the operator must adjust the position of this prism to regulate the amount of optical shadowing and image contrast.

The rays exiting the prism are observed to define two distinct planar wavefronts that meet in the image plane (see Fig. 10-6a). Each front shows localized regions of phase retardation—differential phase retardations—caused by phase objects in the specimen plane. Figure 10-6 (top) shows the reconstructed profiles of the O and E wavefronts in the image plane taken along an axis parallel to the direction of shear with the instrument adjusted to extinction. Each wavefront shows a dip or trough whose width represents the magnified object diameter and whose depth represents the amount of phase retardation ϕ in nm. After combination and interference, the resultant image may be represented as an amplitude plot, from which we deduce that the image of the spherical object shows a dark central interference fringe flanked on either side by regions of brightness. With the background appearing dark, the overall effect is that of a dark-field image.

In practice, a prism setting giving total extinction of the background rays is not used. Rather, the 0th-order interference fringe is displaced to one side of the optic axis of the microscope using the objective prism adjustment screw, an action that introduces a phase displacement between the O- and E-ray wavefronts (Fig. 10-6, bottom). This manipulation is called *introduction of bias retardation*. Since background ray pairs are now differentially retarded and out of phase, they emerge from the objective prism as elliptically polarized waves and partially pass through the analyzer, causing the background to look medium gray. Adding bias retardation now causes the object image to exhibit dark shadows and bright highlights against a medium gray background in regions where there are phase gradients. The amplitude at the edges of objects relative to that of the background depends on whether the O- or E-ray wavefront was phase retarded or phase advanced at the specimen, and is determined by the direction of offset of the interference fringe. On some microscopes bias retardation is introduced by advancing or retracting the objective DIC prism in the light path by turning a positioning screw on the prism holder; on other microscopes containing a $\lambda/4$ plate, the objective DIC prism is fixed, and the bias adjustment is made by rotating the polarizer (Sénarmont method). The amount of displacement between the O and E wavefronts caused by the objective DIC prism is small, usually $<\lambda/10$. Introducing bias retardation makes objects much easier to see, because phase gradients in the specimen are now represented by bright and dark patterns on a gray background. The resultant image exhibits a shadow-cast, three-dimensional, or relieflike appearance that is the distinguishing feature of DIC images and makes objects look like elevations or sunken depressions depending on the orientation of phase gradients. It is important to remember that the relieflike appearance of the specimen corresponds to its phase gradients, not differences in elevation in the specimen, though it may do so if real topological features also correspond to sites of phase gradients.

Alignment of DIC Components

It is important to inspect the appearance of extinction patterns (polarization crosses) and interference fringes in the back aperture of the objective lens to confirm that optical components are in proper alignment and to check for damage such as stressed lens elements,

At extinction

Δ Bias

Σ_1

Σ_2

φ

Δ Object

a

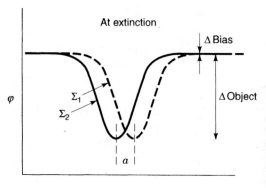

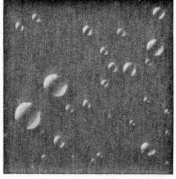

A

x

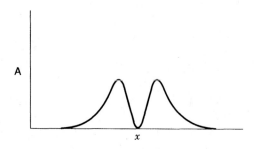

With bias retardation

Σ_1

Δ Bias

φ

Σ_2

Δ Object

a

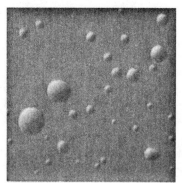

A

x

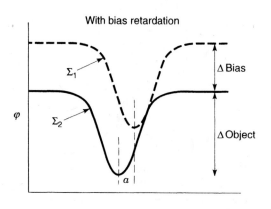

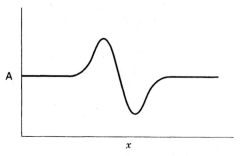

scratches, lint, bubbles, and dirty lens surfaces. Adjustments of DIC optical components are critical to imaging performance, so it is important to recognize misalignments and faults and correct them if necessary. The appearance of the image at different steps of alignment is shown in Figure 10-7, and the operation is performed as follows:

1. Cross the polarizer and analyzer. The polarizer (near the light source) is oriented in an east-west direction as you face the microscope. A mark on the mounting ring of the polarizer indicates its transmission axis. Before adjusting the analyzer, remove all optical components, including the condenser, the objective lens, and DIC prisms. When the analyzer is crossed at 90° with respect to the polarizer, the field looks maximally dark (extinction) when observed through the eyepieces. If the field of view is not dark, move the analyzer in its mounting until the transmission axis is oriented in a north-south direction. If the analyzer is fixed and the polarizer is rotatable, this adjustment is made in the reverse order. When the objective and condenser are inserted (but without the DIC prisms) and the microscope is focused on a blank slide and adjusted for Koehler illumination, the field looks dark in visual mode and a dark extinction cross can be seen in the back aperture of the objective lens with an eyepiece telescope or Bertrand lens. If the polarizer and analyzer are mounted properly, the extinction cross will have straight horizontal and vertical components. There should not be any bright birefringent streaks, which are indicators of strained lenses and inferior performance in DIC.

2. Examine the objective back aperture, with the objective DIC prism in position and the condenser prism removed. A single dark interference fringe extends across the diameter of the back aperture from the northwest to southeast quadrants at a 45° angle. The fringe should be well defined and should run through the middle of the aperture. The objective prism is fixed in some microscope designs, but in others it can be adjusted using a prism positioning screw. The image field as seen through the eyepieces looks bright and featureless.

Figure 10-6

Interference between O and E wavefronts in the image plane. The two views show the DIC prism adjusted for extinction (top) and with the addition of bias retardation (bottom). The pairs of graphs for each condition show the positions of wavefronts (ϕ) and the corresponding amplitudes (A) for profiles taken through an object in the direction of prism-induced shear, which gives the greatest contrast. The x-axis represents the distance x across the object. The graphs indicating the phase shift ϕ show the O and E wavefronts (labeled Σ_1 and Σ_2) in the image plane after passage through the objective DIC prism and analyzer. The dips in the wavefronts represent phase retardations resulting from transit through a phase object. The graphs of amplitude A show the wave resulting from interference between the two original wavefronts. *Objective prism adjusted to extinction:* Notice that under conditions of extinction, the two wavefronts in the top panel are sheared laterally by a distance a along the x-axis, but do not exhibit a phase difference in the regions corresponding to background. These regions have 0 amplitude and appear black in the corresponding intensity plot. *Addition of bias retardation after movement of the objective DIC prism:* The two wavefronts remain sheared by the same amount a, but are now relatively shifted in phase. The corresponding amplitude plot shows a bright edge on the left-hand side and a dark edge on the right-hand side. Moving the DIC prism changes the displacement between the two wavefronts along the y-axis and alters the contrast.

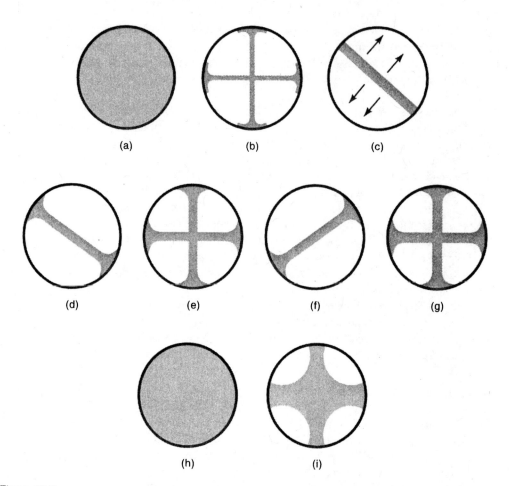

Figure 10-7

Alignment of DIC optical components. The following description is based on alignment of DIC prisms on a Zeiss inverted microscope. The alignment procedure is different on other microscope designs, but the appearance of the expected interference fringes is the same. (a and b) Both DIC prisms removed, crossed polars: Image view (a) reveals a maximally dark field (extinction), while telescope view (b) shows the characteristic extinction cross. (c) Objective DIC prism only; crossed polars; telescope view: A prominent interference fringe is seen running diagonally from northwest to southeast across the field. Adjustment of the prism position with the bias adjustment screw shifts the fringe pattern laterally. For extinction, the fringe is positioned in the center of the field. (d–g) Condenser DIC prism only; crossed polars; telescope view: A prominent interference fringe runs diagonally from northwest to southeast across the field. The patterns seen at successive 90° rotations of the prism are shown, but only the first fringe pattern (d) is correct. Thus, the objective and condenser prism interference fringes must be parallel and overlapping. (h and i) Both DIC prisms; crossed polars; image and telescope views: In image view, the field looks uniformly dark. In telescope view, a dark, diffuse extinction cross is observed. Slight adjustments of the condenser and DIC prisms might be necessary to obtain perfect alignment. The condenser prism is then locked into position. The condenser aperture is stopped down to mask bright peripheral illumination at the objective back aperture.

3. Now inspect the front aperture of the condenser again using a telescope, with the condenser DIC prism in position and the objective prism removed. A single interference fringe running northwest to southeast at a 45° angle should be observed. To see the fringe clearly on some high-NA condensers, it may be necessary to remove the front 1.4 NA cap lens of the condenser. If the fringe is not

well defined or is misoriented, the condenser prism may need to be rotated to optimize the alignment. Usually this adjustment is made by the manufacturer and remains fixed (or can be locked down with a set screw), so it does not need to be altered. If it is not properly displayed, contact the manufacturer and get instructions on how to reset the alignment. The image field as seen through the eyepieces again looks bright and featureless.

4. Mount and focus a specimen (such as buccal epithelial cells) on a slide and set Koehler illumination with both DIC prisms and polarizers in position with the optics set at extinction. The image field looks very dark gray at extinction, while a sharply defined interference fringe or band is seen running in a northeast-southwest direction across the diameter of each refractile object particle (Fig. 10-6, top). The orientation of the interference fringe is parallel to the fringe orientation seen in aperture views of the individual condenser and objective DIC prisms already described, and is the correct orientation for the fringe in the image at extinction. While viewing the image, note that advancing the objective DIC prism (or rotating the analyzer a few degrees to either side of the extinction position) moves the interference fringe bisecting particles or organelles along an axis oriented in a northwest to southeast direction (the shear axis), causing one side of the organelle to look dark and the opposite side to look bright. For a given microscope, only one of these elements is adjusted (prism or polarizer or analyzer) to introduce bias retardation. Adjusting the bias retardation brightens the background, improves image appearance and contrast, and is an essential final step in the adjustment of DIC optics. In addition to the presence of discrete light and dark intensities at opposite edges of each organelle along the shear axis, a broad and indistinct field fringe is sometimes observed, a gradient of light across the entire field of view. With well-designed optics, the field fringe is so broad that the entire image background appears a uniform medium gray. More commonly, some evidence of the fringe remains, so after introducing bias retardation (by adjusting the DIC prism or rotating the analyzer), the field exhibits a shallow gradient of light intensity from one edge to the other.

5. When the back aperture of the objective is examined with an eyepiece telescope with the DIC prism set at extinction, the central region should look dark gray and uniform, but possibly with some brightening at four quadrants at the periphery, giving the appearance of a very broad extinction cross similar to the one observed in the back aperture of a polarizing microscope. The brightening represents an artifact due to partial depolarization of light at lens elements of the condenser and objective. Image contrast can be greatly improved if these regions are masked out by partly closing down the condenser aperture, leaving 75% of the aperture diameter clear. If the optics are in perfect adjustment at extinction, the cross stands upright and is seen to be composed of two broad interference fringes, each bent in the shape of a right angle and meeting in the center of the aperture. On most microscopes the fringe pattern can be adjusted to make the central region of the aperture darker and more uniform. This is done by loosening and rotating the condenser DIC prism a small amount, or by slightly rotating the polarizer or analyzer. Now secure the components, leaving only one for adjusting bias retardation. As a final adjustment and check, move the condenser focus very slightly out of the Koehler position to determine if extinction at the back aperture can be improved still further. However, too great a movement will bring the conjugate interference planes of the prisms too far apart and degrade optical performance.

Image Interpretation

The DIC image has a relieflike quality, exhibiting a shadow-cast effect as if the specimen were a three-dimensional surface illuminated by a low-angle light source. It must be remembered that the shadows and highlights in the shadow-cast image indicate the sign and slope of phase gradients (gradients in optical path length) in the specimen and do not necessarily indicate high or low spots. The direction of optical shear is obvious and is defined by an axis connecting regions having the highest and lowest intensity. Finally, the direction of the apparent shadow casting reverses for structures with refractive indices that are lower and higher than the surrounding medium. Thus, dense nuclei, mitochondria, and lysosomes might have the appearance of raised elevations, while less dense pinocytotic vesicles and lipid droplets look like sunken depressions. The degree of contrast and extent of three-dimensionality depend on the amount of bias retardation between wavefronts imparted by the objective prism. The axis of optical shear cannot be changed by changing a setting on the microscope. However, the orientation of bright and dark edges can be reversed 180° by moving the DIC prism to place the optic axis of the microscope on the other side of the null position of the prism. This has the effect of reversing the relative phase retardation of the O and E wavefronts. Therefore, the only way of changing the shear axis relative to the specimen is to rotate the specimen itself. For certain symmetric specimens such as diatoms, specimen rotation can be used to highlight different features (Fig. 10-8). A precision-rotating specimen stage is very useful in deducing the direction of phase gradients in complex structures.

Figure 10-8

Effect of specimen orientation in DIC microscopy. Since the shear axis is fixed in DIC optics, the specimen itself must be rotated to highlight different features. Notice the differential emphasis of pores and striae in the shell of a diatom, *Amphipleura*, using video-enhanced DIC optics.

Finally, note that the intensity of shadows and highlights is greatest along the direction of the shear axis. If we examine the contrast at the edges of a spherical particle along diameters taken at different azimuths, we observe that the contrast between the particle and the background gradually decreases and reaches zero at 90° along a line defining the axis of the interference fringe. At this position, the irregularities in profiles of the O and E wavefronts are exactly aligned, so subtraction of the two wavefronts cancels out any retardations and gives a positive value that exactly matches that of the background.

The best amount of bias retardation is that giving optimum contrast to the object image and is unique for each object. Since the field of view usually includes many phase objects of different size and refractive index, the best overall bias setting is a compromise. The following guidelines are useful in performing this adjustment:

- The amount of bias retardation required to maximally darken one slope or edge of an object also gives the maximum possible contrast between the object and the background. Thus, for any given object, there is an optimal amount of bias retardation that requires a particular prism setting.

- If a bias retardation is chosen that is greater than the minimum amount required, the contrast will be reduced.

- Thick light-scattering objects may require a higher bias compensation setting in order to obtain extinction of one edge (gradient slope) of the object.

- When the condenser aperture exceeds about 75% of the objective aperture, light scattering in the optical system increases significantly and contrast becomes reduced.

The Use of Compensators in DIC Microscopy

Although the DIC microscope is largely a qualitative instrument, a compensator can be used to manipulate the amount of bias retardation between O- and E-wave pairs more precisely and give more control to adjusting the contrast of specimen details in the image. The action of compensators as contrasting and measuring devices is described in Chapter 9. The compensator is placed in a specially designated slot between the crossed polars and introduces a known amount of retardation.

A full-wave plate or λ plate, such as the red-I plate with a retardation of ~551 nm, can be used to color the image by introducing a spectrum of interference colors. The colors at the edges of objects and their immediate background can be compared using a Michel Lèvy color chart to estimate the magnitude of the optical path difference.

The Sénarmont compensator contains a fixed λ/4 wave plate and a rotating analyzer, and is frequently used with DIC optics to introduce a known amount of bias retardation to a specimen. This might be needed in certain semiquantitative applications or simply as a monitor of DIC optical alignment. Allen (1985) and Inoué (1989) used the technique to introduce a precise amount of retardation to optimize the contrast of microtubules imaged by video-enhanced DIC microscopy. When using this technique, the λ/4 wave plate and analyzer (adjusted for extinction) are inserted into the optical path and the objective DIC prism is adjusted to give extinction. The analyzer is then rotated to give the desired amount of bias retardation and background intensity (for details on Sénarmont compensation, refer to Chapter 9). The degrees of rotation can be noted for future reference.

Comparison of DIC and Phase Contrast Optics

Figure 10-9 shows DIC and phase contrast images of a flattened protozoan, *Acanthamoeba castellanii*. Both images show a conspicuous nucleus and nucleolus, but intensity differences in the phase contrast image show that the nucleoplasm is less dense that the cytoplasm and that the nucleolus is denser than either the nucleoplasm or the cytoplasm; notice too that the cytoplasm is denser than the surrounding medium. Small phase-dark mitochondria and large phase-light vacuoles are conspicuous. Phase halos around the cells indicate significant differences in optical path length compared to the surrounding medium, which is water.

In the DIC image, notice that the bright and dark edges of the nucleus as well as the vacuoles along the shear axis are reversed compared with intensity distributions at the edges of the nucleolus and the whole cell, which is consistent with the nucleoplasm and vacuole being less dense than the cytoplasm. Organelles are also clearly defined, and there is no phase halo. The shadow-cast, three-dimensional appearance is the result of dual-beam interference.

MODULATION CONTRAST MICROSCOPY

Optical methods based on oblique or off-axis illumination provide an alternative to DIC optics for viewing phase gradients in an object. The principal systems are single side-band edge-enhancement (SSEE) microscopy described by Ellis (1978) and modulation contrast microscopy (MCM) described by Hoffman and Gross (1975; see also Hoffman, 1977). For examining tissue culture plates, Carl Zeiss recently introduced Varel optics,

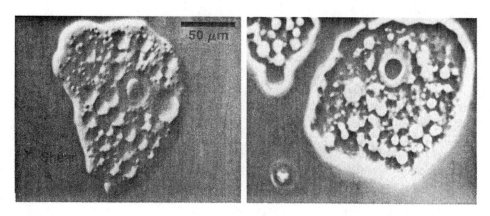

Figure 10-9

Comparison of DIC and phase contrast images of a living soil amoeba, *Acanthamoeba*. Bar = 50 μm. *DIC:* The direction of the shear axis is shown in the micrograph. The cell appears as if illuminated by a grazing incident light source located in the upper left corner. Bright regions at the upper margins of the cell, the nucleolus, and small spherical mitochondria indicate these objects are phase dense (have a higher refractive index) compared with their surround. Bright contrast at the bottom edges of the nucleus and cytoplasmic vacuoles indicates these objects are phase light. *Phase contrast:* Positive phase contrast renders phase-dense and phase-light objects as dark and light contrast features in the image according to their optical path length relative to the background. The cells themselves are surrounded by a bright phase halo, an artifact of the phase contrast optical system. The information content (spatial resolution, detection sensitivity) of the two optical systems is similar.

which uses a related optical system. Like DIC optics, MCM systems produce images that have a three-dimensional or shadow-cast quality, making objects appear as though they were illuminated by a low-angle light source (Fig. 10-10). In both MCM and DIC, brightly illuminated and shadowed edges correspond to *optical path gradients* (phase gradients) of opposite slope in the specimen, but unlike DIC, the MCM system does not require crystalline DIC prisms. Although resolution and detection sensitivity of the Hoffman MCM system are somewhat reduced compared with DIC, the MCM produces superior images at lower magnifications, allows optical sectioning of rounded cell specimens, and offers certain advantages over DIC optics, including the ability to examine cells on birefringent plastic substrates such as cell culture dishes. The Hoffman modulation contrast system is commercially available through Modulation Optics, Inc., Greenvale, New York.

Contrast Methods Using Oblique Illumination

Those who test optical surfaces will already be familiar with the essentials of the schlieren system, which is related to the well-known knife edge test first employed by Leon Foucault in 1859 for measuring the radius of curvature of a lens surface. Toepler later used the method to examine variations in the refractive index of a transparent medium in a sample cell, where inhomogeneities in the medium appear as high contrast

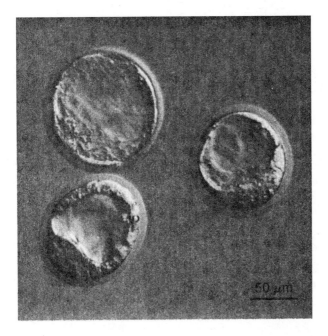

Figure 10-10

Mouse blastocysts, modulation contrast microscopy. As in DIC microscopy, variations in intensity of the image correspond to gradients in optical path length in the specimen. The contrast image is generated by blocking one sideband of the diffracted light. There is no dependence on polarized light and no dual-beam interference mechanism as in DIC microscopy. (Image courtesy of Mahmud Siddiqi and Ann Lawler, Johns Hopkins University.)

streaks (Schlieren in German) (Fig. 10-11). The cell is illuminated with a narrow slit, and the light is refocused with a lens to reform an image of the slit. The sample cell is examined by placing the eye just behind the slit image, while an opaque straight edge is inserted into the focused beam, with the edge aligned with the slit, so as to nearly completely mask the slit and reduce transmission to the eye. Brightness and contrast are modulated by the degree to which the knife edge blocks light from the slit. In forms of schlieren microscopy the optical design is similar. The specimen is illuminated by a slit in an opaque aperture mask placed in the front aperture of the condenser, and an adjustable knife edge located in the back aperture of the objective is used to adjust brightness and contrast. The schlieren image is formed in the following way.

The object field appears evenly illuminated, but phase gradients in the object deflect rays through principles of diffraction, refraction and reflection to regions outside the area of the focused image of the slit, which represents the 0th order or direct light component. The eye sees the object as a relieflike pattern of shadows and highlights. Intensity differences perceived by the eye are due to interference in the image plane between the 0th-order component and a single sideband of the diffracted light component.

Light microscopes using oblique illumination and MCM optics operate by similar principles. *Oblique illumination* can be obtained in a standard transmitted light microscope by selectively illuminating one side of the front lens of the condenser using an opaque mask with an off axis slit in the front aperture position. Alternatively, you can simply rotate the condenser turret until light passing through the condenser iris diaphragm hits one edge of the condenser lens. This condition is analogous to using an illuminating slit as already discussed. The image of the offset condenser aperture is now offset in the conjugate back aperture of the objective lens. Many of the rays diffracted by an object that would be brought to a position peripheral to the offset aperture are blocked by the edge of the lens and become excluded from image formation. In this case, the function of the knife edge at the aperture plane is provided by the edge of the lens itself. While the delivery of light is not well controlled resolution is good because diffracted waves are included on one side of the 0th-order spot in the back focal plane of the objective.

The arrangement of components in a modulation contrast microscope resembles the design of the schlieren optical system described above. The slit and knife edge of the schlieren system occupy conjugate focal planes, and placement of the eye just behind the knife edge allows these planes to function as the condenser and objective aperture

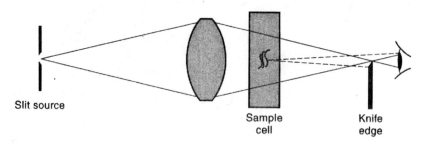

Slit source

Sample cell

Knife edge

Figure 10-11

Optical plan for schlieren optics with off-axis illumination. A knife edge placed close to the eye blocks one sideband of diffracted rays (dotted lines), creating a shadow-cast contrast image of phase gradients in the sample cell.

planes in a microscope. The object and retina define two conjugate field planes of the system. These features are modified in *Hoffman modulation contrast optics* as shown in Figure 10-12. An off-axis slit of some width is mounted in the front aperture of the condenser, while the knife edge at the back aperture of the objective is represented by a modulator plate. The modulator is divided into three asymmetric regions: (1) a nearly opaque section of a circle at the extreme edge of the plate, (2) an adjacent semidarkened rectangle giving 15% transmission, and (3) a large transparent zone that allows 100% transmission. When properly aligned, the image of the condenser slit exactly fills the semitransparent rectangle and produces even, attenuated illumination in the image plane. Sliding the modulator to the right or left exposes a greater or lesser area of the slit and brightens or darkens the background in the image. This right-left shear axis is the same axis that defines the bright and dark contrast regions in the image, but details along a north-south diameter through the object have minimal contrast. To examine contrast

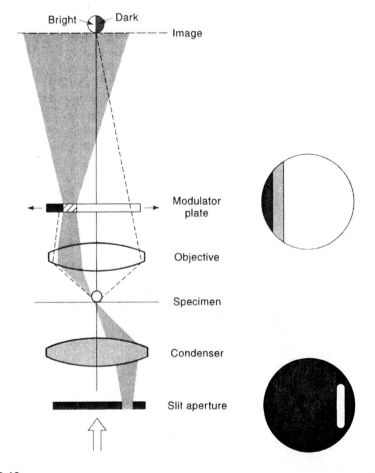

Figure 10-12

Equipment for modulation contrast microscopy. Oblique illumination is provided by an off-axis slit in the condenser aperture. A modulator plate with matching complementary slit in the objective back aperture differentially blocks one sideband of diffracted light. Movement of the plate modulates the transmission of 0th-order light, allowing for regulation of image contrast.

features of the specimen at different azimuths, it is necessary to rotate the specimen on the microscope stage.

Gordon Ellis demonstrated that the mechanism of contrast generation in this and other schlieren-related systems is based on the selective removal of diffracted light on one side of the 0th-order spot in the diffraction plane. This demonstration was made on a schlieren microscope of his own design: the *single sideband edge-enhancement (SSEE) microscope.* As in the Hoffman MCM microscope, there are complementary masks (half-aperture masks) in the front aperture of the condenser and in the back aperture of the objective. Ellis demonstrated (1978) that a transparent object (a water-immersed diffraction grating replica), which is invisible without the masks in a focused bright-field image, becomes visible at focus when the *diffracted light of one sideband is blocked* by the objective aperture mask. In the demonstration, there was no modification of the direct light, so it is possible to conclude that formation of a visible image is strictly a consequence of the change in interference between the diffracted sidebands and the direct light.

Alignment of the Modulation Contrast Microscope

The microscope is first adjusted for Koehler illumination. The condenser slit aperture is mounted in a vertical north-south orientation in the front aperture of the condenser (Fig. 10-13). Slit alignment is performed while viewing the objective aperture plane with an eyepiece telescope or Bertrand lens. The modulator is inserted in a slot near the back aperture of the objective lens and likewise aligned, giving attention that the image of the condenser slit exactly fills the semidarkened rectangle on the plate. A positioning screw on the modulator allows you to slide the plate along an axis perpendicular to the long axis of the condenser slit image. Other versions of MCM include two polarizing elements, one of which is rotated, to vary image brightness. Brightness can also be controlled by moving the modulator plate, but as movement of the plate affects the contrast of phase gradients, this method of control may not be desirable. Since both polarizing elements are located on the same side of the specimen, loss of contrast from birefringent plastic substrates such as culture dishes is not a problem.

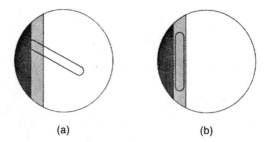

(a)   (b)

Figure 10-13

Alignment of optical components for modulation contrast microscopy. The alignment of the condenser slit aperture and modulator plate is examined using an eyepiece telescope or Bertrand lens. The condenser slit must be perfectly aligned with the rectangular gray area of the modulator plate to properly isolate and control the 0th-order beam. Note the requirement for Koehler illumination for the confocal positioning of the two conjugate aperture planes.

Exercise: DIC Microscopy

- *Review of procedures for aligning optical components for DIC microscopy.* In a sketch, indicate the proper sequence and orientation of the polarizer and analyzer, as well as the condenser and objective DIC prisms. Indicate the pattern and orientation of interference fringes seen with a Bertrand lens when the microscope is set at extinction and when the condenser or objective DIC prism has been removed.

- *Examination of diatom shells by DIC microscopy.* Examine the diatom test plate at 100× with DIC optics and determine if the striae and pores in *Frustulia* and *Amphipleura* can be resolved. Alternatively, examine the myofibrils of striated muscle in the thorax flight muscles of a fruit fly, *Drosophila*. Anesthetize one or two flies, remove the head and abdomen, and place the thorax in a 30–50 μL drop of water between a microscope slide and coverslip, and squash the thorax to disperse the muscle. Blot off excess water, dry, and seal the edges of the coverslip with nail polish to prevent evaporation. Allow to dry thoroughly (!), and clean the glass surfaces before examination. To obtain the highest resolution and contrast for these demanding specimens, oil both the objective and the condenser. Adjust the microscope for Koehler illumination and use the Bertrand lens to check that the condenser diaphragm masks some of the peripheral light, but does not significantly block the aperture. Does the spacing you observe approach the theoretical Abbe limit for resolution in the light microscope?

  Since DIC microscopy uses polarized light, it can be used in combination with a compensator such as a full-wave plate to examine interference colors in phase objects. The high-NA optics available for DIC mean that the specimen can be examined at very high resolution and with optical sectioning. Insert a red-I plate somewhere in the light path between the two crossed polars to observe the effect of optical staining. Be sure you have the correct orientation (axes of the refractive index ellipsoid of the plate at 45° to the axes of the polarizers). Note the colors of the background and of the edge of the shell and uniform portions of the shell itself using a 40× objective. Notice that each edge of the specimen is composed of a double band of interference colors. On the opposite side of the diatom, the same double-color band is observed, only in reversed order. The exact color observed depends on the optical path difference intrinsic to the specimen and the bias retardation introduced by the DIC prism. Although the interference colors indicate optical path differences in the specimen, the DIC microscope is primarily a qualitative instrument.

- *Examination of buccal epithelial cells.* Buccal epithelial cells are excellent specimens for practicing optical contrasting with the DIC prism and for examining the effect of the condenser diaphragm on image contrast and optical sectioning. The cells contain a centrally placed nucleus and have a modified plasma membrane that preserves the angular edges of their original polygonal shape in the epithelium, as well as patterns of multiple parallel ridges and grooves that exhibit an interridge spacing of ~0.4 μm. Gently scrape the underside of your tongue with the edge of a #1.5 coverslip to collect ~10–20 μL of clear saliva and surface cells of the stratified squamous epithelium, and then mount the coverslip on a microscope slide.

1. Examine the preparation by phase contrast and DIC optics using a 40× or 100× oil immersion objective, and compare the quality of the images. Focus through the specimen during observation and notice the clarity of optical sectioning using DIC optics.

2. Carefully examine and test the alignment of polarizers and DIC prisms, as viewed at the back focal plane of the objective lens with a telescope eyepiece or Bertrand lens.

• *Examination of living tissue culture cells.* Obtain a coverslip of tissue culture cells (COS7 cells, U2OS cells, or other flat epithelial cells are ideal). Cultures are prepared 1–2 days beforehand using plastic tissue culture dishes containing #1.5 coverslips presterilized by dipping in ethanol and holding briefly over a flame. Cell organelles that are identifiable in DIC include the nucleus, nucleolus, heterochromatin, lysosomes and peroxysomes, secretion granules, lipid droplets, mitochondria, rough endoplasmic reticulum, Golgi apparatus, centrosomes and centrioles, and stress fibers consisting of bundles of actin filaments. Membrane specializations, including the leading edge, membrane ruffles, filopodia, and microvilli, should also be visible. DIC optics provide high sensitivity of detection and the ability to optically section through the specimen.

1. Carefully remove a coverslip with a pair of fine forceps, making sure not to damage the forceps or wipe off the surface layer of cells. Next clean off and wipe dry (carefully!) the back side of the coverslip with a water-moistened lab tissue, again being careful not to break the coverslip. The cleaned surface should be immaculate and dry.

2. Mount the coverslip with Vaseline spacers as shown in Figure 10-14. Clean off the back surface of the coverslip with a moistened lab tissue. Place the dry coverslip on a clean surface, cell-side-up, and wipe away cells and medium within 1 mm of two opposite edges of the coverslip. Act quickly so that the cells do not dry out. Smear a small bead of Vaseline on the edge of the palm of your hand and drag the edge of the coverslip over the surface to make a small, even ridge of Vaseline on the two dried edges of the coverslip (the side facing the cells). Mount the coverslip on a slide as shown, with the edges containing Vaseline parallel to the long axis of the slide, and gently tap the coverslip to ensure good attachment to the glass slide. Place 1 drop of Hepes-buffered Minimal Essential Medium (HMEM) against one open edge of the coverslip to nearly fill the chamber. Do not overfill, or the coverslip will float loose from the slide and be difficult to focus. Use the edge of a filter paper to wick away excess medium beforehand if necessary. It is better to leave a little airspace than to overfill the chamber. To prevent evaporation, you can seal off the two exposed edges with small drops of immersion oil.

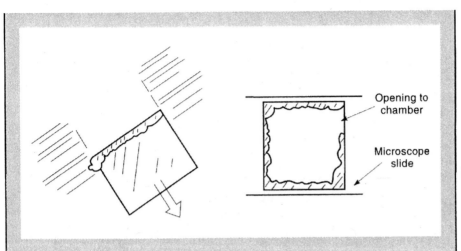

Figure 10-14
A quick mount for the examination of cells grown on coverslips. A thin layer of petroleum jelly is spread evenly across a glass microscope slide or the palm of the hand, and a coverslip is drawn across the substrate sufficient to accumulate a small ridge of jelly on the side of the coverslip facing the cells. The coverslip is mounted on a microscope slide as shown, is gently tapped down to assure solid contact, and the chamber is filled with culture medium.

3. Examine the preparation immediately with a $40\times$ dry objective, but plan to make most of your observations using a $40\times$, $60\times$, or $100\times$ oil immersion lens. Illuminate the cells with monochromatic green light. It is advisable to use UV- and IR-blocking filters. Make a labeled sketch of a well-spread cell showing all the recognizable organelles and structures. Include a scale bar on your drawings.

FLUORESCENCE MICROSCOPY

OVERVIEW

With light microscope optics adjusted for fluorescence microscopy, it is possible to examine the distribution of a single molecular species in a specimen, and under specialized conditions, even detect individual fluorescent molecules. In contrast to other forms of light microscopy based on object-dependent properties of light absorption, optical path differences, phase gradients, and birefringence, fluorescence microscopy allows visualization of specific molecules that fluoresce in the presence of excitatory light. Thus, the amount, intracellular location, and movement of macromolecules, small metabolites, and ions can be studied using this technique. Figure 11-1 shows one such example: the distribution of an enzyme, β-1,4-galactosyltransferase, in a tissue culture cell. Typically, nonfluorescent molecules are tagged with a fluorescent dye or fluorochrome, in order to make them visible. Examples of these are DAPI and the Hoechst dyes used to directly label nuclear DNA or rhodamine-labeled phalloidin used to indirectly label cytoplasmic actin filaments. Alternatively, fluorochrome-labeled antibodies can be used to label fixed, permeabilized cells in a method known as *immunofluorescence microscopy,* the technique used in Figure 11-1. These techniques are commonly used to visualize the distribution of certain proteins in a cell or to make visible specific organelles, filaments, and biochemically distinct membrane regions. A variety of new tagging methods is also employed, including inserting short DNA sequences of known epitopes into the coding sequences of proteins (epitope tagging), constructing protein chimeras with green or red fluorescent proteins (GFP, RFP), and several other methods.

Fluorescence microscopy has gained in popularity ever since Coons (1941) developed methods to conjugate proteins to fluorochromes, and as improvements in optics, thin film technology, and opto-electronics increased the specificity and sensitivity of detection of emitted fluorescent light. Among the most important advances were the application of interference filters and the dichroic mirrors and their incorporation in a versatile epi-illuminator (Ploem, 1967), the introduction of special high-NA objective lenses, and the introduction of sensitive films and electronic imaging devices. Advances continue today in all of these areas. Because of its great specificity and relative ease of use, fluorescence microscopy is the most frequently employed mode of light microscopy used in biomedical research today.

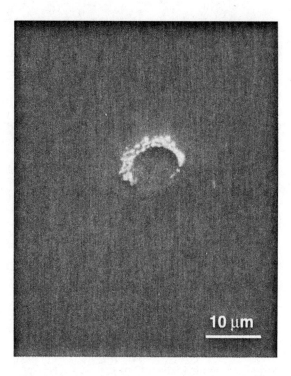

Figure 11-1

Demonstration of specific molecule labeling by immunofluorescence microscopy. Cultured primary endothelial cells from bovine aorta were fixed, extracted, and labeled with an antibody to the enzyme β-1,4-galactosyltransferase and a fluorescein-labeled secondary antibody. Galactosyltransferase is the only molecule labeled, and is observed to be highly enriched in the *trans* cisternae of the Golgi apparatus. Fluorescence microscopy is commonly used to determine the amount, distribution, and dynamics of specific macromolecules in cells. Bar = 10 μm.

Fluorescence microscopes contain special filters and employ a unique method of illumination to produce images of fluorescent light emitted from excited molecules in a specimen. The filters are designed to isolate and manipulate two distinct sets of excitation and fluorescence wavelengths. A band of shorter excitation wavelengths from the illuminator and filters is directed to the specimen, while a band of longer fluorescence wavelengths emitted from the specimen forms an image of the specimen in the image plane. To perform fluorescence microscopy effectively, the microscopist must be able to select the proper fluorochrome, filters, and illuminator for a given application and evaluate the quality of fluorescence signals. In this chapter we discuss the physical basis of fluorescence, the properties of fluorescent dyes, the action of filters comprising a fluorescence filter set, the optical design of epi-illuminators, and the positioning of this equipment in the optical pathway. We also examine important variables that affect image quality and discuss methods for examining fluorescence in living cells.

APPLICATIONS OF FLUORESCENCE MICROSCOPY

Fluorescence microscopy is used extensively to study the intracellular distribution, dynamics, and molecular mechanisms of a large variety of macromolecules and metabolites. While it is impractical to discuss specialized labeling techniques and methods of fluorescence quantitation and analysis, it is important to note some of the principal applications for which fluorescence microscopy is applied. These include:

- *Determination of the intracellular distribution of macromolecules in formed structures such as membranes, cytoskeletal filaments, and chromatin.* Fluorochrome-conjugated metabolites, ligands, and proteins can be used to label membrane channels and ion channels. Target molecules can also be labeled with fluorescent antibodies (immunofluorescence microscopy) or with biotin or epitope tags, or conjugated to fluorescent proteins such as green fluorescent protein (GFP) and other agents. Multicolor labeling is possible, whereby several different molecular species are labeled and viewed simultaneously using dyes that fluoresce at different wavelengths.

- *Study of intracellular dynamics of macromolecules associated with binding dissociation processes and diffusion (fluorescence recovery after photobleaching, or FRAP).* FRAP techniques give the halftime for subunit turnover in a structure, binding constants, and diffusion coefficients.

- *Study of protein nearest neighbors, interaction states, and reaction mechanisms by fluorescence energy transfer or FRET and by fluorescence correlation microscopy.* In FRET, two different fluorochromes are employed, and the excitation of the shorter-wavelength fluorochrome results in the fluorescence of the longer-wavelength fluorochrome if the two moieties come within a molecular distance of one another.

- *Study of dynamics of single tagged molecules or molecular assemblies in vivo using an extremely sensitive technique called total internal reflection fluorescence (TIRF) microscopy.*

- *Determination of intracellular ion concentrations and changes in the concentrations for several ionic species, including H^+, Na^+, K^+, Cl^-, Ca^{2+}, and many other metals.* Ratiometric dyes are used, whose peak fluorescence emission wavelength changes depending on whether the dye is in the free or bound state. The ratio of fluorescence amplitudes gives the ion concentration.

- *Organelle marking experiments using dyes that label specific organelles and cytoskeletal proteins.*

- *Determination of the rates and extents of enzyme reactions using conjugates of fluorochromes whose fluorescence changes due to enzymatic activity.*

- *Study of cell viability and the effects of factors that influence the rate of apoptosis in cells using a combination of dyes that are permeant and impermeant to the plasma membrane.*

- *Examination of cell functions such as endocytosis, exocytosis, signal transduction, and the generation of transmembrane potentials using fluorescent dyes.*

PHYSICAL BASIS OF FLUORESCENCE

Fluorescence is the emission of photons by atoms or molecules whose electrons are transiently stimulated to a higher excitation state by radiant energy from an outside source. It is a beautiful manifestation of the interaction of light with matter and forms the basis for fluorescence microscopy, so we will take a moment to examine the physical basis of the phenomenon.

When a fluorescent molecule absorbs a photon of the appropriate wavelength, an electron is excited to a higher energy state and almost immediately collapses back to its initial ground state. In the process of energy collapse the molecule can release the absorbed energy as a fluorescent photon. Since some energy is lost in the process, the emitted fluorescent photon typically exhibits a lower frequency of vibration and a longer wavelength than the excitatory photon that was absorbed. The situation is depicted graphically in what is known as a *Jablonski diagram* (Fig. 11-2), which shows a series of increasing energy states as a stack of horizontal lines. Each energy level is in

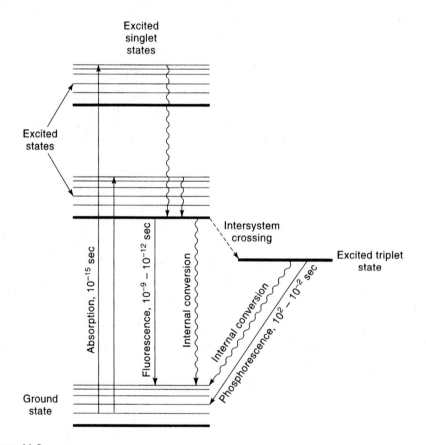

Figure 11-2

Jablonski diagram showing energy levels occupied by an excited electron within a fluorescent molecule (chlorophyll a). Chlorophyll a is unique in absorbing blue and red wavelengths of the visual spectrum. Blue photons are excited to a higher energy level than are red ones (straight upward arrows, left), but the collapse to the ground state by an electron excited by either wavelength can occur through any of the following three pathways: Chlorophyll can give off a photon (fluorescence emission, straight downward pointing arrow); it can release vibrational energy as heat without photon emission (internal conversion, wavy downward pointing arrows); or its electron can enter an excited triplet state (intersystem crossing, dotted downward arrow), which can make the molecule chemically reactive. Electrons in the triplet excited state can return to the ground state through internal conversion or by emission of phosphorescence. Refer to the text for details.

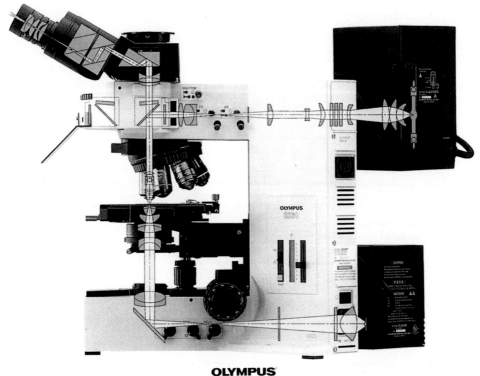

OLYMPUS

Color Plate 4-1. Optical path in the Olympus BX60 upright microscope. The microscope is fitted with a transilluminator (bottom) and epi-illuminator (top) and has infinity-corrected optics. Lenses, filters, and prisms are light blue. Light passing through the objective lens emerges and propagates as a parallel beam of infinite focus, which is collected by an internal tube lens (Telan lens) as an aberration-free image in the real intermediate image plane. The Telan lens is located where the black trinocular headpiece joins the white microscope body. The infinity space between objective and Telan lens allows insertion of multiple optical devices (fluorescence filter sets, waveplate retarders, DIC prisms, analyzer, and others) without altering the magnification of the image. This color plate was provided by Olympus America, Inc.

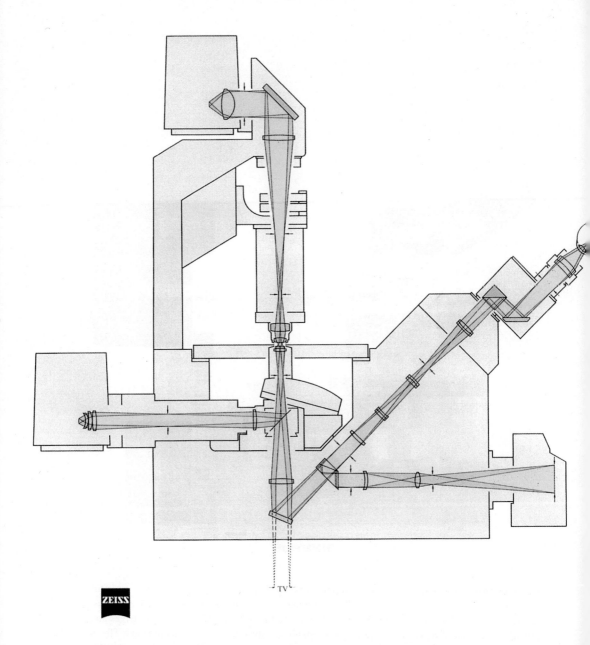

ZEISS

Color Plate 4-2. Optical path in the Zeiss Axiovert-135 inverted microscope. The microscope is fitted with a tran-silluminator (top) and epi-illuminator (bottom) and uses infinity-corrected optics. This plate shows the locations, marked by pairs of arrows, of multiple field planes (full beam diameter, bright yellow) and aperture planes (full beam diameter, dull gold.) Lens, mirror, and prism locations are shown in light blue. In this design, the stage is fixed to the microscope body and the specimen focus dial raises and lowers the objective lens. The black square outline at the site of intersection of the epi-illuminator beam with the microscope axis marks the position where filter sets are inserted for fluorescence microscopy. The identifications of conjugate sets of focal planes are described in Chapter 1. This color plate was provided by Carl Zeiss, Inc.

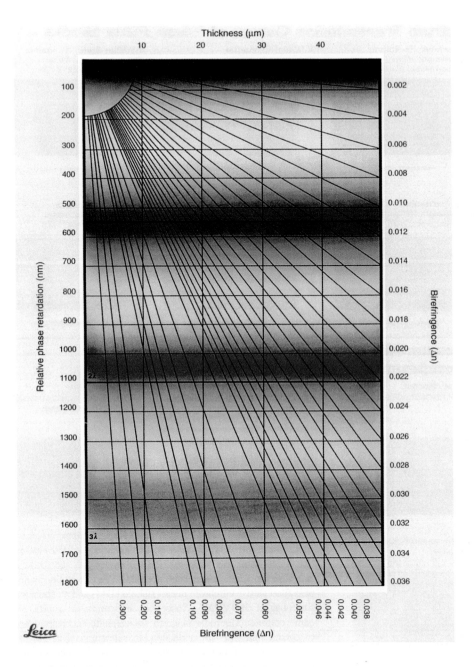

Color Plate 9-1. Michel Lèvy chart showing four orders of the interference color spectrum. Removal of the wavelengths shown on the left edge of the chart through destructive interference yields the indicated interference colors. The chart is used to determine the phase difference between O and E rays for birefringent specimens examined in a polarizing microscope equipped with a 1-plate compensator. The procedure for adjusting the compensator with white light illumination is described in Chapter 9. The Michel Lèvy chart also indicates the refractive index or thickness of a birefringent specimen if one of the two parameters is independently known. In geology, the chart is used to determine the identity, refractive index, or section thickness of birefringent crystals (indicated by the diagonal lines on the chart). Color plate courtesy Leica Microsystems Wetzlar GmbH.

Spectrum Transmission Curves of Nikon Filter Blocks

UV (Ultraviolet) Excitation
Main wavelength: 365 nm
Sample filter block: *UV-2A*

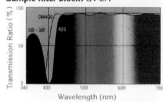

V (Violet) Excitation
Main wavelength: 405 nm
Sample filter block: *V-2A*

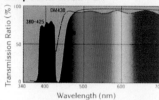

BV (Blue Violet) Excitation
Main wavelength: 436 nm
Sample filter block: *BV-2A*

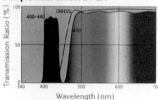

B (Blue) Excitation
Main wavelength: 480 nm
Sample filter block: *B-2E*

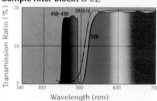

B (Blue) Excitation
Main wavelength: 480 nm
Sample filter block: *B-3A*

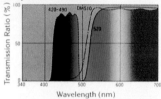

G (Green) Excitation
Main wavelength: 546 nm
Sample filter block: *G-2A*

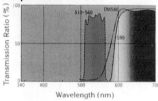

Spectrum Absorption and Emittance of Leading Fluorochromes

Quinacrine Mustard
* V or BV excitation filter can be used.

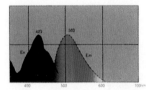

FITC (Fluorescein Iso Thio Cyanate)
* Because of the wide excitation range in the short wavelength side below 500 nm, V or B filters can be used (the B filter is more efficient).

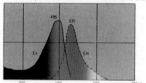

Rhodamine B200
* Offers a wide application range in the G excitation range.

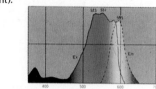

TRITC (Tetramethyl Rhodamine Iso Thio Cyanate)
* G excitation filter can be used.

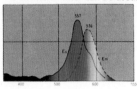

Nikon

Color Plate 11-1. Transmission curves of common fluorescence filter sets. TOP: Filter sets for excitation at UV, violet, blue violet, blue, and green excitation wavelengths are shown. Each set shows the transmission profiles of an excitation bandpass filter (left), a dichroic mirror (labeled DM) and an emission filter (right). BOTTOM: Absorption and emission spectra of some common fluorochromes; the wavelengths corresponding to spectral maxima are indicated. In selecting a filter set to excite fluorescence of a given dye, the excitation bandpass filter must cover the excitation peak of the dye. Likewise, dichroic mirror and emission filter profiles must cover the principal emission peak of the dye. Thus, filter blocks B-2E and B-3A are suitable for examining FITC fluorescence, and block G-2A is suitable for examining the fluorescence of Rhodamine B200 and TRITC. This color plate was provided by The Nikon Corporation, Inc.

turn composed of a number of sub energy levels, which do not concern us here. There are two categories of excited states, characterized by different spin states of the excited electron—the singlet excited state and the triplet excited state. Most commonly, an excited electron occupies an excitation level within the singlet excited state (straight upward pointing arrows), and when it collapses to the ground state, energy can be given up as fluorescence emission (straight downward pointing arrows). Alternatively, energy can be given up as heat (internal conversion), in which case no photon is emitted (wavy downward pointing arrows). When excited above the ground state, there is a probability that an electron can also enter the triplet excited state through a process called intersystem crossing (dotted arrow). The triplet state is important, because molecules with electrons in this state are chemically reactive, which can lead to photobleaching and the production of damaging free radicals (discussed later in this chapter). During fluorescence, the absorption and re-emission events occur nearly simultaneously, the interval being only 10^{-9}–10^{-12} seconds; therefore, fluorescence stops the moment there is no more exciting incident light. The emission process is called *phosphorescence* if the period between excitation and emission is not instantaneous and lasts fractions of a second to minutes. These processes should not be confused with *bioluminescence,* such as that exhibited by firefly luciferase, in which electrons are excited by chemically driven processes rather than by absorbing external radiation.

Molecules that are capable of fluorescing are called *fluorescent molecules, fluorescent dyes,* or *fluorochromes.* If a fluorochrome is conjugated to a large macromolecule (through a chemical reaction or by simple adsorption), the tagged macromolecule is said to contain a *fluorophore,* the chemical moiety capable of producing fluorescence. Fluorochromes exhibit distinct excitation and emission spectra that depend on their atomic structure and electron resonance properties. Fluorescent dyes usually contain several unconjugated double bonds. The spectra for fluorescein-conjugated immunoglobulin (IgG) are shown in Figure 11-3. Molecules absorb light and re-emit photons over a spectrum of wavelengths (the *excitation spectrum*) and exhibit one or more characteristic excitation maxima. Absorption and excitation spectra are distinct but usually overlap, sometimes to the extent that they are nearly indistinguishable. However, for fluorescein and many other dyes, the absorption and excitation spectra are clearly distinct. The widths and locations of the spectral curves are important, particularly when selecting two or more fluorochromes for labeling different molecules within the same specimen.

Re-emission of fluorescent light from excited dye molecules likewise occurs over a broad spectrum of longer wavelengths (the *emission spectrum*) even when excitation is performed with a monochromatic source such as a laser. This is because electrons occupy excited states for various lengths of time during which they give up varying amounts of vibrational energy, some of it as heat, resulting in the re-emission of lower-energy, longer-wavelength photons over a spectrum of wavelengths. Because some energy is given up during the process, the wavelength of a fluorescent photon is usually longer than the wavelength of the photon exciting the molecule. It will be remembered that the energy of a photon is given as $E = hc/\lambda$, where h is Planck's constant, c is the speed of light, and λ is the wavelength. Since the photon energy is reduced during absorption and re-emission, the wavelength increases. The reader is encouraged to review the relationships between the energy, frequency, and wavelength of photons described in Chapter 2.

Finally, the shapes of spectral curves and the peak wavelengths of absorption and emission spectra vary, depending on factors contributing to the chemical environment of the system, including pH, ionic strength, solvent polarity, O_2 concentration, presence of

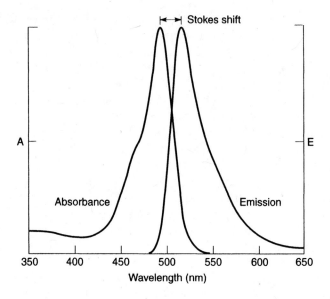

Figure 11-3

Normalized absorption and fluorescence emission spectra of fluorescein-conjugated IgG. Both spectra span a wide range of wavelengths. Fluorescein has an absorption/excitation peak at 492 nm, but is also stimulated by ultraviolet wavelengths. Fluorescein emission has a peak at 520 nm and looks yellow-green to the eye, but actually fluoresces at wavelengths ranging from blue to red. The difference in nanometers between the excitation and emission maxima is called the Stokes shift.

quenching molecules, and others. This fact explains why the fluorescence of a dye such as fluorescein varies depending on whether it is free in solution or conjugated to a protein or other macromolecule.

PROPERTIES OF FLUORESCENT DYES

The excitation and emission spectra of fluorescent molecules are commonly observed to overlap. The difference in wavelength or energy between the excitation and emitted fluorescent photons is called the *Stokes shift*. In practice, the Stokes shift is the difference between the excitation and emission maxima (Fig. 11-3). Depending on the particular fluorescent molecule, the shift can range from just a few to several hundred nanometers. The Stokes shift for fluorescein is ~20 nm, while that for porphyrins is over 200 nm. Dyes exhibiting a large Stokes shift are advantageous in fluorescence microscopy, because the bands of excitation and fluorescence wavelengths are easier to isolate using interference filters. Another important criterion for dye selection is the *molar extinction coefficient,* which describes the potential of a fluorochrome to absorb photon quanta, and is given in units of absorbance (optical density) at a reference wavelength (usually the absorption maximum) under specified conditions. The *quantum efficiency (QE) of fluorescence emission* is the fraction of absorbed photon quanta that is re-emitted by a fluorochrome as fluorescent photons. QE varies greatly between different fluorochromes and for a single fluorochrome under different conditions. For soluble fluorescein dye at alkaline pH, the quantum efficiency can be as high as 0.9—an extremely high value—but for protein-bound fluorescein at neutral pH, the quantum efficiency is typically 0.6–0.3.

Other important characteristics of dyes are their resistance to photobleaching, solubility in aqueous media, and chemical stability. Quenching and photobleaching reduce the amount of fluorescence and are of great practical significance to the microscopist. *Quenching* reduces the quantum yield of a fluorochrome without changing its fluorescence emission spectrum and is caused by interactions with other molecules including other fluorochromes. Conjugation of fluorescein to a protein usually causes a significant reduction in the quantum yield because of charge-transfer interactions with nearby aromatic amino acids. Proteins such as IgG or albumin that are conjugated with 5 or more fluorescein molecules, for example, fluoresce less than when bound to 2–3 molecules, because energy is transferred to nonfluorescent fluorescein dimers. *Photobleaching* refers to the permanent loss of fluorescence by a dye due to photon-induced chemical damage and covalent modification. As previously discussed, photobleaching occurs when a dye molecule, excited to one of its electronic singlet states, transits to a triplet excited state (Fig. 11-2). Molecules in this state are able to undergo complex reactions with other molecules. Reactions with molecular oxygen permanently destroy the fluorochrome and produce singlet oxygen species (free radicals) that can chemically modify other molecules in the cell. Once the fluorochrome is destroyed, it usually does not recover. The rate of photobleaching can be reduced by reducing the excitation or lowering the oxygen concentration. Methods for reducing oxygen concentration as a way to protect live cells are described at the end of the chapter.

Demonstration: Fluorescence of Chlorophyll and Fluorescein

The phenomena and principles of fluorescence can be demonstrated using a filtered alcoholic extract of spinach leaves and an aqueous solution of fluorescein. The procedures for preparing these samples are given in Appendix II.

Although an extract of spinach contains a mixture of chlorophylls a and b, plus carotene and xanthophyll, most of the fluorescence phenomena are due to chlorophyll a. The fluorescence spectra of chlorophyll a, the major chlorophyll component, are shown in Figure 11-4. The figure shows that the absorption/excitation spectrum is broad and distinctly bimodal, with absorption peaks widely separated in the blue and red portions of the visible spectrum. Fluorescence emission at red wavelengths with an emission peak at 670 nm is the same regardless of whether blue or red wavelengths are used to excite the molecule. Notice too the very large Stokes shift from 420 to 670 nm—a difference of 250 nm. Normally the chlorophyll and other pigments in intact chloroplasts act as energy transducers and do not fluoresce; rather, the electrons excited by absorbed photons are transferred to nearby enzyme assemblies (photosystems I and II) for the fixation of carbon dioxide into carbohydrate. However, when chlorophyll is extracted into a soluble form in alcohol, no electron transfer is possible, and absorbed photons are re-emitted as deep red fluorescence.

If the flask of chlorophyll is placed over a bright white light source, a deep emerald green color is seen. Green wavelengths are observed because red and blue wavelengths are selectively absorbed. (Refer to subtraction colors in Chapter 2.)

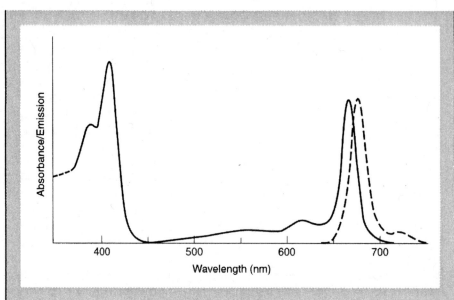

Figure 11-4

Absorption and emission spectra of chlorophyll a. The spectrum for absorption-excitation (solid curve) is unusual in showing two prominent peaks at 420 nm (blue) and 660 nm (red) and a pronounced trough corresponding to green wavelengths of the visible spectrum. The emission curve is shown as a dotted line. It might seem unusual that the major chlorophyll species of plants does not absorb in the green, where the peak of solar radiation occurs. This job is performed by other pigments (chlorophyll b, xanthophyll, and carotene), which transfer captured energy of incident radiation to chlorophylls a and b, thus providing an efficient design for light absorption across parts of the UV and much of the visual spectrum.

Red fluorescence is emitted by the solution, but its presence is masked by the bright green color of the nonabsorbed illuminating wavelengths. If the flask is now illuminated with deep blue wavelengths <450 nm or with long-wave UV (invisible) wavelengths from a black light in a darkened room, the red fluorescence can be easily observed. Chlorophyll demonstrates features important to fluorescence microscopy: the requirement to selectively isolate bands of wavelengths corresponding to the excitation and emission maxima, and the benefits of using molecules that exhibit both a large Stokes shift and high quantum efficiency.

We can examine a 10 μM solution of fluorescein using the same illuminators. Fluorescein appears bright yellow in white light (blue wavelengths are efficiently absorbed, leaving red and green wavelengths, which the eye perceives as yellow). Brilliant yellow-green fluorescence is observed under excitation illumination with the black light. The effect of environmental conditions can be demonstrated with fluorescein by adding a few drops of 10 N sodium hydroxide, which causes a 2-fold increase in quantum efficiency and therefore a dramatic increase in fluorescence. If an opaque mask with a slit 5 cm long by 2–5 mm wide is placed up against the flask on the side facing the observer, and the slit of bright fluorescence is examined in a darkened room at a distance of ~3 meters while holding a holographic (sinusoidal) diffraction grating immediately in front of the eye, the full fluorescence emission spectrum of fluorescein can be examined. It begins with the longest of the blue-green wavelengths, reaches a peak in the green and yellow green, and then tapers off in the orange and red. It is clear that fluorescein re-emits light over a broad spectrum of wavelengths and that the yellow-green fluorescence color is not due to a single spectral line or narrow band of yellow-green wavelengths.

Table 11-1 and Figure 11-5 indicate important properties of fluorescent dyes commonly used in fluorescence microscopy. New classes of dyes such *Alexa dyes* (Molecular Probes, Inc.) and *cyanine dyes* (Amersham-Pharmacia, Inc.) are especially deserving of attention, because they have a high quantum efficiency and offer high resistance to photobleaching. *Green fluorescent protein (GFP)* isolated from the jellyfish *Aequorea victoria,* and its mutated allelic forms, blue, cyan, and yellow fluorescent proteins, are used to produce fluorescent chimeric proteins that can be expressed in living cells, tissues, and whole organisms. The most recent member of the group, DsRed, or *red fluorescent protein,* comes from a marine anemone. The topic of fluorescent dyes and their applications is extensive and complex. Readers wishing to explore the subject in greater depth can begin by reading the chapters by Tsien and other authors in Pawley (1995) and by consulting the *Handbook of Fluorescent Probes and Research Chemicals,* which is available on the Internet at www.probes.com or on hard copy through Molecular Probes, Inc. The spectra and properties of many fluorescent labels including cyanine dyes are available at the Web site of Jackson ImmunoResearch Laboratories, Inc., at www.jacksonimmuno.com, and of GFP and DsRed at the Web site of Clontech Laboratories, Inc., at www.clontech.com. Another excellent reference is available at the website of Omega Optical, Inc at www.omegafilters.com.

It is common practice to label cells with multiple fluorescent dyes to examine different molecules, organelles, or cells in the same preparation using different fluorescence filter sets. As an example, Figure 11-6 shows the excitation and emission spectra of three dyes (DAPI, fluorescein, and rhodamine) that are suitable for specimens illuminated with different spectral regions of a mercury arc lamp. Many factors influence the decision regarding dye selection, including spectral region, Stokes shift, quantum efficiency, solubility, and photostability of the dye; spectral profile of the illuminator; and availability of suitable filter sets. In the example shown here, DAPI and rhodamine are each excited by bright spectral lines of the mercury lamp at 366 and 546 nm, respectively; the excitation of fluorescein is not as favorable, but this is compensated by the fact that fluorescein (or Cy2 or Alexa 488) has a high quantum efficiency.

AUTOFLUORESCENCE OF ENDOGENOUS MOLECULES

All cells contain endogenous metabolites that autofluoresce and contribute background fluorescence to the image. In some cases, these signals are strong enough that they are mistaken for the signals of fluorescently tagged molecules. Some common sources of autofluorescence are the B vitamins, flavins, flavin proteins, and flavin nucleotides (FAD and FMN), reduced pyridine nucleotides (NADH and NADPH), fatty acids, porphyrins, uncoupled cytochromes, lipofuchsin pigments, serotonin, and catecholamines

TABLE 11-1   Excitation and Fluorescence Wavelengths of Some Commonly Used Fluorescent Dyes[a]

Fluorochrome	Excitation Color Band	Mean Absorption Wavelength	Mean Fluorescence Wavelength	Quantum Efficiency	Resistance to Photobleaching
Indo-1 + Ca^{2+}	UV	330	401		
Fura-2 + Ca^{2+}		340	510		
Diaminonaphthylsulphonic acid (DANS)		340	525		
DAPI[b]		345	460		
Amino methylcoumarin (AMCA)		349	448		
Hoechst dye 33258		355	465	√	
Cascade blue		375, 398	424		
Lucifer yellow	Blue	428	540		
Acridine yellow		470	550		
Acridine orange		470	530–650		
$DiOC_6$		484	501		
FM 1-43[d]		473	578		
Fluorescein isothiocyanate (FITC)[c]		492	520	√	
YOYO-1[b]		491	509		
BODIPY fluorescein[c]	Green	503	511	√	√
Fluo-3 + Ca^{2+}		506	526		
Oregon green 514[c]		511	530		
TOTO-1[b]		514	533		
Propidium iodide[b]		520	610		
Tetramethylrhodamine isothiocyanate (TRITC)[c]		540	578	√	√
BODIPY tetramethylrhodamine[c]		542	574		
Carboxy-SNARF-1, acid pH		548	587		
Phycoerythrin-R		565	578	√	√
$DiIC_{18}$, membrane dye[d]		569	565		
Lissamine-rhodamine B	Yellow	575	595		
Texas red[c]		592	610	√	√
Allophycocyanine	Red	621, 650	661	√	
Ultralite T680		656, 675	678		

[a]Most values are from Haugland's *Handbook of Fluorescent Probes and Research Chemicals* (1996): Molecular Probes, Inc., Eugene, Oregon. In most cases, the solvent used was methanol. Absorption maxima are typically close to the peak excitation wavelength.

[b]Dye bound to DNA

[c]Dye bound to protein (IgG)

[d]Dye associated with lipid

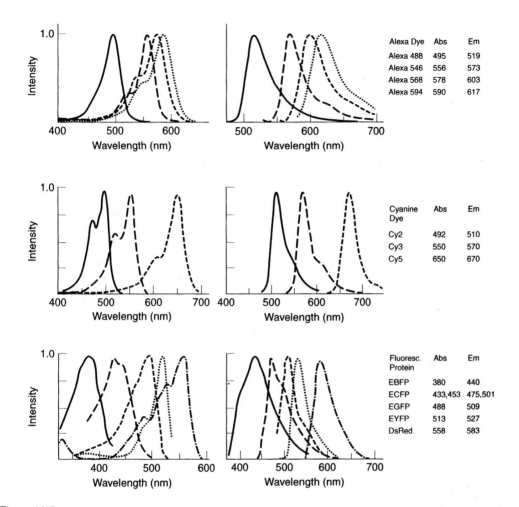

Figure 11-5

Absorption and emission spectra of recently introduced dyes and proteins for fluorescence microscopy. The Alexa series of dyes introduced by Molecular Probes, Inc. (Eugene, Oregon), and the cyanine dyes of Amersham International, Inc. (available from Jackson ImmunoResearch Laboratories, Inc, West Grove, Pennsylvania), are exceptionally photostable, have high quantum efficiency, and are soluble in aqueous media. The GFP series of proteins and DsRed protein, provided as DNA vectors by Clontech, Inc. (Palo Alto, California), are used to construct fluorescent protein chimeras that can be observed in cells after transfection with the engineered vectors. This technique avoids the problem of purifying, tagging, and introducing labeled proteins into cells or having to produce specific antibodies. Fluorescent protein chimeras are also suitable for studies of protein dynamics in living cells. For examples and visualization methods, see Sullivan and Kay (1999).

(see Demonstration and Table 11-2). Background fluorescence emission is greatest when live cells are examined with blue and UV excitation wavelengths. The strength of endogenous autofluorescence depends on the particular metabolite and excitation wavelength being employed and also on the cell type. Macrophages, neurons, and sperm cells exhibit particularly strong autofluorescence. Fixation of cells with aldehydes in preparation for labeling with fluorescently tagged marker molecules may also induce

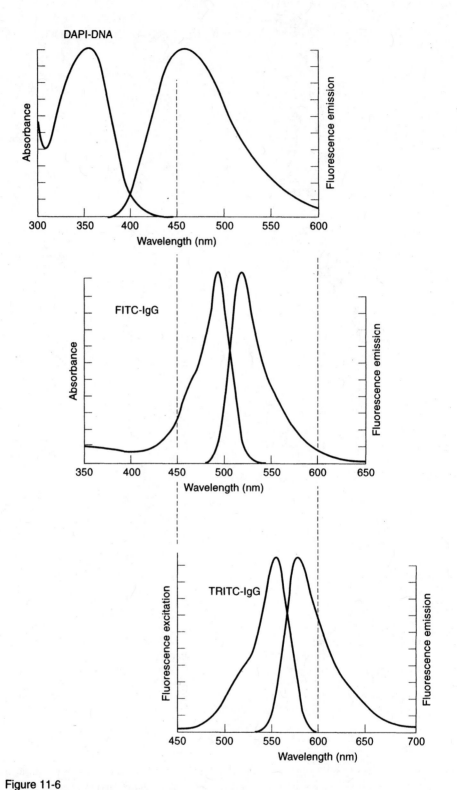

Figure 11-6

Use of multiple fluorescent dyes for examining cells with mercury illumination. The absorption and emission spectra of DAPI, fluorescein, and rhodamine are shown. The absorption spectra of DAPI and rhodamine overlap strong spectral lines of the mercury arc lamp at 366 and 546 nm. Although the spectra of these three dyes partially overlap, it is possible to examine each dye separately using fluorescence filter sets that are specific for each dye.

TABLE 11-2   Fluorescence of Naturally Occurring Substances

Specimen	Substance	Color
Powdered milk	Oxidized riboflavin (lumiflavin)	Blue
Margarine	Fatty acids	Blue
Yeast extract	Oxidized vitamin B_2 (lumiflavin)	Blue
Brain extract	Catecholamines, serotonin	Blue
Yeast on agar plate	Vitamin B_2 (riboflavin)	Green
Liver extract	Vitamin B_2, other B vitamins	Yellow
Carrot extract	β-carotene	Yellow
Butter, milk	Free riboflavin	Yellow
Spinach extract	Chlorophyll a, b	Red
Shells of brown eggs	Porphyrins	Red

unwanted fluorescence, particularly in cell nuclei and organelles. For immunofluorescence studies, aldehyde-induced fluorescence can be diminished by treating fixed samples for 10 minutes with 20 mM sodium borohydride or ammonium chloride. Fortunately, autofluorescent signals are usually low in amplitude. Interference from autofluorescence can sometimes be avoided by simply selecting a longer-wavelength fluorochrome.

Autofluorescence adds to the background signal in a cell and may overlap the signal of a fluorochrome used in a labeling experiment, causing misinterpretation of the distribution pattern of the fluorochrome. After acquiring fluorescence images of labeled specimens, it is therefore important to prepare similar exposures from unlabeled specimens. If necessary, a camera exposure time can be selected that minimizes the autofluorescent contribution, but still allows adequate imaging of the labeled experimental material.

Demonstration: Fluorescence of Biological Materials Under Ultraviolet Light

Fluorescent compounds and metabolites are abundant in living cells and tissues. To become familiar with these signals and recognize them when they occur, examine the fluorescence of naturally occurring compounds in foodstuffs and tissue extracts illuminated with a handheld black light in a darkened room. A list of common foodstuffs and their fluorescence properties is given in Table 11-2. Instructions for preparing certain extracts are given in Appendix II.

ARRANGEMENT OF FILTERS AND THE EPI-ILLUMINATOR IN THE FLUORESCENCE MICROSCOPE

The fluorescence microscope is modified in several important ways in order to obtain fluorescence images that are bright and well defined:

- A bright light source such as a mercury or xenon arc lamp is required because only a narrow band of wavelengths, and consequently a small portion of the illuminator output, is used to excite fluorochromes in the specimen.

- For efficient high-contrast imaging, both the illuminator and objective lens are positioned on the same side of the specimen. In this arrangement, the lamp and light delivery assembly are called an *epi-illuminator*, and the objective lens functions both as the condenser, delivering excitatory light to the specimen, and as the objective lens, collecting fluorescent light and forming an image of the fluorescent object in the image plane.

- *Fluorescence filter sets* containing three essential filters (excitation filter, dichroic mirror, and barrier [or emission] filter) are positioned in the optical path between the epi-illuminator and the objective. This arrangement is shown in Figure 11-7.

- High-NA, oil immersion objectives made of low-fluorescence glass are used to maximize light collection and provide the greatest possible resolution and contrast.

Epi-illumination is made possible by the employment of a dichroic mirror, which is mounted together with exciter and barrier filters as a fluorescence filter set in a filter

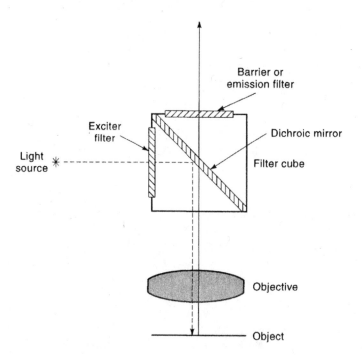

Figure 11-7

Arrangement of filters in a fluorescence filter cube. The diagram shows the orientation of filters in a filter cube in an epi-illuminator for an upright microscope. The excitation beam (dotted line) passes through the exciter and is reflected by the dichroic mirror and directed toward the specimen. The return beam of emitted fluorescence wavelengths (solid line) passes through the dichroic mirror and the emission filter to the eye or camera. Excitation wavelengths reflected at the specimen are reflected by the dichroic mirror back toward the light source. Excitation wavelengths that manage to pass through the dichroic mirror are blocked by the barrier (emission) filter.

cube that is positioned along the optic axis. The first component, the *excitation filter* (or exciter) selectively transmits a band of short wavelengths (relative to the fluorescence wavelengths) for exciting a specific fluorochrome in the specimen. The second component, the *dichroic mirror,* reflects the short wavelength excitation light toward the objective lens and specimen, while transmitting returning long-wave fluorescent light toward the detector. The dichroic mirror also directs any excitation wavelengths reflected by the specimen back toward the exciter filter and illuminator. The third component is the *emission* or *barrier filter,* which transmits the band of fluorescence wavelengths while blocking any residual short excitation wavelengths. The fluorescent wavelengths then form an image on the eye or camera.

Lamps for Fluorescence Excitation

The excitation source can be a band of wavelengths isolated from a mercury or xenon arc illuminator with an interference filter, or it can be monochromatic light from a laser, but illuminators must produce photons of the proper wavelength and at a sufficiently high intensity to be efficient. To obtain a particular wavelength of the desired intensity, it may be necessary to use a certain illuminator—for example, a mercury arc excites DAPI much more effectively than a xenon arc. Therefore, it is important that the microscopist match the illuminator to the fluorescent dye in question.

For fluorescence microscopy, 100 W mercury and 75 W xenon arc lamps are commonly used. Both lamps give bright, continuous emission over the visible spectrum (400–700 nm), but mercury is distinct in containing sharply defined emission lines at 366 (UV), 405, 436, 546, and 578 nm. It is important to reexamine the spectra and characteristics of arc lamps, which are presented in Chapter 3. At 546 nm, mercury and xenon lamps are 10–100× brighter than a 100 W quartz halogen lamp, which is typically used for bright-field microscopy, but is too weak to excite most fluorochromes adequately. The bright emission lines of the mercury arc are useful for exciting certain fluorochromes such as DAPI and Hoechst dye 33258, lucifer yellow, rhodamine, Cy3, and Texas red, but at other spectral regions xenon is brighter and may be more useful (fluorescein, Alexa 488, Fluo3, allophycocyanine, Cy5). Refer to Table 11-1 and Figure 11-5 on fluorescent dyes and the spectra of xenon and mercury light sources in Chapter 3 to confirm this for yourself and to find other well-matched dye-illuminator partners.

The Epi-Illuminator

The epi-illuminator consists of the lamp and its collector lens plus a connector tube fitted with a field stop diaphragm, a relay lens, and slots for additional filters. It is attached to the lamp at one end and is mounted at its other end to the microscope in the vicinity of the fluorescence filter cube. The lamp housing should contain a focusable collector lens to fill the back aperture of the objective as required for Koehler illumination. On most research microscopes, the epi-illuminator is a built-in component of the microscope body. Apart from proper adjustment and focus of the lamp, little other manipulation of the illumination pathway is required. Check that the field stop diaphragm is centered and is opened to provide optimal framing of the specimen. Epi-illumination is much more efficient than transmitted mode (or diascopic) illumination because there is much less background in the fluorescence image.

One of the most important adjustments in fluorescence microscopy is the alignment of the illuminator. Since the amplitude of fluorescence signals depends directly on the amount of excitation, uneven illumination of the object by a misaligned lamp will result in bright and dark regions in the fluorescence image. This is especially detrimental for quantitative work. Lamp alignment is discussed in Chapter 3 and demands attention to two points: (1) centration of the image of the arc and its reflection on the optic axis, and (2) spreading of the illumination beam with the lamp collector lens to fill the back aperture of the objective evenly and homogeneously.

Filters

Filters used to isolate bands of wavelengths in fluorescence microscopy include colored glass filters, thin film interference filters, or a combination of filter types. The filters include long-pass or short-pass edge filters or narrow or broad bandpass filters. The design and action of filters are discussed in Chapter 2; the optical performance of representative filters used for fluorescence microscopy is shown in Figure 11-8 and in Color Plate 11-1. It is essential to understand the properties and value of these indispensable components of the fluorescence microscope. You will need to know how to "read" the absorption and emission spectra of a dye, and the transmission profiles of the filters. This knowledge is required to select the best among several possible filter sets for examining the fluorescence of a certain fluorochrome. You will also need to interpret and explain differences among fluorescence microscope images obtained using different filter sets. In addition, you may need to create new filter sets using components of filter sets already in hand. This will greatly extend the capacity of a limited collection of filters. As you install and remount filters, remember that interference filters are usually mounted in a specific orientation with respect to the illuminator and the specimen in order to obtain optimal performance (see Chapter 2). The transmission profiles of filter sets for fluorescence microscopy can be examined at the Web sites of Chroma, Inc., at www.chroma.com, and of Omega Optical, Inc., at www.omegafilters.com.

The Dichroic Mirror

The dichroic mirror or beam splitter is a special long-pass filter coated with multiple layers of dielectric materials similar to those contained in thin film interference filters, but specially designed for reflection and transmission at certain boundary wavelengths. The mirror is mounted at a 45° angle with respect to the optic axis within a filter cube and faces the light source. At this angle, the dichroic mirror reflects short excitation wavelengths at a 90° angle along the optic axis to the specimen, but transmits long fluorescence wavelengths that are collected by the objective and directed to the image plane. The transition from near total reflection to maximal transmission can be remarkably sharp, occurring over 20–30 nm, allowing the mirrors to act as precise discriminators of excitation and fluorescence wavelengths (Fig. 11-8). (For a description of the design and performance of interference filters, see Chapter 2.) The specifications for a dichroic mirror assume a 45° angle between the axis of the incident beam and the plane surface of the mirror. Dichroic mirrors should be handled with extreme care, because the exposed dielectric layers can be scratched and damaged during handling and cleaning. The cleaning procedure is the same as that used for cleaning other optical surfaces (Chapter 2).

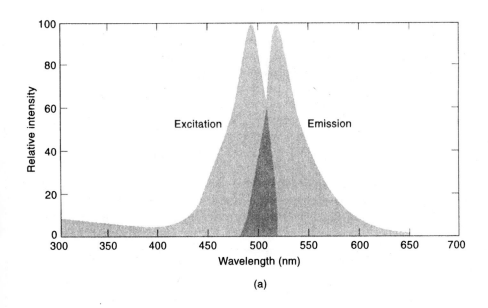

(a)

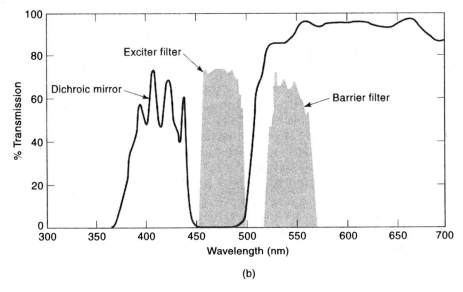

(b)

Figure 11-8

Transmission profiles of filters in a fluorescein filter set. The figure shows (a) the excitation
and emission spectra of fluorescein and (b) the transmission profiles of three filters belonging
to a high-performance filter set (XF100, Omega Optical Company, Inc. Brattleboro, Vermont).
A bandpass exciter filter and a long-pass barrier (or emission) filter (shaded profiles shown in
b) transmit bands of light that occupy the peak regions of the respective excitation and
emission spectra shown in the upper panel. To maintain a distinct separation of these
components, the transmission profiles are not exactly centered on the excitation and emission
maxima of the dye. The dichroic mirror (heavy line) reflects light (100% reflection corresponds
to 0% transmission on the curve) or partially transmits light (80–90%), depending on the
incident wavelength. The pronounced trough in the transmission profile, representing a peak
of reflectance, is used to reflect the band of excitation wavelengths from the exciter filter onto
the specimen. The boundaries between transmitted and reflected bands of wavelengths are
designed to be as steep as possible to assure complete separation of the reflected and
transmitted wavelengths. The pattern of rapidly rising and falling spikes, typical of dichroic
mirror profiles, is known as ringing. The performance of this filter set is extraordinary and
reflects major improvements in thin film technology over the last several years.

Transmission profiles for dichroic mirrors usually show multiple broad peaks and troughs that correspond to bands of wavelengths that experience high transmittance/low reflectance (at peaks) and low transmittance/high reflectance (at troughs) (Fig. 11-8). Filter sets are designed so that the band of excitation wavelengths (high-percent transmission) from the exciter precisely matches a trough in the dichroic (low-percent transmission) so that these wavelengths are reflected to the specimen. Longer fluorescent wavelengths emitted by the specimen must also match the peak to the right of the trough so that they are transmitted to the emission filter and detector. It is important that the transmission, reflectance, and emission characteristics of the exciter and dichroic be closely matched, and that they be appropriate for the absorption and emission maxima of the dye; otherwise, excitation wavelengths can pass through the dichroic and fog the image, or fluorescent wavelengths can be reflected at the dichroic, reducing image brightness. Even when filters and fluorochromes are appropriately matched, performance is usually compromised somewhat if the transmission profiles of the exciter and dichroic overlap. When this happens, some excitation light passes through the dichroic, reflects off the walls of the filter cube, and can be partially transmitted by the emission filter, because the angle of incidence with that filter is oblique. Transmission of unwanted wavelengths through a filter set is called *bleed-through,* and the amount of bleed-through for typical filter sets is generally about 10%. Microscope manufacturers continue to improve fluorescence optical designs to give higher contrast images.

It is important to recognize that in addition to reflecting the excitation band of wavelengths, a dichroic mirror usually reflects bands of wavelengths shorter than the excitation band. Therefore, you cannot always depend on the dichroic filter to block transmission of unwanted short wavelengths, which to a greater or lesser extent always leak through the exciter filter. It is usually wise to insert an additional UV-blocking filter into the beam when examining live cells by fluorescence microscopy.

Advances in thin film coating technology allow for creation of multiple transmission peaks and alternating reflection troughs in a single interference filter or dichroic mirror. When matched appropriately, two filters and a dichroic mirror can be combined to create a *multiple fluorescence filter set* that allows simultaneous excitation and fluorescence transmission of multiple fluorochromes (Fig. 11-9). Multiple-wavelength filters and dichroic mirrors are now commonly employed in research-grade microscopes and confocal fluorescence microscope systems (see Chapter 12). Double, triple, and even quadruple fluorescence filter sets are available for fluorescence microscopy, although these sets are expensive and suffer somewhat from bleed-through—that is, the transmission of fluorescence from one dye through bandwidths intended for other dyes. The clearest multifluorochrome images are obtained by taking separate gray-scale pictures with filter sets optimized for each dye and then combining the images into a single composite color image, either electronically with a computer or in the darkroom.

OBJECTIVE LENSES AND SPATIAL RESOLUTION IN FLUORESCENCE MICROSCOPY

Proper selection of an objective lens is important, especially for imaging dim fluorescent specimens. High-NA, oil immersion plan-fluorite lenses and planapochromatic objective lenses are ideal, because at NA = 1.3 or 1.4 their light-gathering ability is especially high. These lenses give excellent color correction, so different fluorescent wavelengths are brought to the same focus in the focal plane. They are also transparent to UV light—a requirement for examining UV-excitable dyes such as DAPI, Hoechst,

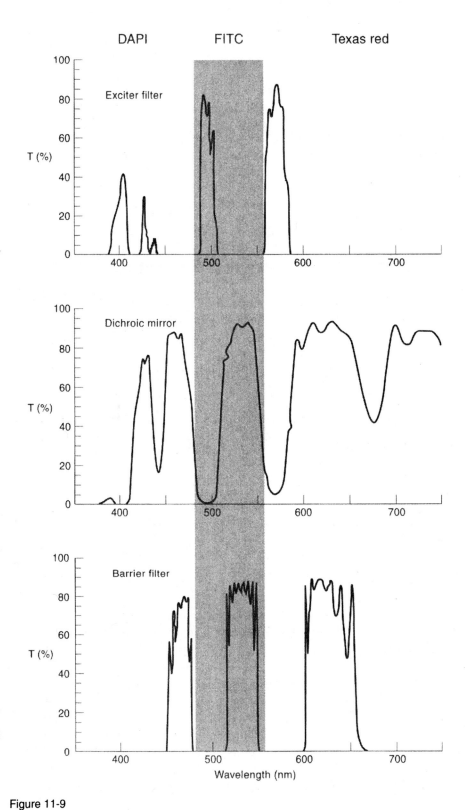

Figure 11-9
Transmission profiles of a triple-band filter set for DAPI, FITC, and Texas red. Each of the three filters contains multiple bandwidths that transmit or reflect three distinct bands of wavelengths simultaneously. The eye or camera sees a multicolor image based on the three dyes. Profiles of the exciter filter (top panel), dichroic filter (middle panel), and emission filter (bottom panel) are shown. The broad vertical band across the drawing distinguishes the spectral regions of the filters for FITC from those for DAPI and Texas red.

and AMC. In addition, they contain low-fluorescence glass—a feature that minimizes background fluorescence and gives high contrast. Since image brightness (photon flux per unit area and time) is proportional to NA^4/M^2, where NA is the numerical aperture and M is the magnification, a $60\times$, 1.4 NA planapochromatic objective is among the brightest objectives and is very well suited for fluorescence imaging.

The spatial resolution d for two noncoherent fluorescent point objects is the same as in bright-field microscopy with incoherent light, and is given as $d = 0.61\lambda/NA$, where λ is the mean wavelength of fluorescent light transmitted by the barrier filter. Resolution, brightness, and other features defining the optical performance of objective lenses are described in Chapter 4.

CAUSES OF HIGH-FLUORESCENCE BACKGROUND

For research microscopes with properly selected fluorescence filter sets, the amount of background fluorescence in the image of a specimen containing a single fluorochrome is usually 15–30% of maximum specimen brightness—not 0% as might be expected. An example of background signal in a cell labeled with a single fluorochrome is shown in Figure 11-10. Because background fluorescence is always present, it is important to take steps to keep the background signal as low as possible.

- Less than ideal performance of filter sets, where transmission and reflectance by interference filters and the dichroic mirror are not 100% and where the transition boundaries between transmission and reflection are not sharply defined, is a major contributor to background signal. These problems are compounded when poorly performing filters are used together in a single filter set.

- Specimen preparation must include complete neutralization of unreacted aldehyde groups and blocking of remaining reactive sites to minimize nonspecific binding of the fluorescent probe. It is also necessary to completely remove unbound fluorochrome by thorough rinsing. However, even if these precautions are followed, background fluorescence from unbound antibody can be high for a mounted coverslip containing labeled cultured cells. This is because the labeling reaction is an equilibrium between bound and free states of the antibody. Even with an equilibrium dissociation constant of 1 nM and a modest concentration of 10^6 antigen-binding sites per cell, a significant fraction of antibody would be expected to dissociate and be free in the mounting medium.

- Reflections and scattering in the optical pathway cause rays to enter a filter at an oblique angle, reducing the filter's transmission/reflection efficiency. One site of concern is the back wall of the filter cube, where excitatory rays that are partially transmitted by the dichroic mirror are reflected and transmitted by the emission filter because they are not incident at an angle perpendicular to the plane of the emission filter. In recent Zeiss designs where the back wall of the filter cube is removed, reflections at this location are removed, and image contrast is improved by 15–20%.

- Dust, fingerprints, and scratches on filters and lens elements scatter significant amounts of light, resulting in an increase in background signal and reduced contrast. Cleaning filters and optics significantly reduces this problem. Interference filters also deteriorate gradually over time due to handling and the presence of water

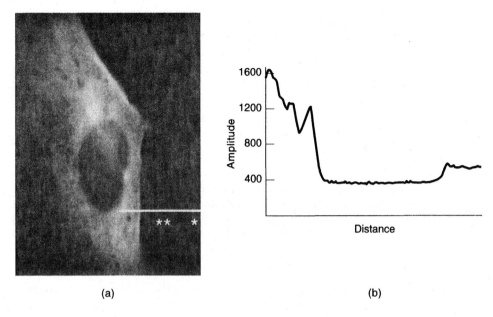

(a)   (b)

Figure 11-10

Comparison of object and background fluorescence in a typical fluorescence light micrograph. (a) A live cultured U2OS cell microinjected with fluorescein-conjugated protein that binds to filaments in the cell. Protein that escaped from the micropipette and bound to the coverslip is indicated with an asterisk. Subsequent retraction of the cell during culture in fresh medium exposed a protected region of coverslip that appears black and is marked with a double asterisk. The horizontal white line (22 μm) represents a row of pixels whose numeric values are shown in (b). (b) Intensity profile of the pixel values under the line in (a). The dark background region is not black at all, but shows an amplitude that is 25% of that at the bright edge of the cell. This is typical of most fluorescent cell and immunofluorescent cell specimens. The causes of background fluorescence are discussed in the text.

vapor and chemicals in the air. Blemishes in old interference filters in the form of microscopic scaling, pinholes, and scratches can scatter significant amounts of light.

- Fluorescence from other sources, including the glass in certain objective lenses, immersion oil, plastic tissue culture dishes, and autofluorescence from the specimen itself contribute to the background. Most microscope manufacturers make objectives of low-fluorescence glass and provide low-fluorescence immersion oil for fluorescence microscopy. The new low-fluorescence immersion oils available from microscope manufacturers and independent sources such as Cargill, Inc., increase contrast significantly and must be employed.

THE PROBLEM OF BLEED-THROUGH WITH MULTIPLY STAINED SPECIMENS

Another problem in imaging double-stained specimens is bleed-through, the crossover of fluorescence signal from one fluorochrome through the filter set of the other fluorochrome. A common example is double labeling with fluorescein and rhodamine using separate filter sets to capture different images of each of the two fluorochromes. Bleed-through can be reduced, but is never completely eliminated, because:

- The excitation spectra of two or more fluorochromes are broad and overlap to a significant extent (Figs. 11-5 and 11-6; Color Plate 11-1). Thus, the excitation of fluorescein at 490 nm also causes rhodamine to fluoresce. The solution is to choose fluorochromes with well-separated excitation spectra so that the excitation peak for one fluorochrome is many wavelengths shorter than the excitation peak for the second fluorochrome.

- The fluorescence spectra of the fluorochromes may overlap, allowing fluorescence from the shorter-wavelength fluorochrome to contribute to the image of the longer-wavelength dye. To minimize the problem, a narrow bandpass filter is used as an emission filter to collect only the peak fluorescence of the lower-wavelength fluorochrome; a long-pass barrier filter is usually used for the longer-wavelength fluorochrome.

- The fluorescence emission of one fluorochrome (fluorescein) may stimulate a second longer-wavelength dye (rhodamine) to fluoresce. Selecting well-separated dyes and assuring that labeling and fluorescence of the dyes are balanced help reduce this problem.

- If the amount of labeling and the intensity of fluorescence of the two fluorochromes are not equally balanced, the brighter signal can overwhelm and penetrate the filter set for the second signal and cause a significant contribution to the image of the second dimmer signal. The intensity of fluorescence from dyes such as fluorescein and rhodamine should be similar and is adjusted according to the amount of dye in the specimen and the type of illumination used. Microscopists using epi-illumination with a mercury arc lamp frequently forget that rhodamine is excited $10\times$ more effectively than fluorescein owing to the bright 546 nm emission line in the mercury arc spectrum.

Even when these factors are controlled, the amount of signal crossover and bleed-through generally remains about 10–15%. In experiments involving double staining, you should always examine single-stained specimens using the filter set for the other fluorochrome to assure that the amount of bleed-through is minimal. In many cases, the main cause of bleed-through is unequal staining by the two dyes. Remember that it is nearly always desirable to return to the lab bench to prepare a proper specimen, rather than to use unequal exposures and image processing on a poorly prepared specimen just to save time.

EXAMINING FLUORESCENT MOLECULES IN LIVING CELLS

Fluorescence microscopy is potentially damaging to living cells, since light sources are intense and rich in damaging UV and IR wavelengths, and because fluorescence filter sets are not totally efficient in removing these unwanted wavelengths. The chief concern is phototoxicity to the cell through absorption of photons by introduced and endogenous fluorophores and the generation of reactive oxygen species, including singlet oxygen ($^1O_2$), superoxide ($O_2 \cdot^-$), hydroxyl radical ($OH\cdot$), various peroxides ($ROOR'$), hydroperoxides ($ROOH$), and others. They react with oxidizable metabolites and components in the cell such as the pyridine nucleotides in nucleic acids, several amino acids, glutathione, lipids, and ascorbate. Among the immediate effects is damage to membrane lipids and proteins, including ion pumps, channels, and gates, which lead to loss of ion balance, loss of transmembrane potential, and rapid cell death. We discussed the consequences of phototoxicity on cell shape and cell behavioral responses in Chapter 3. Briefly, cells cease movement; intracellular organelle trafficking ceases; cells round up and form vacuoles and eventually lyse. Excitable cells such as neurons and free-living amoebae are among the most sensitive to light exposure in the microscope. Just 3 seconds of continuous exposure to the blue excitation wavelengths from a mercury lamp filtered by a standard fluorescein filter set is sufficient to cause *Acanthamoeba* and *Dictyostelium* amoebae to stop moving and round up. The requirement to minimize the exposure of living cells to light can place extreme demands on photography, because short exposures reduce the signal-to-noise ratio and result in grainy images. Control of the exposed dose is especially difficult during the acquisition of time-lapse sequences. The following approaches reduce the effects of phototoxicity:

- Additional UV and IR cutoff filters should be inserted near the illuminator, since the majority of the spectral output of mercury and xenon lamps is, respectively, in the UV and IR regions of the spectrum, and because most fluorescence filter sets are not completely effective in removing these unwanted wavelengths.

- Minimize exposure to light to allow time for dissipation and degradation of free radicals. Phenol red-free basal salt solution is recommended to reduce photon absorption.

- Addition of millimolar concentrations of anti–free radical reagents to live cells and in vitro preparations of purified cell organelles and filaments minimizes free radical damage, because the reagents readily react with free radicals and help spare endogenous molecules. Sodium ascorbate, reduced glutathione, or Trolox (a water-soluble form of vitamin E) at 10 mM concentrations are effective, particularly for in vitro systems.

- To retard the rate of free radical formation, the concentration of dissolved oxygen can be reduced to ~4% of the saturating value for buffer in room air by adding oxygen-depleting enzymes to the medium. Low oxygen concentration is usually not damaging to tissue culture cells—in fact, cells exist at similarly low values of oxygen tension in body tissues. One effective commercial product is Oxyrase (Oxyrase, Inc., Mansfield, Ohio), a preparation of respiratory particles from *Escherichia coli* membranes. Alternatively, an oxygen-scavenging system can be constructed based on a combination of catalase, glucose oxidase, and D-glucose. Cells in well chambers open to the air can be covered with a layer of embryo-grade mineral oil such as that available from Sigma Chemical Company (St. Louis, Missouri) to prevent resaturation of the medium with oxygen.

Exercise: Fluorescence Microscopy of Living Tissue Culture Cells

The purpose of this exercise is to learn how to: Prepare and examine live cells whose organelles are labeled with specific fluorescent dyes; Become familiar with the action of narrow band-pass and long-pass filter sets that are used to examine double-stained specimens; Evaluate and select filter sets based on their transmission profiles; Learn how to balance the intensities of different fluorochromes to minimize bleed-through in inappropriate filter sets. Note: Read the entire text of the exercise before beginning the work. Work in pairs to make the exercise go smoothly.

In this exercise we will stain tissue culture cells with organelle-specific fluorescent dyes to examine the morphology and abundance of organelle systems and to compare the performance characteristics of different filter sets. Many, but not all, of these dyes can be used to stain cells in the living state. Large well-spread cells, such as COS-7, U2OS, CHO, LLCPK cells, or primary cultures of endothelial cells, if available, are appropriate for this exercise. Prepare representative sketches of the labeled cells. Keep a record of your observations, including the time course of the staining response, the stability of fluorescence, and other relevant information. For this exercise pick one or more of the following marker dyes. Instructions for preparing dye solutions are given in Appendix II.

Dye	Cell structure	Filter Set
Acridine Orange	Lysosomes	Fluorescein
Bodipy-Ceramide	Golgi Apparatus	Fluorescein
DAPI, Hoechst 33258	DNA	DAPI
$DiOC_6$	Mitochondria; endoplasmic reticulum	Fluorescein
Ethidium Bromide	DNA	Rhodamine
Lyso-Tracker Green	Lysosomes	Fluorescein
Mito-Tracker Red	Mitochondria	Rhodamine
JC1	Mitochondria	Rhodamine
Rhodamine Phalloidin	Actin Stress Fibers	Fluorescein

Procedure:

Transfer a cell-containing coverslip from a culture dish to a 35 mm dish containing HMEM (MEM culture medium plus 10 mM HEPES pH 7.2 and without serum). Buffering with HEPES (10 mM, pH 7.2) is necessary when handling and examining cultured cells in room air. Serum and phenol red should be omitted because their fluorescence decreases contrast. Remove the culture medium, add 2 mL HMEM and tilt gently to wash the medium over the cells. Pull off the medium with a Pasteur pipette and replace it with 2 mL HMEM containing a specific fluorescent dye at the appropriate concentration. Disperse the dye by gently rocking the culture dish. Incubate for the indicated time, retrieve and save the dye solution for reuse at a later time, rinse the coverslip 3× with 1 mL HMEM to remove any excess unbound dye, wipe off the back of the coverslip with a damp Kimwipe to remove cells and medium, and prepare a Vaseline mount as described in Figure 10-9. Draw off extra medium from the edge of the coverslip chamber with a filter paper and examine the preparations immediately. For each dye employed, describe the degree of staining specificity for organelles and explain why a dye might not appear to stain specifically. Also describe the intracellular abundance and morphological pattern exhibited by each type of labeled organelle. Make sketches and include scale bars on your drawings.

Microscope set-up.   Examine the cells with a 40–100× oil immersion lens using epi-illumination with the 100 W mercury arc lamp. The microscope should also be fitted with phase-contrast or DIC optics so that structural features in the cells can be examined in transmitted light without having to move the specimen or change the lens.

Precautions regarding fixatives.   If formaldehyde or glutaraldehyde is used as a fixative, use ventilation during preparation to prevent fumes from contacting the eyes or skin. Place all fixatives and dyes in a toxic waste container for proper disposal afterwards. Be sure to consult an experienced cell biologist for advice and supervision.

Specific details:

DiOC₆/ER, mitochondria: $DiOC_6$ is a lipophilic dye that stains organelles according to the lipid composition of their membranes. Cells are stained with 2 mL of 2.5 μg/mL $DiOC_6$/HMEM for 1–2 min. Remove the dye and rinse the cells 2× with HMEM. Mount and examine. Can you distinguish the mitochondria from the ER? Does the nucleus also stain? Why? At dilute concentrations of dye (1/10 the indicated concentration) only the mitochondria are stained. Living cells can be treated for 30–60 min and examined the same as fixed cells. With longer labeling times or at higher concentrations of dye, the ER also becomes labeled and is seen as a beautiful lacy network at the periphery of the cell. Make sketches of the stained organelles.

Bodipy-ceramide/Golgi apparatus: Rinse a coverslip containing cells in HMEM; fix with 5 mL 0.5% glutaraldehyde in HMEM for 5 min. Remove the fixative and rinse cells with HMEM several times. Incubate for 30 min at 5°C in 5 μM C_6-bodipy-ceramide/BSA complex in a refrigerator. Ceramide is a lipid that partitions to certain lipid environments in the cell and BSA serves as a carrier for the lipid. Remove and save the ceramide, rinse the cells at 22°C with HMEM containing 3.4 mg/mL delipidated BSA several times over 1 hr, and then mount and examine the cells. (The BSA removes excess ceramide and improves the specificity of labeling.) If things go well, you're in for a treat! Make drawings of the Golgi apparatus from several cells. Are there any indications of staining of pre-Golgi or post-Golgi structures? As an alternative, label living cells and observe the pathway of dye uptake and partitioning from the plasma membrane to endosomes to the Golgi. Wash a coverslip 3× with HMEM. Place the coverslip in 5 μM ceramide/BSA for 15 min at 37°C, then replace the solution with complete culture medium and incubate for an additional 30 min (for endosomes) or 60 min (for Golgi) at 37°C.

Ethidium bromide/nucleus: After rinsing in HMEM, the cells are fixed in 4 mL 0.5% glutaraldehyde in HMEM for 15 min. Remove the fixative, dehydrate the cells in 100% ethanol for 1 min, and rinse in HMEM. Add ethidium bromide to 5 μg/mL (alternatively, DAPI to 0.5 μg/mL) in HMEM, being careful not to contact the skin, and incubate the cells at 22°C for 60 min. Alternatively, stain living cells with Hoechst dye 33258 at 0.5 μg/mL for 15 min at 22°C. Mount and examine. Remove the dye and rinse the cells. Look for reticular patterns caused by euchromatin and heterochromatin in the nuclei. Also look for nucleoli, which are the sites of RNA processing in the nucleus. What do mitotic chromosomes look like? Make sketches.

MitoTracker or LysoTracker/mitochondria, lysosomes: Simply add these potential uptake dyes to 250 nM in the culture medium and incubate the cells at 37°C for 5–10 min. Mount and examine. Carefully discard the staining solution. Are the dyes specific for mitochondria and lysosomes? How many mitochondria/lysosomes are there in a typical cell? What is the distribution of their lengths/diameters? Determine if you can see mitochondria round up and migrate toward the center of the cell, a stress response of the cell to non-ideal environmental conditions and prolonged observation. Make sketches.

Bodipy phalloidin/actin stress fibers: After rinsing in HMEM, fix the cells in 4 mL 0.5% glutaraldehyde in HMEM for 15 min. Dilute the dye 1:40 in the culture medium containing 0.1% Triton ×100 detergent and incubate the cells at 37°C for 30 min. Mount and examine after thorough rinsing in HMEM. Are the filaments you see individual actin filaments or filament bundles? How are stress fibers oriented with respect to the long axis of the cell, cell borders, and vertices that represent points of cell attachment? Estimate the amount of fluorescence associated with stress fibers vs. unorganized filaments subjacent to the plasma membrane. Prepare careful drawings.

Examination of double stained specimens. If the microscope is equipped with two different filter sets for the same dye (for example, for examination of fluorescein, a filter set with a narrow band-pass emission filter and another set with a long-pass emission filter), it is useful to examine specimens double stained with fluorescein and rhodamine.

It is possible to directly examine the transmission profiles of the dichroic mirror-emission filter combination of a given filter set right at the microscope. Just switch the microscope to transmitted white light, mount a diffraction grating on the stage, close the condenser diaphragm down to a minimum, and examine the colors of the first order diffraction spectra with the aid of a Bertrand lens. Examine all of the filter sets to determine the bands of wavelengths transmitted by the sets.

(1) Is there any bleed-through of fluorescence of one dye through the filter set intended for viewing the other dye?

(2) Is the amount of rhodamine bleed-through different for the two types of fluorescein filter sets? Why? In preparing to answer why bleed-through occurs, it is useful to compare transmission profiles of the various filters of the filter sets and to compare them with the excitation and emission

spectra of fluorescein and rhodamine (Figs. 11-6, 11-8 and Color Plate 11-1).

(3) Is rhodamine fluorescence totally excluded from the fluorescein filter set with the narrow band-pass filter? Why? Since the long-pass fluorescein filter set shows transmission of red wavelengths, it should be obvious that rhodamine fluorescence will also be transmitted by this filter set design, making it unsuitable for examination of double stained specimens.

(4) What is the purpose of equipping a microscope with a filter set containing a long-pass emission filter?

(5) Name two steps you could take to reduce the amount of rhodamine bleed-through in a fluorescein filter set.

CONFOCAL LASER SCANNING MICROSCOPY

OVERVIEW

Thick fluorescent specimens such as rounded cells and tissue sections can pose problems for conventional wide-field fluorescence optics, because bright fluorescent signals from objects outside the focal plane increase the background and give low-contrast images. Confocal and deconvolution microscopy solve the problem by rejecting signals from nearby sources above and below the focal plane. In confocal microscopes this is accomplished optically by illuminating the specimen with a focused scanning laser beam (point scanning) and by placing a pinhole aperture in the image plane in front of an electronic photon detector. Both fluorescent specimens and reflective surfaces can be examined using this technique (Fig. 12-1). Confocal images can also be produced using a spinning Nipkow disk that gives tandem scanning with literally thousands of scanning beams. In deconvolution microscopy a standard wide-field fluorescence microscope is used, and the image of an optical section is obtained computationally with a computer that removes out-of-focus light from the image. (Deconvolution microscopy is not covered in this book. For discussion of this topic, the reader is directed to an excellent perspective by Chen et al., as well as early pioneers and developers of the method, David Agard and John Sedat (Pawley, 1995).) The high-contrast images provided by confocal and deconvolution methods can provide clear answers to commonly asked questions about fluorescent microscope specimens: Is a fluorescent signal distributed on a membrane surface or contained throughout the cytoplasm as a soluble factor? Within the limits of resolution of the light microscope, are different fluorescence signals colocalized on the same structure? What is the three-dimensional structure of the specimen?

By using a stepper motor that changes the microscope focus in 100 nm steps along the z-axis, confocal and deconvolution microscopes make it possible to acquire a stack of images or z-series at different focal planes and generate a three-dimensional view of the specimen using computer software. Microscope savants and manufacturers foresee a time when it will become routine for microscopists to acquire z-section stacks of live cells in multiple color channels, with stacks acquired at regular time intervals, so an entire color movie can be constructed showing dynamic events in a cell in three dimensions—a truly valuable and exciting experience! Such sequences are called five-dimensional, because intensity information for every point in x, y, and z dimensions in

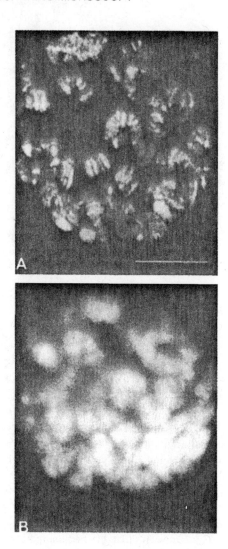

Figure 12-1

Confocal optical section of chromosomes in a *Drosophila* salivary gland nucleus. Nuclei were labeled with chromomycin A3 to show the banding of polytene chromosomes. The confocal image shows a striking pattern of fluorescent bands on the chromosomes, while the wide-field fluorescence image of the same cell looks blurred. Bar = 10 μm. (From White et al., 1987; with permission, Rockefeller University Press)

a specimen is available at various times (as in a time-lapse sequence) and in color. Confocal and deconvolution microscopy already provide this capability.

The principle of confocal imaging was developed and patented in 1957 by Marvin Minsky, who is well-known for work on neural network computing and artificial intelligence, while he was a postdoctoral fellow at Harvard University. In subsequent years, Brakenhoff, Sheppard, Wilson, and many others contributed to the design of a practical working instrument. Amos and White demonstrated the value of confocal imaging for fluorescent biological specimens around the time the first commercial instruments appeared in 1987. Since then, interest in confocal microscopy and improvements in the capacity of confocal imaging have increased at a rapid pace. For a historical perspective and annotated bibliography of this technology, refer to articles by Inoué and by Webb in Pawley (1995).

Confocal microscopes are *integrated electronic microscope systems,* in which the light microscope is part of an electronic imaging system containing an electronic detector or camera, a computer and computer software, and electronic devices for image display, printing, and storage (Fig. 12-2). The integration is so complete that microscopists now frequently refer to their microscopes as *digital* or *video imaging systems* and their activity at the microscope as *electronic imaging.* Electronic microscope systems are revolutionizing the face of research, serving as indispensable tools for documentation and analysis, and facilitating investigations on molecules, cells, and tissues that until now have simply not been possible. This transformation is the result of rapid technological advances in opto-electronics, lasers, fiber optics, thin film dielectric coatings, computers, printers and image storage devices, and image acquisition and processing software. In order to use these imaging systems, we must not only learn principles of light microscopy, but also master electronic imaging and image processing operations. Some of the challenges involve very fundamental concepts, such as how to take a picture. This is especially relevant to confocal microscopy, where there are no camera buttons to push, and even if there were, there is no confocal image to take a picture of! Acquiring an image with an electronic microscope system requires us to use computer software and make decisions that affect the resolution of space, time, and light intensity in the image. Since the ability to make these decisions quickly and with confidence requires training and education, we focus on electronic image acquisition in this and the following chapters.

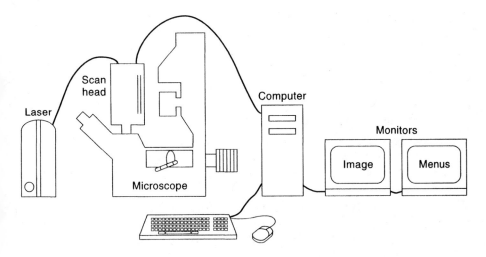

Figure 12-2

Basic components of a confocal laser scanning microscope (CLSM). A laser provides a beam of light that is scanned across the specimen by the scan head under the control of a computer. The scan head also directs fluorescence signals from the specimen to its pinhole and photomultiplier tube (PMT). The computer holds the image in an image memory board until it is processed and displayed on a computer monitor. A second monitor displays software menus for image acquisition and processing. It is important to realize that a confocal image is never generated in the microscope. Instead, the image is built up electronically, point by point and over time, from fluorescent signals received by the PMT and accumulated in the image memory board of the computer.

In this chapter we discuss the optics and essential electronic features of the confocal microscope, review briefly the parameters that are important in acquiring and evaluating an electronic image, and then consider how these parameters are affected by the operating controls of the confocal microscope.

THE OPTICAL PRINCIPLE OF CONFOCAL IMAGING

The confocal microscope is an integrated microscope system consisting of a fluorescence microscope, multiple laser light sources, a confocal box or scan head with optical and electronic equipment, a computer and monitor for display, and software for acquiring, processing, and analyzing images. The scan head generates the photon signals required to reconstruct the confocal image and contains the following devices:

- Inputs from one or more external laser light sources
- Fluorescence filter sets
- A galvanometer-based raster scanning mechanism
- One or more variable pinhole apertures for generating the confocal image
- Photomultiplier tube (PMT) detectors for different fluorescent wavelengths

The general arrangement of components in the scan head is shown in Figure 12-3. A computer converts the voltage fluctuations of the PMTs into digital signals for image display on the computer monitor.

The optical principle of confocal microscopy is shown in Figure 12-4 and is described as follows:

- *Epi-illumination* is used, where the light source and detector are both on the same side of the specimen plane and separated from it by the objective lens, which functions as both a condenser and objective. The components of fluorescence filter sets (exciter filter, dichroic filter, emission filter) perform the same functions as they do in wide-field fluorescence microscopy.
- A laser beam is expanded to fill the back aperture of the objective and forms an intense diffraction-limited spot that is scanned from side to side and from top to bottom over the specimen in a pattern called a *raster.* This procedure is called *point scanning.*
- The heart of confocal optics is the *pinhole aperture,* which accepts fluorescent photons from the illuminated focused spot in the raster, but largely excludes fluorescence signals from objects above and below the focal plane, which, being out of focus, are focused on the pinhole as disks of much larger diameter. Because the size of the disk of an out-of-focus object is spread out over such a large area, only a fraction of light from out-of-focus objects passes through the pinhole. The pinhole also eliminates much of the stray light in the optical system. Examine Figure 12-4 carefully to see how the pinhole blocks out-of-focal-plane signals. The combination of point scanning and the use of a pinhole as a spatial filter at the conjugate image plane are essential for producing the confocal image.
- Fluorescent wavelengths emanating from an excited point in the specimen at any time *t* are collected by the same objective and focused as a small diffraction spot

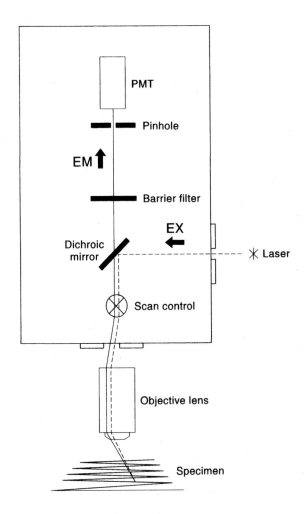

Figure 12-3

Optical pathway in a confocal scan head. A laser beam is reflected by a dichroic mirror onto components of the scan-control mechanism that sweeps the beam in a raster back and forth across the specimen. The objective lens collects fluorescent light, which is descanned at the scan control, transmitted through the dichroic mirror and emission (barrier) filter, and passes through the pinhole to a photomultiplier tube. EX and EM indicate the paths taken by the excitation and fluorescence emission wavelengths.

that just fills the diameter of a variable pinhole aperture placed in front of a *PMT detector* in a plane that corresponds to the image plane in a wide-field fluorescence microscope. The pinhole is optically confocal with and conjugate to the specimen plane. Thus, the PMT does not see an image, but receives a constant stream of changing photon fluxes; the computer in turn sees a constantly changing voltage signal from the PMT, digitizes it, and displays the signal on the monitor.

- To generate an image of an extended specimen, the laser beam is scanned across the object by a raster scanning mechanism that is typically based on two high-speed vibrating mirrors driven by galvanometer motors (Fig. 12-5). One mirror oscillates left-right while the other moves up and down. Fluorescent photons emitted from an excited point in the specimen are collected by the objective. Because the speed of

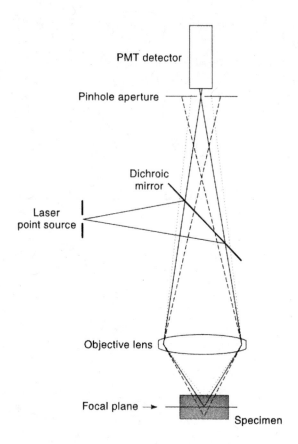

Figure 12-4

The confocal principle in epifluorescence laser scanning microscopy. Excitation wavelengths from a laser point source are confocal with a scanning point in the specimen. Fluorescent wavelengths emitted from a point in the specimen are focused as a confocal point at the detector pinhole. Fluorescent light emitted at points above and below the plane of focus of the objective lens is not confocal with the pinhole and forms extended disks in the plane of the pinhole. Since only a small fraction of light from out-of-focus locations is delivered to the detector, out-of-focus information is largely excluded from the detector and final image. The dichroic mirror and barrier filter (the latter is not shown) perform the same functions as in a wide-field epifluorescence microscope.

the galvanometer mirrors is inconsequential relative to the speed of light, fluorescent light follows the same light path on its return and is brought to the same position on the optic axis as the original exciting laser beam. This process is called *descanning*. The fluorescent light then passes through a dichroic mirror and becomes focused at the confocal pinhole. Because descanning is instantaneous, the image in the pinhole always remains steady and does not move back and forth like the beam in the plane of the specimen; however, the focused spot varies in intensity over time as the spot excites different locations in the specimen.

- Fluctuations in light intensity are converted into a continuously changing voltage (an analogue signal) by the PMT detector. The analogue signal is digitized at regular time intervals by an analogue-to-digital converter to generate pixels (digital picture elements) that are stored in an image frame buffer board and are displayed on

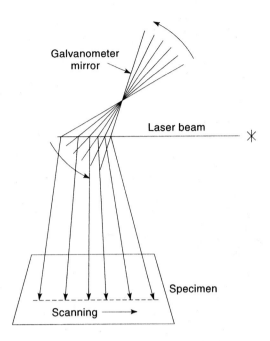

Figure 12-5
The scan-control mechanism in a CLSM. The sketch shows the delivery of the excitatory laser beam to the specimen by one of two galvanometer-driven mirrors that vibrate in mutually perpendicular axes within the confocal scan head. One mirror controls scanning along the x-axis, the other along the y-axis; both motions are coordinated to generate a pattern of a raster on the specimen. The speed and angular extent of deflection of the mirrors can be controlled to regulate the scanning rate and the extent of the scan.

a computer monitor. Thus, a confocal image of an object is reconstructed from pho-ton signals and is displayed by a computer; the confocal image never exists as a real image that can be seen by the eye in the microscope.

- A microscope system of this design is called a *confocal laser scanning microscope (CLSM)*.

Demonstration: Isolation of Focal Plane Signals with a Confocal Pinhole

Warning—Care should be taken not to accidentally reflect the light beam into the eye during this exercise!

To become familiar with confocal optics, construct the confocal reflection optical system shown in Figure 12-6 using common optical components mounted on an I-beam optical bench, a 1–5 mW HeNe laser, and a coin as a specimen. The specimen support plate must be rigid enough so that you can slide a coin or other reflective specimen on its surface without moving the plate. Position the objective lens so that reflections from the coin are refocused at the pinhole aperture (a pinhole

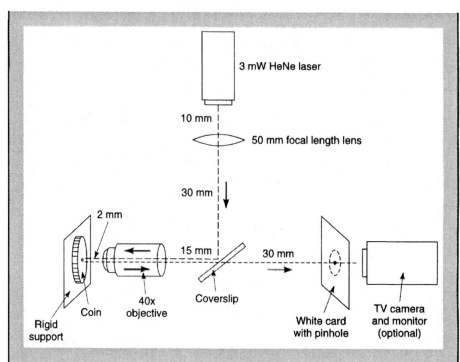

Figure 12-6
Optical bench for demonstrating confocal reflection optics.

pierced in a white card) mounted in the confocal image plane. A white card allows you to see if the laser beam is diffuse or focused at the pinhole. The detector behind the pinhole can be a white screen for visual inspection, or for demonstration before a large group, a video camera connected to a TV monitor. Fluctuations in light intensity reaching the camera are seen as a brightening or darkening on the TV monitor. Note that a large portion of light from the laser passes through the coverslip instead of being reflected toward the objective, and that a large portion of light is reflected back to the laser instead of being transmitted to the pinhole. With a 1–3 mW laser, sufficient light will reach the pinhole.

- Use a reflective specimen such as a coin with a bas relief surface (or some other reflective irregular surface) that can be held up against a rigid but adjustable support. Move the support and coin back and forth along the I-beam until the laser beam appears as a tightly focused spot on the surface of the coin. As the coin is moved on a rigid surface in the specimen plane, adjust the position of the objective (and/or pinhole support) so that reflections from the highest points on the coin are focused as a bright spot at the pinhole and give a bright signal on the camera monitor or screen.

- Reflected light from out-of-focal-plane features (recessed background regions on the coin) forms an expanded 5–10 mm disk at the pinhole rather than a focused point. The decreased transmission of photons through the pin-hole is seen as a darkening on the TV monitor. The rejection of out-of-focal-plane signals is easy to observe.

- If the coin is moved around on its support plate so that it is scanned by the fixed laser beam, the pinhole screen and monitor show bright and dark signals that look like flickering, but an actual image is not observed. This reinforces the idea that the confocal signal is a continuously changing voltage, not an image. In the scan head of an actual confocal system, the voltage signal is digitized at regular intervals into pixel values that are stored in a computer in an image frame buffer board and displayed on a computer monitor.

ADVANTAGES OF CLSM OVER WIDE-FIELD FLUORESCENCE SYSTEMS

The principal gain in imaging performance using CLSM is the ability to optically section through fluorescent objects up to $10-50$ μm thick or to obtain high-contrast images of surface topologies using confocal reflection optics. Contrast and definition are improved, sometimes considerably, due to the reduction in background signal and a greatly improved signal-to-noise ratio. With a stack of images representing different focal planes spaced at regular intervals along the optic axis (the z-axis), the object can be displayed in a variety of ways:

- *Composite or projection view,* where all of the information contained in a stack or series of images taken at a series of focal planes along the z-axis is projected into a single focused image. Consider a three-dimensional object such as a fluorescent neuron with thin extended processes in a tissue section. In a single wide-field fluorescence image, out-of-focus processes appear blurry and indistinct. A single confocal image shows the neuron extensions as fragmented streaks and dots, but a composite view of all of the focal planes contained in a stack or z-series shows the processes as continuous well-defined projections (Fig. 12-7).

- *Three-dimensional views* of an object which can be obtained from a z-stack of confocal images with the help of computer software for image display. Many confocal software programs can display an object at different angular perspectives or rotate the stack about the x-, y-, or z-axis, or a combination of axes, to view the cell in three dimensions. This mode of viewing is valuable for examining complex three-dimensional objects such as columnar epithelial cells and details in tissue sections (Fig. 12-8).

- Transverse xz or yz *cross-sectional views,* which can be generated by most confocal software programs. The object appears as if it had been cut transversely—that is, in a plane oriented parallel to the optic axis.

- *Five-dimensional views* including information in x-, y-, and z-dimensions, in a timed sequence, and in multiple colors. Such sequences can be displayed as a three-dimensional multicolor movie in real time or time-lapse mode. Five-dimensional viewing will become more convenient as computer processing speed and storage capacity continue to improve.

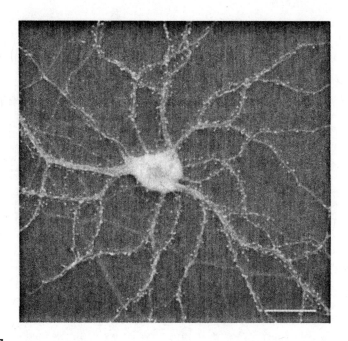

Figure 12-7

Composite or projection view of a neuron in a 10 μm thick section of rat brain tissue. The image is a composite of 20 optical sections obtained by reflection-mode CLSM. The highly branched and extended processes of the single neuron are clearly defined in a single image, even though the cell is highly three-dimensional. A 10 μm section of tissue was fixed, labeled with neuron-specific antibodies and HRP-conjugated secondary antibody, and contrasted with diaminobenzidene. Bar = 20 μm (Image courtesy of Nicole Stricker, Johns Hopkins University.)

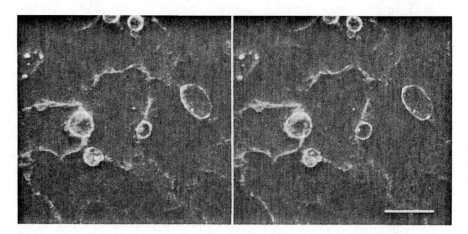

Figure 12-8

Volume view of transformed liver hepatocytes in culture. The two micrographs can be fused into a single three-dimensional image by crossing the eyes or using a stereo viewer. Basilateral and apical membrane surfaces were labeled with specific antibodies and tagged with rhodamine- and fluorescein-conjugated secondary antibodies to distinguish the two membrane systems. The large vesicles (fluorescein signal) at the margins between adjacent cells represent apical membranes that function as bile canaliculi. Bar = 10 μm (Image courtesy of Pamela Tuma, Johns Hopkins University.)

The following additional advantages of CLSM should also be considered:

- Magnification can be adjusted electronically by varying the area scanned by the laser (zoom factor) without having to change the objective lens. Since the display area on the monitor remains unchanged, the image appears magnified or zoomed.

- Digitization by an *analogue-to-digital converter* transforms the continuous voltage signal into discrete digital units that correspond to steps of light intensity, thus allowing you to obtain quantitative measurements of fluorescence intensity.

- There are few restrictions on the objective lens. However, for accurate color focus and proper spatial resolution of light intensities, high-NA plan-fluorite or apochromatic oil immersion objectives should be employed.

- Epi-illumination and point scanning are ideal for *confocal reflection microscopy,* in which the laser beam scans the three-dimensional surface of a reflective object. Fluorescence filter sets are not used for this mode of imaging; rather, the focused spot of the laser is reflected off the surface. Reflections from features lying in the focal plane pass through the confocal pinhole aperture, whereas light from reflections above and below the focal plane is largely excluded just as in confocal fluorescence microscopy.

CRITERIA DEFINING IMAGE QUALITY AND THE PERFORMANCE OF AN ELECTRONIC IMAGING SYSTEM

The quality of a confocal image or any image is determined by four principal factors:

1. Spatial resolution
2. Resolution of light intensity (dynamic range)
3. Signal-to-noise ratio
4. Temporal resolution

Spatial resolution describes the smallest resolvable distance between two points in an image. Resolution between two points in the image plane and along the z-axis depends on the excitation and fluorescence wavelengths and the numerical aperture of the objective lens and settings in the confocal scan head. The numerical aperture of the objective is crucial, since it determines the size of the diffraction-limited scanning spot on the specimen and the size of the focused fluorescent spot at the pinhole. The role of the numerical aperture in determining spatial resolution has already been discussed (see Chapter 6). In wide-field fluorescence optics, spatial resolution is determined by the wavelength of the emitted fluorescent light; in confocal mode, both the excitation and emission wavelengths are important, because the size of the scanning diffraction spot inducing fluorescence in the specimen depends directly on the excitation wavelength. (See Chapter 5 for the dependence of the size of the diffraction spot on the wavelength of light.) Thus, unlike wide-field fluorescence optics, the smallest distance that can be resolved using confocal optics is proportional to $(1/\lambda_1 + 1/\lambda_2)$, and the parameters of wavelength and numerical aperture figure twice into the calculation for spatial resolution (Pawley, 1995; Shotton, 1993; Wilhelm et al., 2000).

In the confocal microscope, spatial resolution also depends on the size of the pinhole aperture at the detector, the zoom factor, and the scan rate, which are adjusted using software that controls the scan head. Decreasing the size of the pinhole reduces the thickness of the focal plane along the z-axis, thereby allowing higher resolution in optical sectioning, which is essential for high-quality projection images and high-resolution three-dimensional viewing. Reducing the size of the pinhole also improves contrast by excluding out-of-focal-plane sources of light. Under certain conditions, the lateral spatial resolution in the xy plane obtainable in a confocal fluorescence microscope can exceed that obtainable with wide-field optics by a factor of ~1.4, a condition sometimes called *superresolution.* Normally the detector pinhole is adjusted to accommodate the full diameter of the diffraction disk. However, if the pinhole is stopped down to about one-quarter of the diffraction spot diameter, the effective spot diameter is slimmed down so that the disk diameter at one-half maximum amplitude is reduced by a factor of ~1.4. Oldenbourg et al. (1993) have shown that image contrast is also improved by stopping down the pinhole.

As given by R. Webb in Pawley (1995), the minimum resolvable distance d between two points in the xy horizontal plane of the confocal microscope can be approximated as:

$$d_{x,y} \approx 0.4\lambda/NA.$$

Resolution along the z-axis is described:

$$d_z \approx 1.4\lambda n/NA^2.$$

The effect of pinhole diameter on section thickness is shown in Figure 12-9. A constricted pinhole in confocal optics increases horizontal resolution in the x,y plane by a factor of ~1.4 but decreases the axial resolution by the same factor. The effects of zoom factor and scan speed on resolution are discussed later in this chapter.

Dynamic range (DR) describes the *resolution of light intensity* in the image and is defined as the number of gray levels that are assigned to an image by the analogue-to-

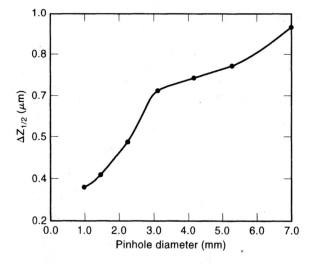

Figure 12-9

Confocal pinhole diameter vs. optical section thickness. The thickness of the optical section through the z-axis of the confocal image (Z) is plotted vs. the diameter of the pinhole aperture in millimeters. Fluorescent beads (0.2 μm) were examined with a Nikon Planapo 60×, 1.4 NA oil immersion lens. (From Pawley, 1995; Kluwer/Penum Publishers, with permission) digital converter. The maximum possible DR is the DR of the detector. The DR of the PMT detector is defined as the ratio of the saturated signal to the detector readout noise, and is calculated in volts or electrons. The DR of a detector is therefore an inherent feature of the imaging system. To achieve the imaging potential of the detector, the operator should try to acquire images that fill its dynamic range, from black (no signal) to white (saturating signal). This is accomplished by adjusting the amplitude range of the photon signal using the electronic gain and offset controls of the PMT. The procedures for adjusting gain and offset are given in this chapter.

Signal-to-noise ratio (S/N) defines the degree of visibility or clarity of an image. It depends directly on the amplitudes of the object and its background, and on the electronic noise of the imaging system. For bright images, S/N is determined mainly by the intensities of the object and background; however, for dim images, the electronic noise of the imaging system becomes a determining factor. Image brightness at the pinhole is affected by many variables, including laser power, fluorochrome density in the specimen, NA of the objective lens, confocal zoom factor, raster scan rate, the choice of fluorescence filter sets, and other factors. Reduced photon throughput and/or noisy performance by the detector results in images that lack contrast and look grainy. Generally, there is little we can do to improve the imaging performance of a detector other than to provide more light. Typically 50–100 photons/s/pixel are detected from a moderate to bright specimen by CLSM, which corresponds to an S/N ratio of 25 for a sensitive confocal system. This value compares with maximum S/N ratios of 100 to several hundred for video and digital CCD camera systems. Thus, the image quality of a bright object in confocal is okay, but not good or excellent. The usual way to improve S/N is to increase the amount of light by reducing the scan rate or opening the pinhole. Alternatively, a number of individual frames can be averaged, which is called *frame averaging* or *Kalman averaging,* because S/N improves in proportion to the square root of the number of averaged frames. We return to the topic of improving S/N later. One of the compromises in obtaining optical slices by confocal imaging is a reduction in photons contributing to the image and therefore to image quality.

Temporal resolution depends on the raster scan rate and the processing rates of the detector, the analogue-to-digital converter, and the computer. Frames are typically captured at a rate of 2 frames/s for a 512×512 pixel image, but rates of 100 frames/s or higher can be acquired for images of limited size.

In summary, optical performance in the confocal microscope is affected by several variables, all of which must be evaluated and controlled by the microscopist. It almost never happens that time, space, and intensity are all resolved optimally for a given combination of specimen and microscope. Generally, it is necessary to make compromises. For example, to optimize resolution of light intensity to obtain smooth-looking images with high signal-to-noise ratio, you might need to decrease the spatial and/or temporal resolution of the image. Recognizing how to balance these parameters requires education and experience and is the key to successful confocal imaging.

ELECTRONIC ADJUSTMENTS AND CONSIDERATIONS FOR CONFOCAL FLUORESCENCE IMAGING

A number of mechanical adjustments, many of which are actuated using the confocal software, affect the brightness and spatial resolution of the confocal image. These include adjustments for:

- Detector pinhole aperture
- Laser scan rate and image size
- Zoom factor
- Laser selection and laser intensity
- Gain and offset of the PMT detector
- Objective lens selection

Pinhole Adjustment

Parameters such as focal plane thickness, spatial resolution, image brightness, and susceptibility to photodamage are affected by the size of the detector pinhole. Since the pinhole is frequently adjusted to control these parameters, the confocal operator needs to become familiar with its effects. The effect of the pinhole diameter on the thickness of the optical section is shown in Figure 12-9. As already described, a small pinhole gives the thinnest optical section. It also improves horizontal resolution and contrast in the image. However, reducing the size of the pinhole also decreases the number of photons reaching the detector. If we compare two images of comparable amplitude made at different pinhole settings, the image corresponding to the smaller pinhole diameter requires a longer exposure for the same signal-to-noise ratio. However, increasing the exposure time can cause significant photobleaching in immunofluorescence specimens or compromise viability in live cell preparations. In contrast, a wide pinhole increases photon flux and reduces photobleaching by requiring less exposure of the specimen to the laser. If we compare images made at different pinhole settings but the same exposure time, we see that the image acquired using a wider pinhole is brighter and contains less noise.

The optimum average pinhole setting corresponds to an aperture giving 50% of maximum intensity. At this position, ~75% of the light of the Airy disk still passes the pinhole, and resolution will be 20% better than obtained by a wide-field fluorescence system. This corresponds to a pinhole diameter that is slightly smaller than the first dark ring in the Airy disk pattern. Generally, higher magnification objectives require larger pinhole settings because they produce larger Airy disks.

Scan rate, Image Size, and Dwell Time

The galvanometer-based scanning mechanism determines the size of the area in the specimen that is covered by the raster and the rate at which the specimen is scanned. These parameters determine, in part, the pixel size in the captured image, the spatial resolution, and the *laser dwell time*—that is, the length of time the laser dwells on the specimen per pixel element in the image. If the amplitude of deflection of the galvanometer mirrors is small and the scan rate is high, a smaller image area can be presented on the monitor at rates up to hundreds of frames per second. High-speed sampling (20–30 frames/s) gives the appearance of a live image on the monitor and is convenient for focusing the specimen. High speed is also useful for monitoring rapid kinetic events in the specimen. In contrast, a large-amplitude deflection of the galvanometer mirrors coupled with a slow scan rate produces a large-size image at high resolution but at a frame rate of 1 frame/s or less. For fixed, stained fluorescent specimens, the capture rate is typically 2 frames/s for an area covered by 512×512 pixels. The scan rate options indicated in the confocal software accommodate a range of scanning speeds and image sizes.

It should be remembered that the scan rate directly affects the rate of photobleaching of the specimen (slower scan rate gives greater bleaching rate) and the signal-to-noise ratio in the image (faster scanning reduces S/N). The scan rate also affects the laser dwell time on a unit imaging area (pixel) in the image. With rapid scanning, the dwell time can be reduced to about $1.0-0.1$ μs/pixel. The significance for live fluorescent specimens is that shorter dwell times reduce the rate of accumulated photodamage.

Zoom Factor

Electronic zoom is used for two reasons: (1) to adjust the sampling period of the laser scanning mechanism to maintain spatial resolution, and (2) to magnify the image for display and printing. Implementing the zoom command while maintaining the same pixel format in the image causes the galvanometer scanners to cover a smaller area on the specimen and reduce the scan rate. Since the zoomed image contains the same number of pixel elements, the image appears magnified on the monitor (Fig. 12-10). However, if significant changes in magnification are required, it is always preferable to use a higher-power objective lens, rather than increase magnification electronically using the zoom, because the higher-magnification, higher-NA objective gives better image resolution and definition. In most cases, zoom control is used to make minor adjustments in

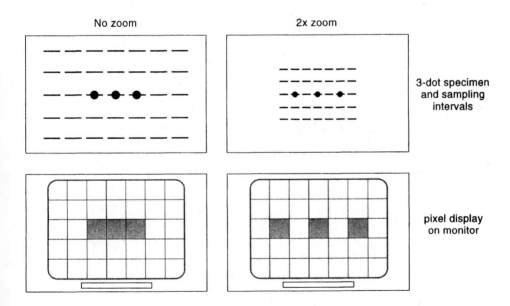

Figure 12-10

Effect of electronic zoom on electronic sampling and image resolution. Increasing the zoom factor reduces the area scanned on the specimen and slows the rate of scanning. The increased number of sampling operations along a comparable length in the specimen increases the spatial resolution and magnifies the display on the monitor. Zoom is typically employed when it is necessary or desirable to preserve the diffraction-limited optics of the objective lens and optical system.

the magnification and to adjust the number of pixels (samples) that cover an object in order to maintain resolution. Because the same amount of laser light is delivered to a smaller area when the zoom is increased, specimens photobleach at a faster rate. The effect of zoom on a biological specimen is shown in Figure 12-11.

To preserve the maximum resolution afforded by the objective lens, you should calculate and employ the required zoom factor. Using the resolution equation given earlier in this chapter, a 100×, 1.35 NA objective lens receiving 520 nm blue-green light is calculated to have a spatial resolution of 0.15 μm. The required sampling interval is calculated as 0.15 μm/2 = 0.075 μm. The factor of 2 means that there are two samples per spatial unit of 0.15 μm, and it is related to the requirements of a sampling theorem for maintaining the spatial resolution provided by the optics. We discuss this topic in greater detail in Chapter 13. The sampling period of any given zoom setting is usually indicated in a statistics table that accompanies an image. If the sampling period is not indicated on the monitor or in an image statistics menu, the diameter of the field of view can be adjusted with the zoom command to obtain the same result. The correct size is determined as the desired sampling period times the number of pixels across one edge of the image. For an image format of 512 × 512 pixels, we calculate 512 × 0.075 μm = 40 μm as the required image diameter.

Zoom settings above and below the optimum value are called *oversampling* and *undersampling*. Oversampled images (zoom too high) are a bit smoother and easier to work with in image processing, but the rate of photobleaching is greater. Undersampling (zoom too low) degrades the spatial resolution provided by the objective lens, but reduces the rate of bleaching. Undersampling is sometimes used in live cell imaging to optimize viability. For low-magnification, low-NA objective lenses, the Nyquist sam-

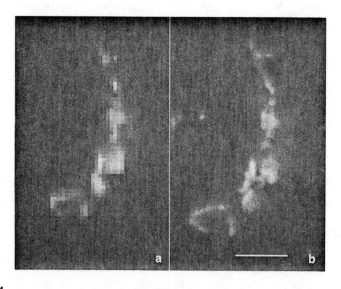

Figure 12-11

Effect of zoom on confocal image resolution. Immunofluorescence micrograph of a Golgi apparatus in a tissue culture cell. (a) Image acquired without zoom. (b) Image acquired at 9× electronic zoom. The image was acquired on a Noran confocal microscope with an Olympus 100×, 1.3 NA oil immersion lens. The nonzoomed image was magnified to about the same final magnification as the zoomed image. Bar = 2 μm. (Image courtesy of Michael Delannoy and Emily Corse, Johns Hopkins University.)

pling factor may need to be as high as 3–4 (instead of 2) to maintain the resolution provided by the objective.

Laser Selection and Laser Intensity

Confocal microscopes employ one or more types of lasers to obtain the wavelengths required for fluorochrome excitation, the most common wavelengths being 365, 488, 515, 568, 633, and 648 nm (see Table 12-1). The laser beam is expanded by a lens to fill the back aperture of the objective to give the minimum-size diffraction spot on the specimen. Because the back aperture must be uniformly illuminated across its diameter, and because the light intensity profile across a laser beam is Gaussian, the beam is expanded considerably so that the portion filling the lens is more uniform. The power of lasers fitted by manufacturers for confocal illumination (range, 5–50 mW) is chosen based on the factor of light loss from beam spreading and the intensity of light required to give close to saturating excitation of the appropriate fluorochromes in the focused beam in the specimen.

Laser power is adjusted using the laser's power-control dial, or additionally with an *acousto-optical tunable filter (AOTF)*. For lasers that emit at several wavelengths, the AOTF can also be used to select a specific laser line for excitation. Normally lasers are operated at their midpower range during acquisition to prolong the lifetime of the laser. Many lasers can also be placed in a standby (low-power) mode, at which the amplitude of laser light is minimal. In cases where specimens are subject to photobleaching or biological damage, laser power is reduced to a point where fluorescence emission is still adequate for acquiring an image. Since the intensity of the focused laser beam can damage the eye, confocal systems contain an *interlock device* that prevents the operator from seeing the laser through the eyepieces during laser scanning or during visual inspection of the sample with a mercury arc lamp.

Gain and Offset Settings of the PMT Detector

The gain and offset controls of the PMT are used to adjust the light intensities in the image to match the dynamic range of the detector (Fig. 12-12). These adjustments assure that the maximum number of gray levels is included in the output signal of the

TABLE 12-1   Lasers Employed in Confocal Microscopy

	Wavelength (nm)			
Laser type	UV	Blue	Green	Red
Argon	351–364	457, 488	514	
Helium/cadmium	322	442		
Krypton/argon		488	568	647
Green helium/neon			543	
Red helium/neon				633
Red laser diode				638

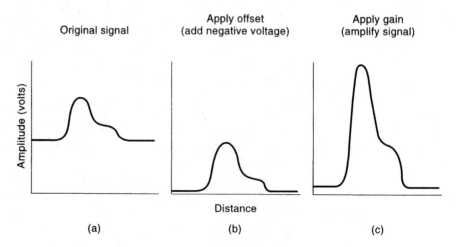

Figure 12-12

Adjustment of gain and offset. Gain and offset are electronic controls used to stretch the available photon signals from an object to fill the dynamic range of the detector in order to include the maximum number of gray levels, from white to black, in the captured image. Offset adds a voltage (positive or negative) to the output signal so that the lowest signals just approach the threshold for detection on the PMT (black level). Gain amplifies the signal by multiplying the output voltage from the PMT by a constant factor prior to its digitization at the analogue-to-digital converter and is increased until the maximum signal values just approach saturation. Offset should be applied first before adjusting the gain.

PMT. After digitization in the computer, the photon signal is displayed as shades of gray ranging from black (no signal) to white (saturating signal) on the computer monitor. PMTs with a dynamic range of 10 or 12 bits have 2^{10} (1024) or 2^{12} (4096) gray levels, respectively, which is also the number of gray levels in the respective image files in the computer. Setting the gain and offset of the PMT should not be confused with adjusting the contrast and brightness of the image during image processing. Processing can stretch existing pixel values to fill the black-to-white display range on the monitor, but can never create new gray levels. Thus, when a poorly captured image containing only 50 out of a possible 4000 gray levels is stretched from black to white during image processing, the resulting image looks grainy. Sometimes it is not possible to fill all of the gray levels during an acquisition, but ordinarily this should be the goal. We will discuss the gray-level ranges of images in Chapters 13 through 16.

Offset is an electronic adjustment that adds a positive or negative voltage to the signal so that a selected background signal corresponds to a PMT output of ~0 volts (black). *Gain* is an electronic adjustment that amplifies the input signal by a voltage multiplication process that causes it to be assigned to a higher gray-level value. Generally, offset should be set first, followed by adjustment of the gain. Increasing the gain beyond its optimal setting can make an image look somewhat grainy, but it is sometimes desirable to capture the maximum number of gray levels, even if the gain must be increased somewhat to do so. In practice, these adjustments are often made using a special color display function to depict pixel values on the monitor. In one commonly used function, the saturating pixels are shown in red, black-level pixels are shown in blue, and all intermediate gray levels are shown in shades of gray. When properly adjusted, a few red and blue pixels will be seen in the image, indicating that the full dynamic range of the PMT is being utilized. Figure 12-12 shows the effect of gain and offset on image contrast.

Objective Lens Selection

The quality of the confocal image is critically dependent on the optical performance of the objective lens. The lens should be well corrected for the principal lens aberrations discussed in Chapter 4. Different color wavelengths must be focused precisely in the confocal pinhole, and spherical and off-axis aberrations such as coma and astigmatism must also be corrected to a high degree. This is because the laser must be focused to a well-defined diffraction spot at locations in the specimen that are off the optic axis, and fluorescent wavelengths must be collected and focused to a spot in the pinhole. As we now also appreciate, the NA of the objective is important, because it determines the spatial resolution and brightness of the image and the depth of focus in the specimen. The role of the NA in determining spatial resolution has already been discussed (Chapters 3 and 6). Image brightness (photon flux per area per unit time) is also determined by the NA and is of great practical importance for confocal microscopy. Since the amount of light reaching the photomultiplier tube is usually small, we are frequently forced to operate the confocal system at high electronic gain, which tends to produce noisy images. The gain can be reduced and the image quality improved by providing a brighter diffraction spot at the pinhole with a high-NA objective. Since image brightness is proportional to NA^4/Mag^2 (Chapters 4 and 6), you should select the lowest magnification commensurate with the resolution requirements, and within that constraint, choose the highest available NA. For these reasons, a well-corrected plan-neofluorite or plan-apochromatic objective of 60–100× magnification is ideal for the examination of cellular details.

In addition to the confocal pinhole, the objective lens plays an equally important role in determining the depth of field in the specimen and the depth of focus in the image, and therefore the thickness of a confocal image. The thickness d of the focal plane along the z-axis in the specimen is defined as λ/NA^2 and is determined by the wavelength λ and the numerical aperture NA of the objective lens (Chapter 6). Since in epi-illumination mode the objective focuses the laser spot on the specimen, high-NA lenses reduce the thickness dimension of the diffraction spot for laser excitation and reduce the focal plane thickness at the exit pinhole. Reducing the size of the pinhole diaphragm limits the thickness of the confocal optical section, but only to the limit set by the microscope optics. Therefore, the combination of high-NA lenses and small pinhole apertures generates the thinnest optical sections. The effects of the mechanical adjustments on image brightness and spatial resolution are summarized in Table 12-2.

PHOTOBLEACHING

Confocal specimens are subject to photobleaching from repetitive exposure to the intense beam of the laser. The rate and extent of bleaching are generally greater than for specimens examined with a wide-field fluorescence microscope and can make it difficult to obtain suitable images of weakly fluorescent specimens. If an immunolabeled specimen appears dim and its image looks unacceptable due to improper staining, a new specimen should be prepared. In other cases, dyes with lower rates of photobleaching

TABLE 12-2   Effect of Increasing Confocal Parameters on Image Intensity and Spatial Resolution

Parameter	Effect
↑ Pinhole diameter	↑ Intensity
	↓ Spatial resolution
↑ Zoom	↑ Intensity
	↑ Spatial resolution
↑ Scan rate	↓ Intensity
	↑ Temporal resolution
↑ Objective NA	↑ Intensity
	↑ Spatial resolution

and antifade reagents (mounting medium with components that reduce the rate of photobleaching) can be used. Cyanine dyes (Amersham International, marketed by Jackson ImmunoResearch Laboratories, West Grove, Pennsylvania) or Alexa dyes and new antifade reagents such as SlowFade and ProLong (Molecular Probes, Eugene, Oregon) are suitable reagents.

The microscope operator can reduce the rate of photobleaching through one or more of the following methods:

- Reduce the laser power.
- Insert neutral density filters into the excitation beam.
- Scan at a faster rate.
- Lower the zoom factor (scan over a larger area).
- Widen the pinhole (allows shorter exposure time).
- Use a long-pass emission filter to capture as much fluorescent light as possible.
- Reduce the number of frames used for frame averaging.

For z-series intended for three-dimensional viewing, the operator can sometimes collect the stack of images from front to back so that more distant details, made dimmer by photobleaching from repetitive scanning, match the viewer's expectations that fluorescent signals from more distant locations will appear fainter in the three-dimensional view.

GENERAL PROCEDURE FOR ACQUIRING A CONFOCAL IMAGE

The sequence of steps used to acquire an image with a confocal laser scanning microscope usually proceeds as follows:

- *Prepare for acquisition.*
  1. Focus the specimen in visual mode with the appropriate filter set.

2. With software, pick the appropriate laser line and filters (dichroic mirror, emission filter), and select the PMTs to be used for confocal imaging.

3. With software, indicate the appropriate objective magnification and NA.

4. Begin by using a neutral density (attenuation) filter of moderate density (10%T); alternatively reduce the laser power.

5. Select the pixel number format of the image; a good place to begin is 512×512 pixels.

6. For purposes of evaluating and adjusting the initial raw image, use a live (single-frame) display at the slowest scan rate.

7. Before shifting to confocal mode, be sure that the laser safety interlock mechanism is working.

8. Begin live scanning.

- *Make on-the-fly adjustments of the pinhole and the zoom factor.*

    1. Adjust the pinhole diameter for each objective lens. Typically the pinhole is set at or slightly below the diameter of the Airy disk. The recommended pinhole size is usually given in the software or posted on a card by the microscope.

    2. Increase the zoom factor to match the pixel size of the scanning mechanism to the required spatial resolution or the resolution of the optics. Remember that increasing the zoom decreases the diameter of the field of view and increases the rate of photobleaching.

- *Make interactive adjustments of the gain and offset.*

    1. Set the offset (dark or black level). If the image intensity is bright, it may be necessary to employ a neutral density filter. Examine a dark background region on the specimen slide. Increase the amount of offset slowly and notice on progressive scans how the screen display exhibits darker shades of gray. When the image background approaches a black color, the offset is set properly. This adjustment is easier to perform using a color display function (look-up table or LUT) where minimum intensifier output (black) is shown as blue, saturating pixels are shown as red, and intermediate values are shown as shades of gray. Adjust the offset to 5% above the background so that there will be no true black in the image. If the offset is insufficient, darker parts of an image appear medium gray, and the full range of possible gray levels is not included in the image. If too much offset is applied, darker regions in the image are displayed as black, and they can never be retrieved by subsequent image processing. Ideally the background should be set a few levels higher than true black (digital value $= 0$).

    2. Set the gain. Increasing the gain amplifies the light intensity and increases the image contrast. Increase the gain until you begin to approach saturation of bright highlights in the image. The use of a color display function makes this adjustment easier to set. If the gain is too high, brighter parts of the image become saturated and appear white, and the resolution of these parts can never be retrieved by subsequent image processing. If the gain is too low, brighter parts of the image appear gray, and the number of gray levels is limited. If the proper gain setting is a medium or low position (a strong fluorescent specimen), you may use a slow or medium scan speed to acquire an image; if the gain must be set at a high setting (a weak specimen), a fast scan speed should be used. A

rule of thumb is to scan slowly enough that the image stands out above the background by a small amount.

- *Acquire the final image.* Use frame averaging or frame accumulation. The number of frames to be averaged depends on the noise in the image, but numbers ranging from 4 to 16 are typical. Remember that the factor improvement in S/N is the square root of the number of averaged frames or the square root of the factor by which the exposure time has been increased.

- *Adjust the step size for a z-series.* For z-series, the step size of the stepper motor controlling the stage or objective is adjusted to give a sample period of 2 steps per diffraction disk radius in the z-dimension (more than this if the sample is thick and/or high spatial resolution is not required).

- *Review the following general points:*

  1. Use the highest numerical aperture lens available (NA 1.3–1.4) to reduce focal plane thickness and increase image brightness.

  2. Room lights must be off to see gray levels on the monitor properly.

  3. For bright images, use an ND filter so that offset and gain can be used at their midrange settings.

  4. Excitation of the specimen at high laser power maximizes fluorescence emission, but causes rapid bleaching and reduces image quality.

  5. ND filters are usually employed even if the resulting image is somewhat noisy to reduce the rate of photobleaching. Noise in dimmer images can be removed by frame averaging or frame accumulation. Noise in the PMT output, made worse by use of a high gain setting, can be significant even in a single scan of a bright image. Thus, frame averaging (Kalman averaging) of a moderate-quality image nearly always produces a better result.

  6. Remember that electronic zoom is related to the sampling size and affects spatial resolution.

  7. A freshly acquired (raw) image has up to 10–12 bit intensity resolution in computer RAM; however, saving an image may reduce its resolution to 8 bits, and the original dynamic range will be permanently lost; therefore, images must be optimized for brightness/contrast before saving. It is often good practice to histogram stretch an image before saving it as byte so that image data are distributed over the maximum number (256) of gray levels.

  8. Laser alignment is critical. If the laser spot does not fill the back aperture of the objective, the scan spot in the object plane is wider and image resolution is reduced.

TWO-PHOTON AND MULTIPHOTON LASER SCANNING MICROSCOPY

A fluorescent molecule can be induced to fluoresce by the absorption of a single photon of the proper excitation wavelength. However, since the energy E of a photon is described as

$$E = h\nu,$$

where h is Planck's constant and ν is the frequency of vibration, the same energy can be provided by two photons having half the frequency (twice the wavelength) of a single excitatory photon if these wavelengths are absorbed by the fluorochrome simultaneously. Thus, illumination with intense 800 nm light can induce fluorescence by a 2-photon mechanism in a fluorophore that is normally excited by 400 nm light in a single photon mechanism. The same 800 nm wavelength could be used to induce fluorescence by a 3-photon mechanism in a different fluorochrome that absorbs 267 nm light. Fluorescence from 2-photon or multiphoton excitation is not commonly observed, because 2 or more photons must be absorbed by the fluorochrome simultaneously, a condition requiring very intense illumination. However, the application of pulsed infrared lasers in microscopy has led to the development of the *2-photon laser-scanning microscope (TPLSM)*. Like the CLSM, the TPLSM uses a laser, a raster scanning mechanism, and an objective lens for point scanning of the specimen. But TPLSM has unique characteristics and offers several distinct advantages:

- Excitation wavelengths in the near IR are provided by a pulsed IR laser such as a titanium:sapphire laser. Excitation occurs in the near UV, causing fluorescence emission in the blue portion of the visual spectrum.

- Fluorochromes in the focal plane are differentially excited because with 2P excitation the efficiency of fluorescence excitation depends on the square of the light intensity, not directly on intensity as is the case for 1P excitation. Thus, by properly adjusting the amplitude of laser excitation, it is possible to differentially excite fluorochromes within the focal plane. The laser power, pulse duration, pulse frequency, and wavelength are critical variables for determining 2P excitation. Figure 12-13 shows the difference between 2P- and 1P-induced fluorescence in a transparent cuvette containing a fluorescent dye.

- Differential fluorescence excitation in the focal plane of the specimen is the hallmark feature of TPLSM and contrasts with CLSM, where fluorescence emission occurs across the entire thickness of a specimen excited by the scanning laser beam. Thus, in TPLSM, where there is no photon signal from out-of-focal-plane sources, the S/N in the image is improved. Because the fluorescence emission occurs only in the focal plane, the rate of photobleaching is also considerably reduced for the specimen as a whole.

- The use of near-IR wavelengths for excitation permits examination of relatively thick specimens. This is because cells and tissues absorb near-IR light poorly, so cellular photodamage is minimized, and because light scattering within the specimen is reduced at longer wavelengths. The depth of penetration can be up to 0.5 mm for some tissues, 10 times the penetration depth of a CSLM.

- A confocal detector pinhole is not required, because there is no out-of-focal-plane fluorescence that must be excluded. Nevertheless, emitted fluorescent wavelengths can be collected in the usual way and delivered by the galvanometer mirrors to the pinhole and detector as in CSLM, but the efficiency of detection is significantly reduced. This method is called *descanned detection*. A more efficient method of detection involves placing the detector near the specimen and outside the normal fluorescence pathway *(external detection)*. Several other detection methods are also possible.

Despite the potential advantages, there are still practical limitations and problems that remain to be solved. Since the titanium:sapphire laser presently used for TPSLM

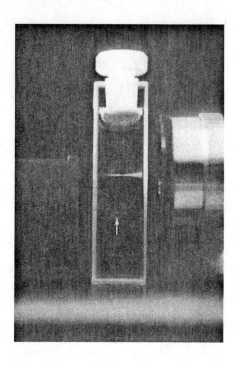

Figure 12-13

Excitation of fluorescence in conventional and 2P confocal microscopy. A cuvette of fluorescent dye is illuminated by the focused beams of a visible and an infrared laser. In conventional single-photon excitation (above in photo) the laser beam excites molecules along the entire path; in 2P illumination with an IR laser (below in photo and marked with an arrow), the laser beam excites a minute spot. Above a certain density of photons in the focused beam, frequency doubling occurs and fluorochromes emit fluorescent light, but below the critical density no 2P excitation occurs. By controlling the laser power, excitation can be induced in a volume that is close to the thickness of an optical section obtained using a confocal pinhole. (With permission, Imaging Sciences, Inc.)

has a peak efficiency around 800 nm (range, 750–1050 nm), for 2P excitation this has meant that only UV-excitable fluorochromes, such as NAD(P)H, serotonin, Indo-1, can be examined. Fluorochromes excited by longer wavelengths such as fluorescein and GFP have been more difficult targets, and rhodamine and red light–excitable dyes have remained largely out of reach. Other technical matters still under study include:

- Local heating from absorption of IR light by water at high laser power.
- Phototoxicity from long-wavelength IR excitation and short-wavelength fluorescence emission.
- Development of new fluorochromes better suited for 2P and multiphoton excitation.
- Improvements in laser performance allowing 2P excitation of fluorochromes at longer wavelengths; application of an optical parametric oscillator pumped by a titanium:sapphire laser and the use of neodymium lasers allow efficient 2P excitation at a wide range of UV and visible wavelengths.

Excellent examples of the application of 2-photon imaging in neuroscience applications are given by Denk and others in Yuste et al. (2000).

CONFOCAL IMAGING WITH A SPINNING NIPKOW DISK

Instead of illuminating the object by raster scanning using a single spot, it is possible to scan the specimen with thousands of points simultaneously using a spinning *Nipkow disk* (Fig. 12-14). A Nipkow disk contains thousands of minute pinholes arranged in rows of outwardly spiraling tracks. The arrangement and spacing of the pinholes is such that every point in the specimen receives the same amount of illumination from the rotating disk. There are substantial advantages inherent to this design:

- The returned fluorescent light can generate a real confocal image that can be seen by the eye or recorded on a camera, so no PMT-based imaging system is required.

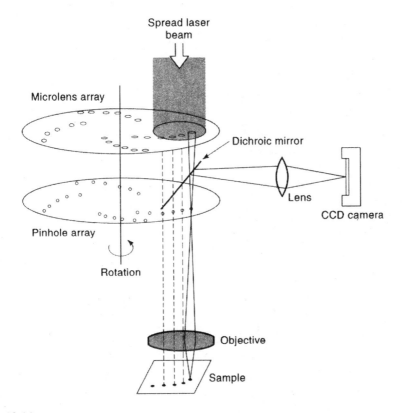

Figure 12-14

Tandem scanning confocal microscopy using a spinning Nipkow disk. The Yokogawa design features two disks each with >20,000 pinholes that rotate as a single unified piece around a central axis. The upper disk is fitted with microlenses that focus incident rays on a pinhole in the second disk. The pinholes of the disk are confocal with the specimen and the surface of an electronic imager such as a charge-coupled device (CCD) camera. A fixed dichroic mirror positioned in between the rotating disks transmits excitatory wavelengths from the laser while reflecting fluorescent wavelengths to the camera.

Direct viewing of the confocal image through an eyepiece increases the speed, ease, and efficiency of specimen positioning, focus, and examination.

- With a high-dynamic-range CCD camera, the gray scale range is extended to thousands of levels, and light intensity can be quantitated with higher resolution and accuracy. S/N is increased and images look considerably smoother and less noisy.

- Time resolution is greatly improved, because the rotating disk provides 360 frames/s (3 ms/frame) as opposed to 12–36 frames/s for single-point scanning confocals.

- The reduction in sampling time reduces the rate of photobleaching and the amount of phototoxicity.

- The very rapid scan rate (12.5 μs) results in very efficient fluorescence emission. Among the benefits is improved detection of fine structural details represented by clusters of only a few fluorescent molecules.

- Since there are no adjustments for the pinhole, scan speed, frame size, or zoom factor, the Nipkow confocal is considerably easier to use from the point of view of microscope-operating software.

The pinholes in the Nipkow disk perform the same function as the pinhole in a single-point scanning system in generating a confocal image of an optical section. Incorporation of a z-axis stepper motor and three-dimensional imaging software also permits volume viewing and single-plane composite or projected views. Early Nipkow disk imaging systems had the disadvantage of very low light transmission, but a recently introduced double-disk design by Yokogawa Electronics, Inc., (Tokyo) remedies the problem by including microlenses that greatly boost light-gathering efficiency and transmission. The Yokogawa-Nipkow system has been thoroughly reviewed in a recent publication by Inoué and Inoué (2000). The popularity of modified Nipkow disk confocal systems is certain to increase in the future. The disadvantages of this system are that the optical section is somewhat thicker than with a conventional confocal with a stopped-down pinhole and that the microlenses of the disk do not transmit UV light, which precludes the use of UV-excitable dyes.

Exercise: Effect of Confocal Variables on Image Quality

This exercise should be performed during or after a training session on the use of a confocal microscope by an experienced user.

- The following adjustments can be used to brighten a confocal image. Examine the effects of each manipulation on image quality beyond the effect on brightness:

    1. Remove a neutral density filter or increase the laser power.

    2. Open the pinhole diaphragm.

    3. Increase the PMT gain.

    4. Decrease the PMT offset.

5. Reduce the scan speed.

6. Increase the zoom factor.

- Examine the effect of pinhole diameter on the thickness of the optical section.

- Compare the effects of built-in confocal filter sets with bandpass vs. long-pass emission filters using a specimen labeled with multiple fluorochromes. Note the effects on brightness/quality of the image and bleed-through of non-specific fluorochromes.

VIDEO MICROSCOPY

OVERVIEW

Video imaging was one of the first electronic imaging technologies to be widely applied in light microscopy and remains the system of choice for many biomedical applications. Video camera systems produce excellent images (Fig. 13-1), are easy to use, and allow convenient storage and subsequent playback from VCR tapes. In this chapter we examine the technology and application of video microscopy, which has its origins in commercial broadcast television. A video system using a video camera and a television monitor for display can be attached to printers and VCRs and can be interfaced with digital image processors and computers. Given the growing interest in digital imaging systems based on charge-coupled device (CCD) detectors, video cameras may seem like a thing of the past. In fact, video microscopy is the best solution for many microscope specimens. For comprehensive, in-depth coverage of the topic, the following are excellent references: *Video Microscopy: The Fundamentals* by Inoué and Spring (1997), *Video Microscopy* edited by Sluder and Wolf (1998), and *Electronic Light Microscopy* edited by Shotton (1993).

APPLICATIONS AND SPECIMENS SUITABLE FOR VIDEO

Some of the unique benefits of video microscopy include:

- High temporal resolution at 30 frames per second.
- Continuous recording for hours at a time. Most computer-based digital systems do not allow this because of limited acquisition speed and limited image storage capacity.
- Rapid screening of microscope specimens. Finding regions suitable for detailed study is convenient in video, especially evident when objects are indistinct and must be electronically enhanced in order to see them.
- Time-lapse documentation. Time-lapse video recording allows continuous monitoring of changes in shape and light intensity. Video is also commonly used for

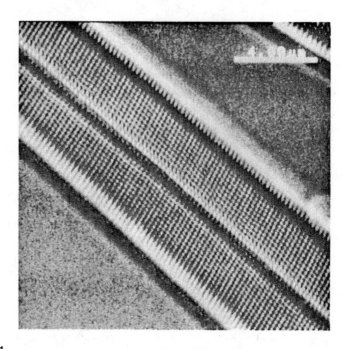

Figure 13-1

Video-enhanced DIC image of the diatom *Amphipleura.* The spacings between pores and between striae in the silicaceous shell (~0.20 and 0.24 μm, respectively) are very clearly defined even though they are at the theoretical limit of resolution. Microscopy: Zeiss DIC optics with a 100×, 1.3 NA Plan Neofluar oil immersion objective and 1.3 NA oiled condenser; illumination with the 405 nm violet line of a 100 W mercury arc lamp; a Hamamatsu Newvicon video camera and Argus 10 digital image processor for frame averaging, background subtraction, and digital contrast adjustment. The distortion that is visible is caused by the method used for background subtraction. Bar = 4 μm.

time-lapse studies such as changes in cell shape and movement (membrane ruffling, mitosis, cytokinesis) and cell and tissue migration in embryonic development, wound healing, and chemotaxis.

- Multipurpose function of VCR tape as an experimental log and data archive. A microphone can be attached to the VCR, allowing you to record observations to the audio channel of the tape.

- Convenience and ease of use of video cameras.

CONFIGURATION OF A VIDEO CAMERA SYSTEM

Video microscope systems such as the one shown in Figure 13-2 usually include the following components:

- Video camera with a power supply and camera control unit or camera electronics unit

- Digital image processor for on-the-fly image processing to reduce noise and improve image contrast and signal-to-noise ratio

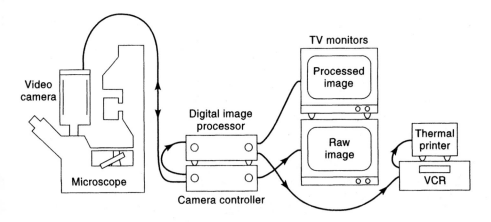

Figure 13-2

The principal components of a video microscope system. The video camera is mounted on a microscope and connected to a camera controller (camera electronics unit or CEU), that displays the raw image on a TV monitor. In this system, the raw image is also sent to a digital image processor, which sends the image to a VCR, a thermal printer, and a second TV monitor for displaying the final processed image.

- TV monitor, VCR, and a thermal printer or freeze-frame device to display and record the live video image
- Interface to computer for additional processing and printing

It is easy to string together or "daisy chain" several items of video equipment with coaxial cables, provided that the last item in a series of connected video devices such as a TV is properly terminated at 75 ohm impedance to avoid back-reflection and distortion of the video signal. This termination feature is usually built into video components and is either activated automatically or set with a special switch on the back of the final unit. Termination should not be activated on any of the intervening devices. Video equipment is sometimes sensitive to magnetic fields, which can generate patterns of electrical interference or even damage expensive integrated circuits. Therefore, video equipment should be located away from nearby magnetic field sources such as electrical conduits, junction boxes in nearby walls, and especially power supplies of arc lamps. (See precautions for using power supplies in Chapter 3.)

In video imaging, differences in light intensity in the image are transmitted by the camera as an *analogue signal*—that is, as a signal with continuous changes in voltage amplitude, which is displayed on a TV screen. Thus, the light intensity at a particular location on the surface of a video camera and on the TV screen corresponds to a unique voltage amplitude. The raster scanning period is fast enough that the image display on the TV appears to be live, occurring continuously and in real time. A video image is qualitative in nature, but can be digitized for purposes of extracting photon measurements and performing image processing with an on-line digital image processor. The processor converts the analogue signal to digital format, performs its processing operations digitally, and then reconverts the signal back to analogue mode before display on the TV monitor. Alternatively, the analogue signal can be sent to a special frame grabber board in a computer, which parses the continuous voltage signal into pixels with digital gray-level values for display on a computer monitor. Although the TV operates at a fixed rate of 30 frames/s, it is possible to slow down the rate at which the video signal is sampled and recorded using a time-lapse VCR.

TYPES OF VIDEO CAMERAS

Video cameras use a video electron tube or a charge-coupled device (CCD) for detecting photons comprising the image. In video, both types of detectors are operated as analogue devices, which means that they generate a continuously varying voltage signal. Regardless of the camera type, the voltage signal is imprinted at regular time intervals with specific voltage pulses for synchronization and blanking, generating a so-called composite video signal that is required for display on a TV monitor. Composite video signals conform to one of several internationally recognized formats: EIA standards in North America, parts of South America, and Japan, and CCIR standards elsewhere. Common EIA video formats include RS-170 (black-and-white broadcast TV), RS-330 (black-and-white closed-circuit TV), and NTSC (color broadcast TV). The components making up a closed-circuit TV system must all be of the same type (EIA, CCIR) and supplied with the correct line frequency. Digital cameras that interface with computers encode signals in other formats.

A *video tube camera* contains a glass-enclosed electron tube. The microscope image is focused on a target plate at the front end of the tube, where photons alter the conductance of the target by causing local displacements of electrical charge. The charge is replenished by a scanning electron beam generated at the rear end of the tube, which is focused into a small spot and scanned across the target surface by magnetic deflection coils in a zigzag pattern of lines called a *raster*. The scanning electron beam generates a current that varies with the conductance in the target, and the information is processed and sent to a TV or VCR as a continuously varying voltage (analogue signal). Registration pulses are added to the ends of each raster line as described. A schematic diagram of a video camera tube is shown in Figure 13-3. For display on the TV, two sequential raster scanned *fields* are *interlaced*—in the sense of interdigitating the spread fingers of one hand with the fingers of the other—thereby giving a single *frame*. The display rate is constant at 30 frames/s. The design of a video tube and a description of the scanning mechanism are given in Figure 13-4.

The *CCD detector* is a slab of silicon that is patterned by narrow transparent strips of conducting materials into square or rectangular units called *pixels* (picture elements) that are typically 5–15 μm on a side. The signatures of photons hitting the CCD surface are absorbed locally in the silicon matrix. There is no scanning by an electron beam to read off the image from the CCD. Instead, charge packets accumulated in the pixels are advanced to an amplifier at one corner of the chip by a timed series of voltage pulses passing through the strips on the surface of the chip so that the pixels are read off serially, row by row, and one pixel at a time. In video, it is common to see so-called interline or progressive scan CCD designs, because these chips can be read out at a faster rate than a standard full-frame CCD chip. The design and mechanism of operation of CCDs are described in Chapter 14.

There is great flexibility in how the charge packets stored in the CCD are read off and displayed. The electron count can be digitally encoded for processing in a computer and displayed on a computer monitor, or sent directly as an analogue video signal to a TV monitor. Many cameras can be operated in both modes simultaneously: continuous video display for live search and study, and computer-based acquisition of single frames

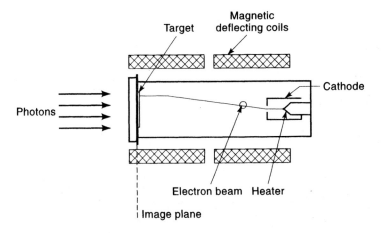

Figure 13-3

Schematic diagram of a vidicon camera tube. Incident photons strike a photosensitive layer of the target electrode, which reduces electrical resistance in the target surface. A focused electron beam generated at the cathode at the back end of the tube is scanned in a raster across the target under the control of magnetic deflection coils and generates a current, whose strength varies according to the local resistance in the target. The current passes through a transducer, which produces a continuously variable voltage (an analogue signal) and is sent to a TV monitor for display. For tubes of the RS-170 format, the target is scanned in a raster 60 times/s in two spatially distinct, interdigitating fields containing 262.5 lines. While many of the commercial cameras are based on the EIA RS-170–type format (525 lines/frame, 60 field scans/s, 2:1 interlace), video tube cameras with higher numbers of scan lines (1025, 1225, or 2000 lines) are available for custom closed-circuit systems. The reader should refer to *Video Microscopy* by Inoué and Spring (1997) for more details, including discussions of the standards used for the new generation of high-definition television systems.

or sequences for analysis and image processing. Communication between the video camera and the computer is made possible by frame grabbers and image buffer boards that are installed in the computer. For video rate display, there is also flexibility in how the CCD device is read out. As an example, the following designs are used by Video Scope International, Ltd., (Sterling, Virginia) for producing EIA video formatted signals for display on TV for a video camera containing a full-frame transfer CCD with 1134 (horizontal) × 486 (vertical) rectangular pixels.

- In *normal mode,* the chip is read 60 times/s, and two fields are interlaced 2:1 to create one frame. Of the 486 lines contained in an image, alternate horizontal rows are saved and the remainder are discarded to create a field containing 243 lines. The subsequent field produces the additional 243 lines to complete one frame. Thus, 30 interlaced frames are generated per second. These are imprinted with synchronization and blanking pulses in EIA format, and are displayed on a 486-line TV display in the normal way.

- In *low-light mode,* a pseudo-interlace pattern is used to generate a brighter image. The CCD is read 60 times a second, and vertically oriented pixel pairs in adjacent line pairs are summed to create a field containing 243 lines. In the subsequent exposure, alternate pairs of lines are summed to create the second field.

- In *dual-field mode,* a single 1/30 second exposure is used to make both frames.

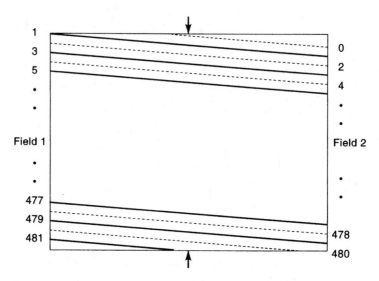

Figure 13-4

Raster pattern used for video image acquisition and display. The format shown is used for EIA RS-170 (black and white broadcast format in North America, Japan, and elsewhere) and related formats. An electron beam scans the target plate in a video tube in two spatially distinct rasters called fields. The set of even-numbered lines is scanned followed by the set of odd-numbered lines, which together number 525, of which ~480 lines or more are used for displaying the image on the TV monitor. The balance of the 525 raster lines is used for blanking and synchronization. Each raster is called a field, and the two interdigitated (interlaced) fields are called a frame, whose beginning and end points are indicated with arrows. The raster scan rate and 2:1 interlacing mechanism were specifically designed to eliminate the perception of flickering and downward sweep in the image, and provide the desired level of vertical and horizontal resolution.

For computer display it is common to see progressive-scan CCD readouts, where no interlacing is used, and the pixels are displayed on the monitor in the order that they are read off from the chip—that is, one row at a time from top to bottom.

For higher-end video cameras based on CCD devices, the distinction between analogue and digital handling of image signals is becoming increasingly difficult to make. For example, CCD cameras with video output to a TV are controlled by a computer, which can also grab and display a video frame, process the image, and extract photometrically accurate data regarding light intensity. In addition, digital image processors with analogue-to-digital and digital-to-analogue converters are now routinely incorporated in video circuitry for the speed and convenience of displaying a live processed image. The sensitivity of various kinds of analogue video and digital camera systems to wavelength and luminosity is shown in Figure 13-5 and 13-6.

ELECTRONIC CAMERA CONTROLS

The camera controller of a video camera system usually has the following electronic adjustments:

- *Gain.* The gain and offset controls of the camera are used to adjust the amplitude of the voltage signal so that bright highlights and dark shadows in the image span the full

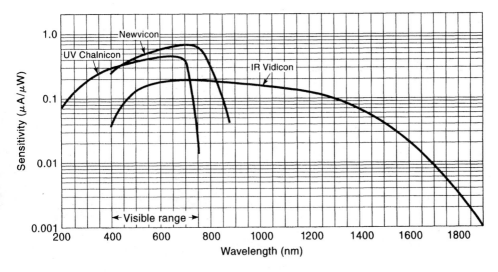

Figure 13-5

The spectral responses of three commonly used video tubes. Video camera tubes vary in their sensitivity to wavelength and luminance, depending on the composition of the target. The Newvicon tube is sensitive across the visual spectrum from 400 to >850 nm and has a peak sensitivity at ~700 nm. Other tubes are relatively more sensitive to UV (UV-Chalnicon) and IR (IR Vidicon) wavelengths. Video CCD cameras also have peak sensitivity in the middle of the visual spectrum.

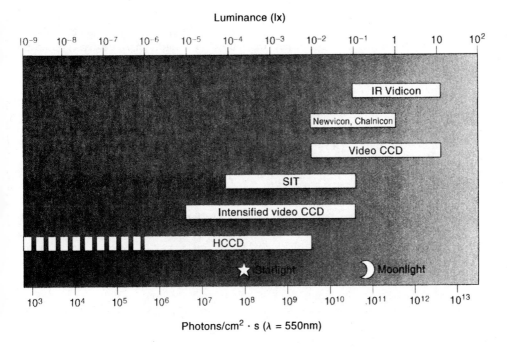

Figure 13-6

Sensitivity comparison chart of detectors used in video cameras. The plot shows the sensitivity ranges of various tube and CCD video cameras, a silicon-intensifier target (SIT) camera, an intensified video camera, and a high-performance CCD (HCCD) camera.

range of available gray levels from black to white. Given that there is adequate illumination, it is usually desirable to utilize the full range of the camera and make minor adjustments for brightness and contrast later (if necessary) using an image processor. If the camera's signal is not adjusted properly or spans a limited range of amplitudes, it may not be possible to adjust the gray-level range subsequently using the processor. Increasing the gain or contrast control amplifies the video signal (multiplies the signal by a constant) and brings the bright features close to the white or saturated value on the monitor. Because the signal is amplified by multiplication, the voltage range corresponding to the black and white end points in the image is increased. For example, a signal ranging from 0 to 10 multiplied by 10 becomes 0 to 100, and the voltage range is increased 10-fold. A sketch showing the effects of gain and offset adjustments on the PMTs of a confocal microscope is shown in Figure 12-12.

- *Offset.* The offset or *black level* is a brightness control that is used to set dark features in the image close to the black level on the monitor. Offset works differently from gain. In this adjustment a positive or negative voltage is added to the image signal to adjust the amplitude of the darkest features in an image to a voltage corresponding to black. Offset changes the amplitude of the entire voltage signal, but since it is added to the signal, it does not change the voltage difference between high- and low-voltage amplitudes contained in the original signal. For example, a signal ranging from 0 to 10 volts modified with an offset of +5 volts becomes 5 to 15 volts, but the difference remains 10 volts as it was originally.

  Both gain and offset affect the contrast of the video image. *Video contrast C_V* is defined as

$$C_V = A\,|\,I_S - I_B\,|/(I_B + I_V),$$

where A is the amplitude factor for gain, and I_V is the intensity corresponding to the voltage offset, and I_S and I_B are the intensity of the signal and background as described in the definition of image contrast in Chapter 2. From the equation it is clear that both gain and offset determine image contrast. We return to the effect of these controls later in the chapter.

- *Gamma (γ).* In the ideal case for an electronic camera, there is a linear relation between the intensity of incident light on the camera and the strength of the camera's output signal. This relationship is described by the γ function,

$$i/i_D = (I/I_D)^\gamma,$$

where i and I are the output current and corresponding light intensity for a given signal, i_D and I_D are the dark current and its corresponding intensity, and γ is an exponent. Thus, in a plot of the log of the output current of the camera vs. the log of the light intensity on the camera target, γ is the slope of the curve. If γ has a value of 1, the relationship is linear (Fig. 13-7). Most video cameras have a γ close to this value. A γ setting <1 emphasizes dark features in the image (makes them brighter and more visible), and has the effect of flattening the image and decreasing contrast between bright and midgray values. A γ setting >1 emphasizes bright features and increases contrast. For the use of γ scaling in image processing, see Chapter 15.

- *Auto-gain control.* This feature automatically adjusts the gain to maintain image brightness. It is very convenient when monitoring fluorescent specimens that are

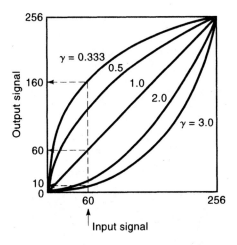

Figure 13-7

Gamma (γ) function of an imaging device. Ideally, a plot showing the log of output signal vs. the log of input light intensity is linear with a γ value of 1. γ values <1 differentially emphasize dark gray values (display dark values over a greater range of gray levels), whereas values of γ >1 emphasize bright pixel values. γ can also be adjusted during image processing to create similar effects.

subject to photobleaching. As the specimen fades during observation, auto-gain control (AGC) maintains the signal strength, although for extensively bleached specimens the S/N ratio deteriorates significantly, causing the image to appear grainy.

- *Enhance.* This control shifts the camera into a high gain mode to increase its sensitivity.

- *Invert.* This control inverts the voltage signal, causing bright features to look black against a bright background, and vice versa.

- *Shading correction.* A bias voltage gradient is applied horizontally and/or vertically so that light intensity gradients in the image caused by uneven illumination from the lamp or optics can be offset and eliminated.

Demonstration: Procedure for Adjusting the Light Intensity of the Video Camera and TV Monitor

Place a linear gray-scale strip ranging from black to white on a medium gray background and illuminate it with a variable-intensity lamp. A lens-mounted video camera, secured to a tripod or ring stand, is directed at the card, and the image is displayed on a TV monitor. The gain and offset dials of the camera and monitor are set to their midrange positions before adjustments are made. To adjust the video image correctly, proceed through the following steps:

1. Adjust the light intensity of the lamp so that it falls within the optimal operating range of the camera. Some cameras include indicator lights to

show when the light intensity is inadequate, properly adjusted, or too bright (saturating). At the microscope, the lamp power supply is adjusted or neutral density filters are used to make this adjustment. If the illumination intensity is not in the optimal range, the camera will not be able to utilize its full range of gray levels to represent the image.

2. Adjust the camera offset and gain (offset first, gain second) to regulate the black and white levels in the image. It is important that the camera, not the TV monitor, be used to optimize the gray-level settings, because the TV adjustments do not affect the signal sent to the VCR or printer. An in-line image processor capable of giving numeric values for features in the image is useful for making gain and offset adjustments. Alternatively, a dedicated signal analysis unit, such as the RasterScope available from Dage-MTI Inc. (Michigan City, Indiana) or an oscilloscope can be used. In the absence of these quantitative aids, the adjustments must be made by eye.

3. Adjust the brightness and contrast of the TV monitor to optimally display the captured image. Adjustments to the monitor affect our impression of image quality, but monitor adjustments do not influence the signal recorded on a VCR or sent to a computer. The TV's adjustment dials for contrast and brightness act in the same way as the gain and offset dials on the camera controller. The brightness (offset) is adjusted first to distinguish black from the darkest grays in the image, followed by the adjustment of contrast (gain) to bright image features to a value close to white, but not so bright as to reach saturation. Since video monitors usually exhibit a γ of 2 or more, bright features tend to be differentially enhanced while darker gray features are buried in a black background. It is therefore important to examine the image of a white-to-black gray scale and adjust the contrast and brightness settings so that all of the gray levels are represented equally on the monitor. Repeat the operation, using a specimen of buccal epithelial cells examined with a 40× lens in phase contrast optics.

4. Adjust the vertical hold to correct the registration of raster lines. The two fields comprising a video frame are traced separately on the monitor. Periodically, you should check that the two fields are interlaced properly—that is, that the spacing between raster lines on the monitor is uniform. To do this, hold a magnifier up next to the screen and bring the raster lines into sharp focus. Adjust the spacing of the lines with the vertical hold control on the TV monitor until the spacing is uniform. Be sure to check different areas on the monitor to obtain the best setting.

For a detailed description of the procedures used to adjust a video camera and monitor, see the writings of Sluder and Hinchcliffe in *Video Microscopy* edited by Sluder and Wolf (1998).

VIDEO ENHANCEMENT OF IMAGE CONTRAST

The camera controls for gain and offset can be used to increase the dynamic range and contrast of low-contrast signals, even signals invisible to the eye (Allen, 1985; Allen et al., 1981a,b; Inoué, 1981). As discussed in the previous chapter and shown in Figure 12-10, increasing the gain amplifies the signal, while offset is used to add or subtract a voltage from the signal to bring background features close to the black level of the image. In this way, a very narrow range of the video signal can be stretched to fill the available gray levels ranging from black to white. Frame averaging, background subtraction, and further contrast adjustments are made using an in-line digital image processor to produce a high-contrast, high-quality image. This procedure, known as *video enhancement of image contrast,* was introduced and popularized by Shinya Inoué, Robert Allen, and others, and has been used to view objects such as lamellapodia and endoplasmic reticulum in living cells. The technique is so sensitive that it is possible to image purified microtubules, which have been shown to retard a wavelength of green light by only $\lambda/300$ (Fig. 13-8)! The procedure for performing video enhancement of image contrast is as follows:

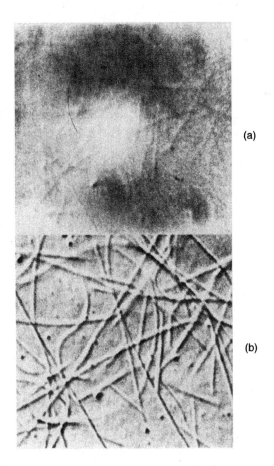

(a)

(b)

Figure 13-8

Video-enhanced DIC contrast image of microtubule polymers in vitro. (a) Raw image before image processing. (b) Processed image after frame averaging, background subtraction, and contrast adjustment. Microtubule polymers on a coverslip were illuminated with the 546 nm green line of a 100 W mercury arc lamp and examined with Zeiss DIC optics using a 100×, 1.3 NA oil immersion objective lens. Video images were obtained with a Hamamatsu Newvicon video camera and processed with an Argus 10 digital image processor. The video image was recorded on a Sony thermal printer that performs dithering to remove raster lines from the video display. Field diameter, 30μm.

- Focus the object and adjust the offset and gain settings as described until the object just becomes visible. This unprocessed image is called the *raw image.* If the image of the object looks indistinct and has very low contrast, even after some adjustments to the gain and offset, the raw image may appear mottled and blotchy (Fig. 13-8a). The blotchy appearance is due to optical faults, imperfect lens surfaces, and dust and scratches on the optics. In addition, uneven illumination can produce prominent gradients in light intensity across the image. These are removed by background subtraction.

- Initiate background subtraction with live frame averaging. For moving objects, a frame average of 4–8 works well. Move the specimen laterally with the stage controls to expose an adjacent blank region on the specimen slide, or if adjacent regions contain objects or debris, defocus the microscope slightly. Upon starting background subtraction, the averaged background image is held in memory in a frame buffer and is subtracted from each new image frame acquired by the camera. Because the background image is subtracted from itself, the TV screen initially looks perfectly blank. Return to focus and relocate the specimen, which should now be visible with remarkable clarity and contrast. The image is now called a *processed image.*

- Make final adjustments to the contrast using the contrast menus of the digital image processor.

- The enhanced images can be examined live on TV or videotaped or printed.

The gain and offset adjustments associated with electronic enhancement can also be used to improve the spatial resolution in a video image. If the image of two overlapping diffraction spots is adjusted so that the amplitude difference between the trough between the two peaks and the peaks themselves ranges from black to white, contrast is greatly improved (Fig. 13-9; see also Inoué, 1989).

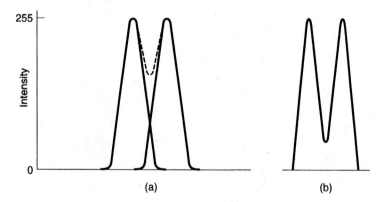

Figure 13-9

Improvement of spatial resolution by video enhancement. (a) According to Rayleigh's criterion for spatial resolution, two identical overlapping diffraction spots corresponding to two point sources of light (solid curves) can be resolved when the diffraction peak of one spot overlies the first minimum of the other. The summation of the individual profiles is depicted by the dotted line and represents the intensity profile seen by the eye. At this distance of separation, the dip in intensity between the two peaks can just be perceived by the eye. (b) Resolution can be improved in video microscopy by adjusting the offset to bring the intensity at the dip close to 0 and readjusting the gain to bring the maxima to saturation.

CRITERIA USED TO DEFINE VIDEO IMAGING PERFORMANCE

The imaging performance of a video camera or any electronic camera is defined in terms of the following parameters: spatial resolution, resolution of light intensity, temporal resolution, and signal-to-noise ratio. These terms were first introduced in Chapter 12. Table 13-1 lists these criteria and provides data for various electronic imaging systems including video.

Spatial Resolution

Vertical resolution is determined by the number of lines in the raster (Fig. 13-10). For many closed-circuit TV cameras, there are up to 486 display lines. If the image of a fine horizontal pattern of black and white stripes was always in perfect register with the raster scanning the image, 486 alternating black and white lines could be resolved. However, usually the match is not perfect, and the 486-line resolution is not observed. Although 486 lines are displayed, the average vertical resolution is therefore estimated as $0.7 \times 486 = 340$ lines.

With respect to *horizontal resolution,* the electronics are designed to resolve 800 black and white lines (400 line pairs) of a black and white square wave pattern, thus matching the vertical resolution of 340–486 lines (Fig. 13-11). To achieve this, the response time of the camera electronics must be fast enough to resolve signal frequencies ranging from 10 MHz (400 cycles/s) down to 30 Hz (1 cycle/s). The range spanned by these spatial frequencies defines the *bandwidth* of the system. At 10 MHz sampling frequency, the response time (rise time) of the camera is very short (~35 ns).

In tube cameras, the diameter of the scanning electron beam at the target plate is approximately 35 μm. To preserve the spatial resolution of the objective lens, the optical magnification must be sufficient to give two 35 μm sampling lines (70 μm) per diffraction spot radius on the target plate. (The requirement for 2-fold greater sampling frequency [raster spacing] than the spatial frequency limit in the image is related to the

TABLE 13-1   Comparisons of Video and Digital Imaging Systems

Imaging System	Video Tube	Video CCD	Video ICCD	Digital CCD
Camera/company	Newvicon video Hamamatsu	Interline CCD Hamamatsu	Interline CCD + Gen II intensifier Hamamatsu	Cooled CCD KAF1400 chip Roper Scientific
Intensity range (photons/cm²/s)	10^{10}–10^{12}	10^{10}–10^{13}	10^{7}–10^{11}	10^{6}–10^{10}
Wavelength (range/max. sens.)	400–875/700	400–950/550	400–850/520	400–1100/700
Dynamic range (min./max.)	80–100	100–300	100	4095
S/N ratio	180	80	80	212
Display rate (frames/s)	30	30	30	1.5 (max. at full frame)
Resolution at face plate (samples/cm)	470	600	600	1343

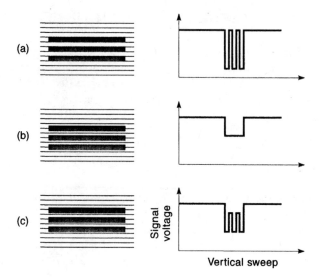

<u>Figure 13-10</u>
Vertical resolution in video. For a test pattern of 3 horizontal black bars whose width is equal to that of a raster line and whose spacing is equal to 2 lines, the resolution and contrast of the lines on the monitor depend on the position of the bars with respect to the raster scan lines of the camera. The raster lines are the white horizontal stripes, and their width is determined by the diameter of the scanning electron beam. (a) With bars perfectly aligned with the raster, maximum resolution and contrast are obtained. (b) With bars straddling two raster lines, there is no resolution, and the region occupied by the bars appears a medium gray. (c) With the bars at an intermediate location (the usual case), the lines are resolved, but the contrast is reduced, the bars appearing medium gray. For this reason, the average vertical resolution is usually calculated as 0.7 of the maximum number of display lines of the camera, or 486 × 0.7 = 340 lines.

Nyquist sampling theorem for preserving signal resolution and is described in greater detail later in this chapter.) For a 100×, 1.4 NA oil immersion objective, the diffraction spot radius is 24 μm (0.24 μm × 100) at 546 nm. Thus, an additional magnification of 2.9× is required for proper sampling (70/24 = 2.9). This can be obtained by inserting a 4× *relay lens* (also called a *TV lens*). In CCD video cameras with small 5–15 μm diameter pixels, the magnification requirement is considerably reduced (40–125× magnification). In this case, a 100× objective would be sufficient to provide the necessary magnification, and a 4× TV projection lens would not be required.

Temporal Resolution

Time resolution of video is generally fixed at the camera scan rate of 30 frames/s (60 fields/s). This is very fast compared to the typical rate of full-frame digital CCD cameras of ~1 frame/s.

Resolution of Light Intensity (Dynamic Range)

The saturated signal divided by the noise signal gives the camera's dynamic range. The intrascenic dynamic range of most vidicon-type tube cameras is 70:1 to 100:1. The

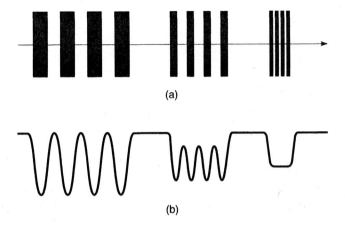

(a)

(b)

Figure 13-11

Horizontal resolution in video. (a) Three patterns of line spacings imaged on the target of a video camera and the path of a focused electron beam (a raster scan line) scanning the target, and (b) a plot of the signal amplitude vs. time corresponding to the image shown in (a). At the coarse spacing, the period of the pattern is resolved, but the shape of the bars is not, the signal being a sinusoidal wave; however, the signal spans the full amplitude (from 0 to 0.7 volts) and thus gives maximal contrast. At the intermediate spacing, the spatial period is still resolved, but the contrast is reduced because the camera does not respond quickly enough to register the true intensity values of the bars and spacings. At the fine line spacing there is no resolution, and all that can be recorded and seen is a constant medium-amplitude signal. The drop in image contrast at higher spatial frequencies is related to the system's modulation transfer function (MTF), which is discussed at the end of the chapter.

lower end of the dynamic range is limited by amplifier noise, and the upper end by signal saturation. The usable light range, however, can be considerably greater: For certain video cameras with a rather limited intrascenic dynamic range of 50:1, the usable light range may be as great as 10,000:1. A dynamic range of 70 to 100:1 is good but not great. It corresponds to 70–100 gray levels (6–7 bit resolution in digital format), about equal to that of the human eye, but less than 8 bit display devices such as computer monitors and dye sublimation printers. This is considerably less than the 10 bit resolution of film (about 1000 gray levels) and the 12–16 bit resolution of high-performance CCD cameras (4096–65,000 gray levels). The effect of reduced dynamic range becomes noticeable when substantial image processing is performed and especially when a subset of gray levels in an image is stretched to fill the entire display range from black to white, in which case the image may look grainy.

Signal-to-Noise Ratio

The signal-to-noise ratio of an image is calculated as the ratio of the object signal (total object signal minus the underlying background) to the standard deviation of the background signal. The higher the S/N ratio becomes, the greater the clarity, contrast, and apparent smoothness of the image. The S/N ratio in video is measured in decibels (dB), where 1 dB = 20 log S/N. For bright objects, video cameras can exhibit an S/N ratio of up to 50 dB (S/N = 316). An image with an S/N ratio of 50 would not show graininess or snow on the TV monitor. In contrast, an S/N ratio of 20 would be noticeably grainy and an S/N ratio <10 might be considered barely acceptable. The S/N ratio is also a statistical parameter that expresses the confidence level at which fluctuations of a certain percent can be detected. The meaning of this statistic is discussed in greater detail in Chapter 14.

ALIASING

Aliasing is the phenomenon whereby periodic structures in an object are not faithfully represented in the object image, but instead by a false period that can be mistaken for being real. Aliasing (from the Latin *alius,* meaning another, other, or different) is an imaging artifact and is a property inherent to all detectors with periodic structures such as raster lines or pixels. We commonly observe aliasing on television. This occurs when there is insufficient magnification produced by the lens of a TV camera focused on periodic structures such as the pattern of pinstripes in an announcer's shirt, bricks in the wall of a house, or seats in an empty stadium. Under these conditions the images of these objects exhibit false patterns of wide bands. An example of aliasing in a diatom imaged by video microscopy is shown in Figure 13-12.

Aliasing has its origins in signal sampling theory and can be described mathematically. If the image of a specimen with periodic structure (a sine wave target or the periodic pattern of pores in a diatom) is reconstructed from sample points that occur at a much higher frequency than detail in the specimen, it is possible to reconstruct and display an accurate image of the specimen. However, at the point where the sampling frequency is less than twice the specimen frequency, it is no longer possible to represent the signal accurately, and a false period (aliasing) is observed. Below this sampling limit, the display of periodic elements in the image actually increases as the sampling frequency becomes further reduced. The factor of 2 requirement for oversampling is

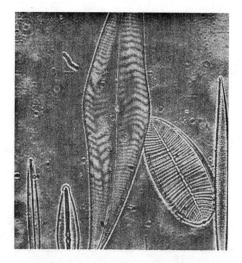

Figure 13-12

Aliasing in a video image of a diatom. The spacing period between rows of pores in the image becomes false (aliased) when the magnification is too low to allow proper sampling of the image on the faceplate of the camera. *Pleurosigma,* DIC microscopy.

called the *Nyquist criterion*. (See Castleman in Shotton (1993) and Inoué and Spring (1997) for theory and details.)

DIGITAL IMAGE PROCESSORS

A digital image processor (digital signal processor) is a valuable tool for increasing the contrast, smoothness, and signal-to-noise ratio of the raw video image and for performing complex operations such as background subtraction and enhancement of image contrast. This unit (a dedicated computer) is positioned between the camera and the TV monitor and other recording devices. The processor converts the raw image into digital form by an A/D converter, performs image processing operations digitally, and reconverts the processed signal back to analogue form through a D/A converter. The final processed image can be displayed on a TV monitor or sent to other devices such as a thermal printer, VCR, or computer, which print, record, or receive the processed analogue signal. Available image processing operations are displayed in a menu on the TV screen and are activated by clicking on them with a computer mouse. The current processors are quite sophisticated and perform many of the functions contained in computer software programs used for processing and analyzing images from digital cameras. Although they are expensive, image processors are indispensable for many video applications. The operations performed by a high-end digital image processor include the following:

Live: Displays the live raw image.

Real-time background subtraction: A background image containing image artifacts such as dust, scratches, and uneven illumination is captured and stored in memory. The background frame is subtracted from each incoming raw image frame, and the corrected background-subtracted frame is displayed in real time (30 frames/s) on the TV monitor.

Frame averaging: Improves the S/N ratio and smoothness of the image. The designated number of frames to be averaged is held in a frame buffer, and the average is computed and displayed at video rate. At each calculation/display cycle, a new frame enters the buffer and the oldest frame is discarded.

Frame accumulation: Accumulates the signals of the designated number of frames into a single image frame to produce a clearer image (higher S/N ratio).

Subtraction: Does a simple subtraction of the input image from a memorized image.

CCD on-chip integration: Integrates the image for the designated number of frame equivalents instead of accumulating or averaging individual 1/30 second frames. This procedure gives a clearer picture of low-light specimens.

Background subtraction plus accumulation: Combines the functions of background subtraction and frame accumulation.

Sequential subtraction: Removes those features that remain stationary in a stack with a designated number of frames. Only moving objects are displayed.

Rapid transition isolation: Removes image features from the processed image that remain stationary or do not change in intensity, leaving an image representing features that exhibit rapid change.

Spatial filtering: Enhances high or low spatial frequency information. Masks composed of various pixel matrices are used for edge enhancement, shadow enhancement, sharpening, or smoothing of the image.

Image trace: Traces the trajectories of moving objects.

Maximum density trace: Traces the pathways of moving objects that are brighter than the background.

Minimum density trace: Traces the pathways of moving objects that are darker than the background.

The image display commands of the processor are also versatile and include contrast adjustment with image histogram display, with adjustments for setting the white and black end points; γ adjustment for differential adjustment of bright and dark features; real-time edge sharpening; split-screen display; pseudocolor; superimposition of a real-time image over an image stored in memory; real-time zooming; superimposed display of scale bar, typed comment, and timer/clock.

Image processors also provide image analysis features for measuring lengths of traced lines, areas, particle counts, interpoint distances, and velocities. Intensity analysis commands allow you to measure intensities of points and average intensities within areas; you can also obtain intensity profiles along a line through an image, an image histogram of a designated region of interest, and a three-dimensional profile of intensity levels within a defined region of interest.

IMAGE INTENSIFIERS

Although the temporal resolution of video cameras is excellent, their sensitivity to light is limited. This is because their quantum efficiency is only moderate (20–30%), the exposure time is short (fixed at $\sim$33 ms), and because the camera electronic noise associated with rapid read rates is high. An image processor placed in-line between the camera and TV monitor can accumulate or average the signals from hundreds of frames for display on the TV, but electronic noise limits its ability to detect faint signals. Even with frame averaging and high gain, the detection limit (as judged from a noisy, barely acceptable image) corresponds roughly to an object of moderate fluorescence intensity as seen through the eyepieces of a fluorescence microscope. Faint fluorescent objects cannot be detected at all. The solution is to use a low-light-level camera or a camera coupled to an image intensifier.

Silicon intensifier target (SIT) and *intensifier silicon-intensifier target (ISIT)* cameras have 10- to 100-fold higher sensitivity to light than vidicon tube and video CCD cameras, and have been employed for low-light fluorescence microscopy for many years. SIT cameras contain a low-light-level vidicon tube that is modified such that photoelectrons generated and emitted at the photocathode at the front end of the tube are focused and accelerated onto a silicon diode *target* that responds more efficiently to the incident electrons than does the standard vidicon tube. ISIT cameras, which can detect even dimmer objects, contain a Gen I–type intensifier in front of the SIT tube. These designs give improved sensitivity and greater dynamic range at low light levels. However, these cameras exhibit a lag in response time that results in the appearance of comet tails for rapidly moving objects, a falloff in resolution from the center to the periphery of the image, gain instability, and significant geometric distortion and shading, which make them less desirable for quantitative applications.

A better solution for low-light imaging is the insertion of a high-resolution, high-gain image intensifier between the microscope and the camera. Image intensifiers increase the intensity of dim fluorescent specimens and are named Gen I, II, III, or IV according to their design and performance specifications. Gen II and Gen III intensifiers give an electronic amplification (gain) of 10^4–10^5 and generate up to 30- to 50-fold increases in the light intensity delivered to a camera that is coupled to the intensifier with an intervening lens. The target at the front end of the intensifier tube receives incident photons and emits electrons, which are accelerated onto and through a *microchannel plate* to further amplify the number of electrons. The electrons excite a phosphor screen at the exit end of the microchannel plate, generating the intensified image. The screen's image is then projected onto the video camera with a relay lens. Alternatively, the intensifier's phosphor screen can be directly coupled by optically transparent cement to the detector (such as a CCD chip) by a large-diameter fiber optic bundle. Details on the design and performance of intensifiers are presented in *Video Microscopy* by Inoué and Spring (1997). There are several points worth noting about intensifier performance:

- Although the electronic gain of a Gen II intensifier is very large (10^4–10^5), the increase in light intensity delivered to a lens-coupled camera is usually considerably less, typically 10- to 30-fold. This decrease occurs because the practical working range of the electronic gain is considerably less than the potential maximum and because of the low efficiency of signal transfer from the intensifier phosphor screen to the relay lens and camera detector. Since intensifiers are usually operated at 10–50% of maximum voltage to preserve their life, the observed increase in light amplification by a gen II intensifier is 10- to 30-fold under typical operating conditions. Despite the losses, a 10- to 30-fold increase in light intensity can be extremely valuable.

- Image fidelity can be altered by the microchannel plate, which, due to the hexagonal packing of its capillaries, contributes a hexagonal chicken-wire pattern to bright images. Newer Gen III intensifier designs minimize this problem by using smaller capillaries.

- Increasing the intensifier gain changes the range of sensitivity of the intensifier, but because the intensities of all image points are amplified by a constant factor (gain), the intrascenic dynamic range (the range within a single image view) remains the same. However, changing the gain increases the interscenic dynamic range considerably (the total dynamic range represented by images obtained under a range of lighting conditions).

Despite some limitations, image intensifiers allow video detection and display of fluorescent objects that are at or below the limit of vision while looking in the microscope. Certain Gen III intensifiers are capable of showing individual GFP or fluorescein molecules at the video rate of 30 frames/s! Intensifier-coupled cameras are employed for examining single-molecule dynamics using total internal reflection fluorescence (TIRF) microscopy. For further orientation, see Spring (1990).

VCRS

Because of limitations in reading and writing to magnetic tape medium, images recorded on a video cassette recorder or VCR have reduced spatial resolution compared to that inherent to the original video signal. The signal response and bandwidth of a

VCR is limited to about 4–5 MHz (as opposed to 10–15 MHz for the camera), which reduces the horizontal resolution (number of lines per picture height) from 800 black and white lines for the camera down to about 320 (4/10 × 800) lines for a home VCR. Once written to tape, the original 10 MHz resolution of the camera cannot be recovered. Nevertheless, VCRs are very convenient image storage devices and are commonly employed. In cases where the full resolution of the camera must be maintained, live images can be captured by a frame grabber in a computer or sent to a freeze-frame device for recording on film. Two VCR formats are commonly available:

VHS (video home system)	320 display lines (240 lines for color)
S-VHS (super VHS)	400 display lines

Clearly the S-VHS system offers greater resolution and is the system of choice for video microscope recording.

SYSTEMS ANALYSIS OF A VIDEO IMAGING SYSTEM

Another way to consider the bandwidth requirement of ~10 MHz for a video camera is to examine the *modulation transfer function (MTF)*. The MTF describes the percent modulation or change in contrast that occurs between the input and output signals for signal-handling devices (microscope optics, radios, electronic circuits). For video imaging systems, this includes the video camera, processor, VCR, and monitor. The *system MTF* refers to the modulation that occurs for a series of connected devices. As seen in Figure 13-13, as the spatial frequency in object details increases (object spacing decreases), the percent modulation (the contrast) in the object signal decreases. At the so-called cut-off frequency (the limit of spatial resolution), the contrast becomes 0. For a given input signal, the spatial resolution can be maintained by increasing the magnification in accordance with the frequency characteristics of the camera or electronic device.

It is easy to calculate the cut-off frequency for your own video microscope system. For a 40×/0.95 NA objective in green light ($\lambda = 546$ nm), the cut-off spatial frequency f_c is

$$f_c = 2\text{NA}_{obj}/\lambda = 3480 \text{ cycles/mm,}$$

or ~0.29 μm/cycle. This frequency is similar to the spatial frequency calculated from Rayleigh's resolution limit d discussed in Chapter 1, where $d = 0.61\lambda/\text{NA} = 0.35$ μm.

The cutoff frequency f_H in the horizontal dimension for a video camera is defined as

$$f_H = M_{objective} \, M_{relay} \, N_H/1.2D,$$

where M represents the magnifications of the objective and relay lenses, N is the lines of horizontal or vertical resolution at $N_H = 800$ and $N_V = 480$, and D is the diagonal measurement of the target in the video electron tube in mm. For the 40× objective lens and with no projection or relay lens (relay lens magnification = 1×), a horizontal resolution of 800, and $D = 15.875$ mm, $f_H = 1680$ cycles/mm. This is considerably less than the 3480 cycles/mm calculated for the microscope optics alone and shows that there is con-

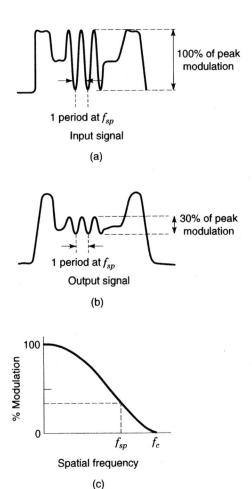

1 period at f_{sp}

Input signal

(a)

30% of peak
modulation

1 period at f_{sp}

Output signal

(b)

% Modulation

100

0

f_{sp} f_c

Spatial frequency

(c)

Figure 13-13

Modulation transfer function. In examining an object, an imaging device such as the eye, microscope optics, or an electronic camera modulates (changes) the contrast of the image depending on the spatial frequencies of object details and the electronic properties of the detector. The relationship between percent modulation and spatial frequency is known as the modulation transfer function (MTF). (a) An input signal consisting of a sinusoidal pattern with modulation at 100%. (b) Output signal of the object pattern showing 30% modulation. The modulation transfer is (30/100)% or 30%. (c) Graph of the MTF plotted as percent modulation vs. spatial frequency. The spatial frequency at which the percent modulation or contrast becomes 0 is called the cut-off frequency and indicates the limit of spatial resolution. The 30% modulation at spatial frequency f_{sp} is indicated. The cut off frequency is indicated f_c.

siderable loss of spatial resolution when the video camera is used in this configuration. However, the spatial resolution can be preserved by adding a 4× TV lens to the system (Fig. 13-14). The MTF then becomes 4 × 1680 = 6720 cycles/mm, which is greater than the cut-off frequency of the optics.

Likewise, we can calculate the effects of the system MTF on the display performance of a VHS VCR whose horizontal resolution limit is typically 320 lines per picture height. Replacing 320 for the value of 800 in the example, we calculate that the

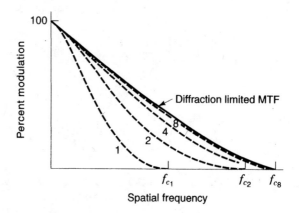

Figure 13-14

Effect of magnification in preserving system MTF. With adequate magnification through the insertion of a relay lens between the microscope and camera, the MTF can be improved up to the limit set by the optics. MTF curves for a system including various 1–8× relay lenses are shown. Because added magnification reduces the diameter of the field of view, a compromise between magnification and MTF performance is sometimes required.

320-line cut-off for the VCR corresponds to 672 cycles/mm in the specimen plane, or only 20% of the optical resolution of 3480 cycles/mm.

DAISY CHAINING A NUMBER OF SIGNAL-HANDLING DEVICES

Linking multiple video components together in a series is called daisy chaining. For example, a video camera can be connected to a digital signal processor, which is in turn connected to a VCR. Although daisy-chained equipment is convenient, it is important to know that the MTF decreases at each signal-handling device. As discussed, the response factor of camera electronics is described in terms of the rise time required for the signal to rise to peak value (the response frequency). When several electronic components are linked together, such as a video camera plus VCR, the overall system rise time is lengthened, and system MTF and spatial resolution are reduced. Thus, two cascaded circuits with the same bandwidth and rise time are found to reduce the horizontal resolution of the system by 30% compared to each circuit individually. The relationship is described in terms of system bandwidth, BW, and is given as

$$1/\text{BW of system} = \sqrt{\sum 1/(\text{BW each component})^2}.$$

For a video camera (10 MHz, 800 black and white lines) coupled to a VCR (5 MHz, 400 lines) the system bandwidth is reduced to 4.47 MHz = 360 TV lines. Another way of stating this is to say that the system MTF for a given frequency f is the product of the values of percent modulation for each of the components in the system:

$$\% \text{ modulation of system} = a\% \times b\% \times c\%.$$

The solution for maintaining resolution is to add a magnifying relay lens (a so-called TV lens) with a magnification range of $2-8\times$ to the imaging system. As seen in the graph of system MTF, a video microscope containing an $8\times$ relay lens gives close to theoretical resolution. The negative side of using a relay lens is that the field of view is reduced; image brightness on the camera also decreases in proportion to the square of the magnification difference.

Exercise: Contrast Adjustment and Time-Lapse Recording with a Video Camera

The objective is to prepare a time-lapse video recording of live cells either in phase contrast mode or in fluorescence mode after staining with a fluorescent dye. The exercise will demonstrate the relative ease of video recording and allow you to experience how time-lapse video imaging can provide valuable information regarding cell motility and the dynamics of cellular and intracellular organelle movements. The exercise will also demonstrate the value and ease of performing video enhancement of low-contrast specimens, but it will also show how electronic recording equipment can potentially degrade image quality (described in quantitative terms as the modulation transfer function, or MTF).

- Prepare the specimen and obtain a live image on the monitor.
  1. For this exercise and for use in the laboratory, it is convenient to prepare a cart containing the video camera, TV monitors, VCR, and other equipment shown in Figure 13-2. The cart can be positioned near the microscope for video recording, but stored in an out-of-the-way place when not in use. On the video cart, turn on the camera and camera controller (video tube camera or video CCD camera), the digital image processor (Hamamatsu Argus-20, Dage DIP, or similar unit), two TV sets for display of the raw and processed images, the thermal printer, and the time-lapse S-VHS VCR.
  2. Mount a coverslip of living cultured cells with Vaseline as described in Figure 10-9. Use cell types such as COS-7, U2OS, or another cell type that has a flat, extended morphology so that organelles can be clearly distinguished and identified. Alternatively, examine a culture of protozoa such as *Acanthamoeba* or *Dictyostelium*, as these cells move relatively quickly and give a high contrast image. Examine the specimen with a $100\times$, 1.3 NA oil immersion objective, and adjust the microscope for Koehler illumination using phase contrast or fluorescence optics. Examine the back focal plane with the focusing telescope eyepiece to check that the oil fills the space between the lens surface and coverslip, that there are no air bubbles trapped in the oil, and in the case of phase contrast, that there is precise alignment of the condenser annulus and objective phase ring. The built-in halogen lamp is suitable for phase contrast, while the epi-illuminator/mercury lamp is required for fluorescence.

3. A raw image of the cells will appear on both TV screens. One screen displays the raw, unprocessed image while the other displays a processed image from the image processor. If the screen of the raw image monitor is black, the light is too dim or the connections and/or circuits are not set correctly; if the screen is white, the camera is oversaturated. Reduce the light intensity to the camera until a normal image appears on the screen. Do no oversaturate the camera! Most camera controllers have a light that indicates when illumination intensity for the camera is too dim, in the optimal working range, or saturating.

4. Special UV and IR cut filters should be placed in the illumination pathway to protect the cells from photodamage. A green interference bandpass filter centered at 546 nm should also be used to further protect the cells and improve image contrast. Brightness and contrast of the raw image are adjusted using the offset and gain controls on the camera controller before further adjustments are made with the digital image processor.

- Adjust the digital image processor for frame averaging, background subtraction, and digital contrast adjustment.

  1. Position the specimen and use the gain/offset dials on the camera controller to optimize the raw image. Adjust the offset first followed by the gain.

  2. Click/select the background subtraction function on the image processor.

  3. Select the frame number = 4, 8, or 16 for frame averaging. This setting selects the number of video frames that will be used for frame averaging. Image quality improves dramatically because the S/N ratio increases as the square root of the number of frames averaged. For dim, noisy images, frame averaging can make a tremendous improvement, although averaging reduces the temporal resolution and may not be suitable for rapidly moving objects. (Another way to see the same effect is to select the operation called averaging. Simply select the number of frames and hit the start button to see the effect.)

  4. Find a clear area of background next to and in the same focal plane as the cell. If there are no clear areas, defocus the specimen by a small amount. Click "Background" to store the background image in the frame buffer of the processor. Then press "Start" to initiate subtraction. The screen will turn blank and medium gray. Reposition the cell and refocus if necessary. You are now performing a simultaneous operation of frame averaging and background subtraction. This operation increases the S/N ratio and smoothness of the image and removes images of dust and optical blemishes in the microscope-camera system.

  5. After the camera's offset and gain controls have been optimized, you may make further improvements in brightness/contrast in the processed image using the digital contrast adjustments on the processor. Be sure the image is being processed for frame averaging and background subtraction. Now click on the "Enhance" button on the processor to make further adjust-

ments to brightness and contrast. It may not be necessary to use these adjustments for low-magnification images of whole cells.

6. For convenience, you may also wish to use the processor to place on the image a scale bar for magnification and a clock for monitoring time.

- Adjust the time-lapse VCR.

1. Adjust the time-lapse VCR for time-lapse mode using the +/− keys. The normal play setting is 2 (2 hours to play a 120 min VCR tape at real-time rate of 30 frames/s). To show a 1 hour recording period in 2 minutes, we need a 30-fold compression factor for time-lapse recording. Thus, we need to record 30× slower than the normal speed setting of 2, which corresponds to a record setting of 60. Hit "Record" on the VCR and be sure the record button lights up. Check that the cell remains in focus during the recording period. On a record sheet, enter the start and stop footage of the recording, objective lens and TV lens magnifications, and any other pertinent information.

2. Play back your recorded sequence. To play the sequence at an accelerated rate, adjust the +/− keys to return to 2 hr play rate. If necessary, adjust the tracking knob on the VCR to remove image flicker.

- Questions

1. What new details or information regarding cell dynamics can you provide using time-lapse recording methods?

2. What is the horizontal cutoff frequency defining the spatial resolution for the combined microscope/camera system? Refer to the text for preparing an answer.

3. Does the addition of a VCR to the imaging system alter the system MTF? If so, how, and by how much?

4. What can be done to minimize degradation in the MTF before recording on the VCR?

DIGITAL CCD MICROSCOPY

OVERVIEW

Digital CCD cameras contain a charge-coupled device or CCD, a photon detector that is divided up into thousands or millions of picture elements or pixels, which store the information from incident photons comprising the microscope image. A computer displays the reconstructed image on a monitor. The light-sensitivity, dynamic range, spatial resolution, and speed of a scientific grade CCD camera are extraordinary. The efficiency of light collection is so great, that a CCD image, compared to a film image of comparable S/N ratio, would be rated at a film speed of ~100,000 ASA. Because they give a linear response over a large range of light intensities, CCD cameras can function as imaging spectrophotometers, producing tens to hundreds of times better resolution of light intensity than video or film cameras. They also have a spatial resolution comparable to film (Figure 14-1). Although scientific grade CCD cameras perform these functions well, they work at *slower rates* than video cameras (~1–10 frames/sec at full resolution), and for this reason, are called slow scan cameras, although recent designs acquire "full frame" images at close to the video rate of 25 or 30 frames/s. One of the greatest attractions of the CCD camera is the ability to see the photographic results instantly, allowing the user to evaluate images and proceed more quickly and efficiently, without having to develop film or make prints in the darkroom. Digital files can also be directly incorporated into manuscripts and used for electronic presentations at conferences.

The combination of microscope and CCD camera together with computer with imaging software defines what is called a *digital imaging system* (Figure 14-2). Although these systems are still relatively expensive, their versatility and convenience have greatly stimulated the use of light microscopy in research (Hiraoka et al. (1987). To use the equipment properly, training and practice are needed to master several software-dependent procedures for image acquisition, processing, analysis, display, and printing, topics that we cover in this and the following two chapters. In this chapter we examine the principles involved in CCD operation, the parameters the user must consider for obtaining high quality images, and the criteria that are important for comparing the performance of different CCD cameras.

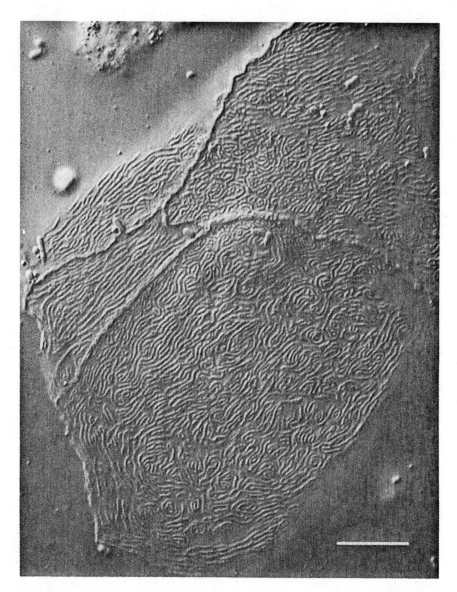

Figure 14-1

Spatial resolution of a CCD detector. This DIC image of a buccal epithelial cell was recorded on a 1.4 megapixel CCD camera having a pixel size of 6.8 μm. Since the unit diffraction spot radius projected on the CCD by a 100×, 1.3 NA objective lens is 26 μm, full optical resolution is retained. The resolution is comparable to that obtained by Kodak TMax100 film. The spacing between ridges on the surface of the cell is approximately 0.4 μm. Bar = 5 μm.

THE CHARGE-COUPLED DEVICE (CCD IMAGER)

The CCD device is located in a hermetically sealed chamber covered with a transparent glass or quartz window. The CCD is mounted on a block that is backed by Peltier cooling elements that reduce the so-called thermal noise in the photoelectron signal. The camera also contains a low noise amplifier, an ADC, and electronics for controlling the

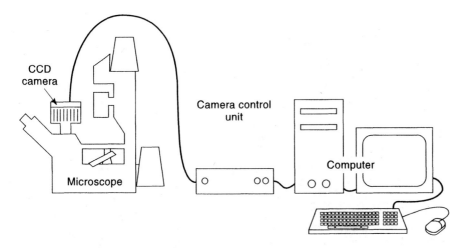

Figure 14-2
Components of a digital CCD imaging system. A CCD camera is mounted on a light microscope. A separate power supply and camera control unit connected to the camera communicates with a computer.

readout of photoelectrons from the face of the CCD. A sketch showing the arrangement of these components is provided in Figure 14-3. Readers will find detailed descriptions of CCD operation and design in volumes by Holst (1996) and Buil (1991).

A *CCD chip* or *imager* is composed of a thin wafer of silicon, a semiconductor material capable of trapping and holding photon-induced electron/hole pairs (Fig. 14-4). The silicon surface is covered with an orthogonal gridwork of narrow transparent strips that carry a voltage, thereby defining thousands or millions of square picture elements or *pixels* in the silicon matrix. The pixels function as light-sensing elements called *photodiodes* that act as potential wells for storing charge carriers derived from incident photons (one electron/hole pair per absorbed photon). The charge carriers are usually called photoelectrons. Photoelectrons can be accumulated and stored for long periods of time until they are read from the chip by the camera electronics. The peak *quantum efficiency (QE)*, the percent of incident photons resulting in photoelectrons, is very high (40–90%) and varies depending on the incident wavelength and electronics design of the chip. Pixels range from 4 to 25 μm on a side and have a typical holding capacity, or full well capacity, of ~1000 electrons/μm^2 when the camera is used in the multipin phase (MPP) mode, which reduces the spillover of saturated pixels into neighboring pixels, a phenomenon called *blooming*. (Despite its convenience in controlling the behavior of saturated pixels in the image, MPP operation reduces the potential full well capacity of the pixels by about 50%. Accordingly, some CCD cameras used for low-light applications do not use this mode.) Therefore, a 6.8 μm pixel in a MPP-operated CCD can hold ~45,000 electrons.

The face of a CCD in a full-frame CCD camera contains thousands of pixels that make up the *parallel register,* the imaging surface that accumulates and stores photoelectrons (Fig. 14-5). Since the image is focused directly on the surface of the CCD, there is a point-for-point correspondence between the pixels representing the image on the chip and pixels on the computer monitor where the picture is displayed and viewed. After an exposure, a timed sequence of voltage potentials moves across the strips on the

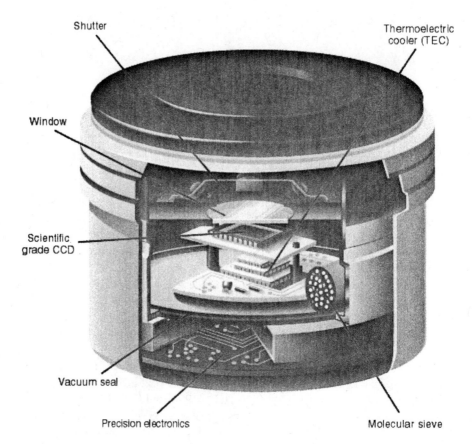

Shutter

Thermoelectric
cooler (TEC)

Window

Scientific
grade CCD

Vacuum seal

Precision electronics

Molecular sieve

Figure 14-3

CCD camera architecture. The CCD detector is mounted in a hermetically sealed chamber charged with dry nitrogen gas or containing a vacuum. A transparent faceplate or window at the front of the chamber is located just in front of the CCD, which is mounted on a thermoelectrically cooled block to reduce thermal noise in the image signal. The cooling element is backed by a fan or by circulating water to maintain the CCD at a constant temperature of $-25°C$ to $-40°C$. The body of the camera contains cooling vanes to help dissipate heat. Depending on the type of CCD, an electromechanical shutter may be situated in front of the faceplate. The camera head also contains several electronic components: a preamplifier (on the CCD chip itself) to boost the signal read from the CCD, an analogue-to-digital converter, and circuits for controlling CCD readout. (Drawing from Roper Scientific, Inc., with permission)

CCD surface, causing all of the electron charge packets stored in the pixels in the parallel register to be transferred one row at a time toward a single row of pixels along one edge of the chip called the *serial register,* from which they are moved one pixel at a time to an *on-chip preamplifier* and the ADC (Fig. 14-6). After the serial register is emptied, the entire parallel register is advanced by one row to fill the serial register, the process repeating until the entire parallel register is emptied. The function of the on-chip preamplifier is to magnify the signal and transmit it as a variable voltage for a short distance along a video cable to an ADC, which converts the signal into the 0 and 1 binary code of the computer. For a 12 bit camera, the assignment of each pixel ranges from 0 to 4095

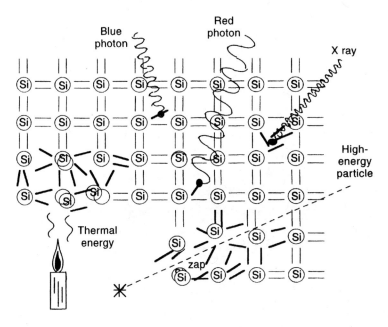

Figure 14-4

Silicon as a photon-sensitive substrate in a CCD imager. The sketch shows the effect of incident photons of various wavelengths on the silicon matrix of a CCD. Incident photons interact with the silicon, breaking covalent bonds between the silicon atoms and generating electrons and electron-deficient sites called electron holes. A voltage potential applied across the CCD holds the accumulating photoelectrons in the silicon matrix until they are read off from the chip and digitized. Red photons penetrate deeper into the matrix than blue photons, accounting for the relative insensitivity of silicon to blue light. High-energy X rays and cosmic rays disrupt many bonds and generate large saturating signals; typically, there are a few cosmic ray hits on the CCD surface per minute. Thermal energy, represented by the candle, also disrupts bonds and generates electrons (thermal noise) that cannot be distinguished from photoelectron counts; however, the problem can be reduced significantly by cooling the CCD to very low temperatures. After the electron charge packets are read off from the CCD surface, the structure of the silicon matrix is restored and the CCD is ready for another exposure. (Sketch from Roper Scientific, Inc., with permission)

steps (12 bit imaging gives $2^{12} = 4096$ possible gray levels). Each step is called an *analogue-to-digital unit (ADU)*.

To fully appreciate the sophistication of the technology, let us review the sequence of events involved in taking a picture with a full-frame CCD camera:

- The camera shutter opens and pixels accumulate photoelectrons.
- The shutter closes, and pixels are moved one row at a time off the parallel register by voltages applied to the strips on the CCD in a pattern and at a rate determined by timers or clocks in the camera electronics. Each row at the end of the parallel register is transferred to a special row of pixels called the serial register.
- Pixels are transferred one pixel at a time down the serial register to an on-chip preamplifier. The amplifier boosts the electron signal and generates an analogue voltage output.

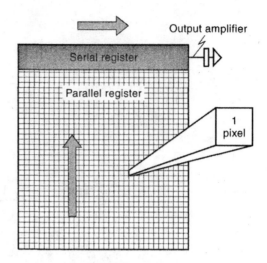

Figure 14-5

Surface design of a full-frame CCD. The majority of the surface area is occupied by an imaging area known as the parallel register, an array of thousands of pixels that are arranged in rows and columns with respect to the output edge of the chip near the serial register. The serial register is composed of a single row of pixels and contains the same number of pixels as the row at the output edge of the parallel register. Photoelectron counts contained in the pixels are transferred serially, from pixel to pixel, until they reach an amplifier, which sends an amplified voltage signal to a nearby ADC.

- An ADC assigns a digital code for each pixel depending on the amplitude of the signal (0–4095 ADUs for a 12 bit system).
- Pixel values are stored in a frame buffer in the computer.
- The process repeats until all 1000+ rows of pixels of the parallel register are emptied.
- For a 1 megapixel CCD chip processed at 2 bytes/pixel, 2 Mbytes are stored in the computer; the image is displayed in an 8 bit (256 gray-level) format on the monitor.
- The CCD chip is cleared (reread without digitization) to remove residual electrons prior to the next exposure.
- The total time for readout and display is ~0.5 s for a 1 Mpixel camera operating at 5 MHz.

Considering the large number of operations, it is remarkable that greater than a million pixels can be transferred across the chip, assigned a gray-scale value ranging from 0 to 4095, and stored in a frame buffer in the computer in less than a second! The accuracy of transfer of electron packets over thousands of pixels is also nearly error-free. The quantitative nature of a digital CCD image is shown in Figure 14-7.

Another important feature in the camera head is the elements associated with cooling. As described below, CCD images are degraded by electron noises, the most serious of which are due to heat and the electronic readout of the camera. *Thermal noise* refers to the generation of electrons from the kinetic vibrations of silicon atoms in the CCD substrate. Thermal electrons cannot be distinguished from photoelectrons and thus contribute noise to the image. Cooling is essential for all scientific-grade CCD cameras, the

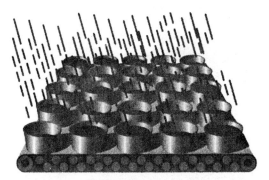

CCD operation
integration of photo-induced charge

CCD operation
parallel shift - 1 row

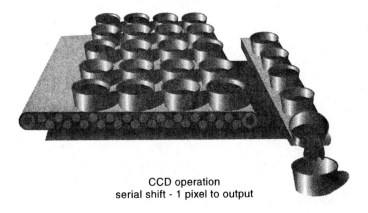

CCD operation
serial shift - 1 pixel to output

Figure 14-6

Sketch showing the concept of serial readout of a CCD. Pixels comprising the parallel and serial registers of a CCD are represented as buckets containing varying amounts of signal (water) being transported on two conveyor belts operating in stepwise fashion. When the serial register is empty, the row of buckets at the edge of the parallel register transfers its contents to the single row of empty buckets on the serial register, which moves its single row of buckets to a measuring station (the amplifier and ADC) at the end of the conveyor. (Sketch from Roper Scientific, Inc., with permission)

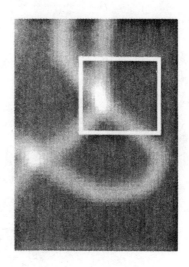

```
 63  84 119 172 219 225 182 135  79  51  36  24  23  19  15  -1  14  14   8   0  -4   7  18
 78  84 104 170 223 224 196 118  84  49  36  28  16  11  25   4  15  13   8  -4   9  11   7
 61  80 115 153 209 204 170 113  73  46  41  29   9  17  11  11   0  12  -2   2   2   3  23
 72  98 121 160 190 207 178 116  68  50  29  22  17  19   7  27  15   9  -3  -4  14   5   8
 64  90 132 167 210 214 180 115  71  37  36  31  13  15   9   8  15   6   0   5 -14   4  12
 75  93 124 169 216 229 196 107  71  56  19  18  22  24   7   5  15  11   8  -1  12   6   7
 97  87 128 193 210 225 193 111  85  47  27  27  21  12   5   2  -1   4   1  -3   7   2 -10
103 108 134 180 201 233 185 115  55  38  26  25  15  20  18   6   2   2   1   4  -3 -13   0
142 132 161 216 238 223 160  90  59  45  17  10   9  13  10  11   4  -9   5   2   7   0   5
172 162 175 231 239 238 155  88  48  28  24  17  15  13   0  14   0  11  -3   4   9   0 -10
226 219 230 260 265 236 161  92  43  31  31  11   5  11   7  13  19   9  18 -11  -9  -2   8
234 247 256 302 311 253 174  97  48  27  12  15   7   7   0  16   8   5   3  -4   0  -6   4
260 263 297 346 349 303 196 126  65  27  30  24   3   6   7   1  12   3   9   0  -2 -13   2
244 293 340 388 399 321 223 130  74  29  24  30  17   4   3  11   0   8   7  -3  -2  -2  -2
209 273 359 423 436 365 264 141  80  57  32  45  13   3  18   8  -7   0  -6   4  -1  -2  -3
176 253 342 430 443 394 291 161  86  59  37  23  18   5   0   7   8  11   1  -3  13  -5  -2
152 218 311 425 470 420 325 208 111  66  52  29  28   9   4   7   8   4  -7  11 -18 -13  -2
129 199 294 413 469 441 384 257 148 111  69  34  20  20   6   3  15   4  -2  -6  -3 -10   9
140 206 294 385 439 442 365 310 223 157 114  76  45  28   9  21   5  15  -4 -13   0  -5  -1
173 233 309 354 392 375 333 303 261 214 135  92  51  47  18  12  13  12  20  -9   4   1  15
221 278 300 321 306 293 286 279 250 231 184 142 108  67  41  18  13   5   8  -8   0   7   5
267 302 291 244 228 211 201 215 241 227 205 184 136 110  68  51  26  11   8   3   0   8  -3
284 279 257 202 133 129 137 151 183 213 209 188 187 155 109  69  49  26  25   8   8  18  -4
275 248 191 143  95  85  87  98 122 166 184 192 206 194 176 135  98  50  44  19  21   0   1
```

Figure 14-7

Image view and corresponding text file of two fluorescent microtubule filaments. Each pixel value (described in analogue-to-digital units or ADUs) indicates the number of incident photoelectrons during the exposure. Notice that where the two filaments cross, the pixel value becomes doubled. The CCD camera used here is capable of resolving up to 4096 gray levels within a single image frame.

benefit being about a 10-fold reduction in the number of thermoelectrons for every 20°C decrease in temperature. Thermal noise can be reduced significantly by cooling the CCD down to −20°C or lower using a stack of 2–3 *Peltier thermoelectric cooling devices.* In the presence of a current, a Peltier bimetallic strip becomes cold on one side and hot on the other. The cold surface is mounted so that it is in indirect physical contact with the CCD, while the heat on the other surface is removed by a fan or backed by a circulating liquid cooling system. Astronomical cameras used for hour-long exposures are cooled with liquid nitrogen. For biological specimens, acceptable images can be obtained from exposures lasting just a few seconds or less from CCD cameras cooled to 0–10°C, but for higher-quality, lower-noise images, deeper cooling to −25°C to −40°C is required. We will deal with the noise components in a CCD image later in the chapter.

CCD ARCHITECTURES

CCD imagers come in three basic designs, which are shown in Figure 14-8 and described as follows:

- *Full-frame CCD.* In this design, whose readout procedure was described in the preceding section, every pixel of the CCD surface contributes to the image. Exposures are usually controlled by an electromechanical shutter, since the imaging surface must be protected from incident light during readout of the CCD. Full-frame CCD imagers are used for specimens requiring high dynamic range images and where the time resolution can be on the order of a second or longer. The fastest frame rates in so-called subarray mode are ~10 frames/s and are limited by an electromechanical shutter.

- *Frame-transfer CCD.* Cameras with frame-transfer CCDs are fast because exposure and readout occur simultaneously. One-half of an elongated rectangular CCD chip is masked at one end by an opaque cover (an aluminum coating on one-half of the surface of the CCD), which is used as a storage buffer. After an exposure, all of the pixels in the image half of the chip are transferred to pixels on the storage side in 1 ms; the storage array is read out while the image array is being exposed for the next picture. No camera shutter is needed because the time required for transfer from the imaging area to the masked area (~1 ms) is only a fraction of the time needed for an exposure. A disadvantage is that only one-half of the potential imaging surface of the chip can be used for imaging. Cameras of this type are chosen for monitoring rapid kinetic processes such as in dye ratio imaging, where it is important to maintain high spatial resolution and dynamic range.

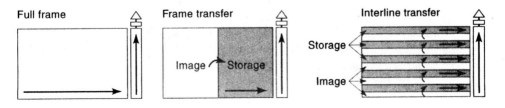

Figure 14-8

Types of CCD designs. (a) Full-frame CCD. (b) Frame-transfer CCD. (c) Interline-transfer CCD.

- *Interline transfer CCD.* In this design, columns of pixels in the parallel register alternate between imaging rows and masked storage-transfer rows, resulting in a pattern of stripes across the entire CCD. Interline transfer CCDs are used in camcorders and video cameras because they can be read out at video rate and provide high-quality images of brightly illuminated objects. Improvements in camera electronics and significant reductions in camera read noise have made this design popular for low-light applications, including dim fluorescent specimens. After an exposure, all of the pixels in the image columns are transferred by one step and in less than 1 ms to pixels in the adjacent masked storage columns; the storage array is read out while the image array is being exposed for the next picture. Cameras with interline CCDs that are read out serially from top to bottom are sometimes called *progressive scan cameras,* a reference to one of the readout modes of video CCD cameras. This type of CCD is also very fast and does not require a shutter to control the exposure, but the dynamic range may be somewhat reduced. The latest interline detectors include very small pixels and microlenses that cover pairs of image and storage pixels so that photons incident on the storage pixels are included in the photoelectron parcels in the imaging pixels.

Note: Interline CCDs for Biomedical Imaging

Because of their speed, lack of requirement for a shutter, and lower wavelength sensitivity (desirable for UV-excitable fluorescent dyes and GFP), present-day interline CCD imagers give superior performance for cameras used for digital microscopy. In earlier interline CCD designs of the type used in video cameras and camcorders, the alternating masked columns of pixels and small pixel size meant reduced light sensitivity and reduced spatial resolution in the image; further, high camera read noise meant reduced dynamic range, with signals suitable only for 8–10-bit ADCs. The situation has completely changed in recent years with the incorporation of microlenses on the surface of the CCD and the inclusion of low-noise camera electronics.

Microlenses are placed on the CCD surface in such a pattern that each microlens straddles a pair of masked and imaging pixels so that light that would ordinarily be lost on the masked pixel is collected and delivered to the imaging pixel area. Thus, by using small pixels and microlenses, the spatial resolution and light-collecting efficiency are now similar to those of a full-frame CCD. The microlens technology also permits extending the wavelength sensitivity into the blue and UV portions of the spectrum, making the interline CCD highly desirable for applications involving short wavelengths, including UV light. The most recent interline chips incorporate nonabsorbing materials in the gating structures on the surface of the chip, boosting their quantum efficiency to close to 80% for green light, a 50% increase over previous designs.

Since the active pixel area of a typical interline CCD is one-third that of a 6.8 μm square pixel on a full-frame CCD, the linear full well capacity in photoelectrons is reduced to about 15,000 electrons ($6.8^2 \times 0.33 \times 1000$). Ordinarily, this would result in a reduced dynamic range for the camera, but electronic improvements have also cut the camera read noise in half, down to as low as 5–8 electrons, so the dynamic range remains comparable to that of 12 bit full-frame CCD cameras. Electronic improvements have also resulted in increased speed of image acquisition, so 12 bit megapixel images can be acquired at 20 MHz or faster rates, 3–4 times the rate of existing full-frame CCDs of comparable size.

ANALOGUE AND DIGITAL CCD CAMERAS

As discussed in the previous chapter, CCDs can be used as photodetectors in analogue equipment such as camcorders and video cameras. Digital and video CCD cameras differ in the way the photoelectron signal is processed and displayed. In digital systems, quanta of photoelectrons stored in the pixels are sent as an analogue voltage signal over a short distance to an *analogue-to-digital converter (ADC)* that changes the signal into the binary code of the computer. The converter is contained in the camera head, or in the *camera control unit,* or on a card in the computer. In video cameras there is no digitization step, and the signal remains in analogue format as a variable voltage signal. Video cameras also add synchronization pulses to the signal to generate the composite video signal required by the raster scanning mechanism in the TV monitor. Because a video CCD imager is read at a faster rate (30 frames/s or greater), read noise is higher, causing the dynamic range and S/N of a video image to be lower than in images produced by a slow-scan digital CCD camera. Finally, video signals are commonly recorded on a VCR tape, while digital images are usually stored as image files on the hard drive of a computer. Sometimes the distinction between digital and video cameras becomes blurred, since some CCD cameras contain dual video and digital output, and because video and digital signals can be easily interconverted.

CAMERA ACQUISITION PARAMETERS AFFECTING CCD READOUT AND IMAGE QUALITY

Readout Rate

Fast readout rates are needed for applications requiring high temporal resolution. At very high readout rates such as those approaching a video rate of 30 frames/s, the image appears live and does not flicker on the computer monitor, and with the addition of sub-array readout and binning, acquisition at rates of hundreds of frames/s is possible. Most scientific CCD cameras can be adjusted to operate at different readout rates ranging from 0.1 to 10 MHz, the processing speed of the ADC and camera electronics (1 MHz = 10^6 byte processing operations/s). However, high readout speeds increase the level of noise in the image. Various noise components are always present in the pixels comprising an electronic image, among which readout noise is one of the major sources. Accordingly, low-intensity images with low pixel values should be read out at slower rates to reduce noise and maintain an acceptable S/N ratio and image quality.

Subarray Readout

It is possible to define a small subset of pixels on the CCD (a *subarray*) corresponding to only a portion of the full image area for acquisition and display on the monitor. The subarray is defined by entering the boundaries of the region in the acquisition software or by defining the *region of interest (ROI)* with a mouse on the computer screen. Subarray readout is fast because the unwanted pixels are not processed by the ADC and are discarded. Image files, particularly image sequences of time-lapse acquisitions, are also correspondingly smaller and more manageable.

Binning

Binning is the combining or pooling together of photoelectrons of adjacent pixels on the CCD to form electronic superpixels (Fig. 14-9). The pooling of photoelectrons occurs in the serial register of the CCD during readout. A superpixel that is 2 × 2 pixels contains the combined photoelectron content of 4 physical pixels, but is processed by the camera and amplifier as a single pixel. Binning reduces spatial resolution, but it offers the following advantages:

- Faster acquisition of image sequences (if the rate of acquisition is limited by camera processing speed)
- Smaller size of image files on the computer
- Shorter exposure time to obtain the same image brightness (a major benefit for live cell imaging)
- Improved S/N ratio for the same exposure time

Gain

Increasing the electronic gain reduces the number of photoelectrons that are assigned per gray level, allowing a given signal to fill a larger range of gray levels (Fig. 14-10). For example, for a gain setting of 10 electrons per gray level, 1000 electrons corresponds to 100 gray levels. If the gain is increased 4-fold, there are now 2.5 electrons per gray level, and 4000 gray levels are obtained. Note the difference between gain adjustment with a digital CCD camera and gain adjustment using a PMT or a vidicon tube,

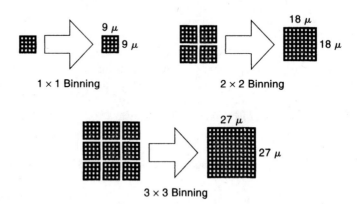

1 × 1 Binning   2 × 2 Binning

3 × 3 Binning

Figure 14-9

Binning. Photoelectrons contained in the unit pixel elements of the CCD are combined during CCD readout to form electronic superpixels. Binning is set by the user in software and is controlled by the timing pulses that drive the pixel parcels during the time of readout in the serial register on the CCD. Since photoelectrons are pooled, camera exposures can be shortened in proportion to the number of pixels included in a superpixel. Binning also reduces the image file size on the computer and allows more rapid frame rates, but spatial resolution is reduced compared to that available in an unbinned image. (Sketch from Roper Scientific, Inc. with permission)

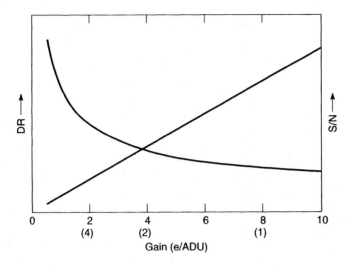

Figure 14-10

The effect of gain on dynamic range and signal-to-noise ratio. As gain is increased, the number of electrons/ADU decreases. For example, for a gain factor of 1×, 2×, and 4×, the number of electrons/ADU is typically 8, 4, and 2 electrons/ADU, respectively. By convention, 1× gain is usually defined as the saturating number of electrons per pixel divided by the read noise of the camera. For images exposed to give a constant number of accumulated electrons, increasing the gain causes the dynamic range (number of gray level steps) to increase exponentially. While having a large number of gray levels can be beneficial, notice that as the gain increases, the S/N ratio, the measure of signal clarity, decreases. For images exposed to give a constant number of ADUs, increasing the gain decreases the S/N.

where the signal voltage is amplified by multiplication by a constant factor. Gain is usually applied when there are a limited number of photons and it is desirable to utilize a large number of gray levels. The disadvantage of increasing the gain is a corresponding decrease in the accuracy of digitization; at high gain, the noise from inaccurate digitization can cause images to look grainy. There is a limit to how much gain can be applied, because at very high gain, image quality deteriorates. Nevertheless, by increasing the gain, the exposure time can be reduced, while retaining a large number of gray levels.

IMAGING PERFORMANCE OF A CCD DETECTOR

Image quality can be described in terms of four quantifiable criteria: resolution of time (sampling rate), resolution of space (ability to capture fine details without seeing pixels), resolution of light intensity (number of gray-level steps or dynamic range), and signal-to-noise ratio (clarity and visibility of object signals in the image). As we will see, it frequently occurs that not all four criteria can be optimized simultaneously in a single image or image sequence. For example, to obtain a timed sequence of a live fluorescent specimen, it may be necessary to reduce the total exposure time to avoid photobleaching and phototoxicity. This can be accomplished by exposing the specimen less often (loss of temporal resolution), binning the image (loss of spatial resolution), and/or by applying a high gain (reduction in dynamic range and S/N ratio). Alternatively, to maximize dynamic range in a single image requiring a short exposure time, you could apply binning or increase the gain. The ability to perform digital imaging efficiently and knowledgeably requires that the user become completely familiar with these terms and gain experience in the art of balancing camera acquisition parameters in order to optimize criteria of high priority. Although mentioned in previous chapters, we will now examine these terms in greater detail and in the context of using a digital CCD camera. This topic is also addressed in Chapter 15.

Temporal Resolution

Full-frame slow-scan CCD cameras do not perform high-resolution imaging at video rates. Generally the readout time is ~1 second per frame, though this depends on the processing speed (given in MHz) of the camera electronics. Since exposures of bright specimens can be made as short as 10 ms using judicious subarray selection and binning, cameras with electromechanical shutters can acquire images of limited size and resolution at rates up to ~10 frames/s. Interline and frame-transfer CCD cameras have no shutters and operate at faster speeds. The latest interline transfer cameras now provide full-frame 12 bit imaging at close to video rates.

Spatial Resolution and Image Dimensions

The spatial resolution of a CCD is determined by the pixel size and can be excellent. The pixels in cameras used for biological imaging are usually smaller than developed silver grains in typical camera film (10 μm). Even with 2- to 3-fold magnification during printing on a dye sublimation printer, the pixels comprising the picture are essentially invisible. With such small detector elements, CCD cameras usually meet the Nyquist criterion for image sampling, thus preserving optical resolution and avoiding aliasing (see Chapter 13 on aliasing and the Nyquist criterion). For example, for a chip with 6.8 μm pixels, there are ~4 pixels per diffraction spot radius produced by a 100×, 1.3 NA objective lens (0.25 μm × 100 magnifications ÷ 6.8 μm/pixel = 3.7 pixels/radius). This is double the Nyquist limit, so spatial resolution is very good. Even with binning at 2 × 2, the Nyquist sampling criterion is very nearly satisfied. The reader should refer to Figure 14-11, which compares CCD pixel dimensions with diffraction spot radii made by different objective lenses. Give this matter serious attention when selecting a CCD camera. The figure shows that a CCD with 10 μm or smaller pixels is ideal for images produced by high-magnification, high-NA, oil immersion lenses as would typically be encountered in fluorescence microscopy.

The number of pixels along the length and width dimensions of the CCD is also important when considering the quality and size of a print made with a dye sublimation printer. A megapixel CCD with 1,000 pixels along one edge of the CCD gives a 3.3 inch print when printed on a 300 pixel-per-inch dye sublimation printer set at a printer magnification of 1×. Pixelation in the image is not observed for printer magnifications under 2–3× (10 inches on a side) which is fine for most microscope imaging applications.

Quantum Efficiency and Spectral Range

Quantum efficiency refers to the efficiency of photon-to-electron conversion in the CCD, whereas spectral range refers to the wavelengths that can be detected. Standard

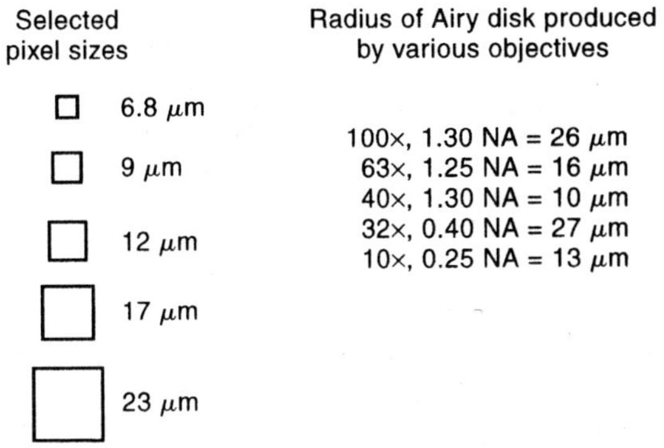

Selected
pixel sizes

Radius of Airy disk produced
by various objectives

☐   6.8 μm

100×, 1.30 NA = 26 μm

☐   9 μm

63×, 1.25 NA = 16 μm
40×, 1.30 NA = 10 μm
32×, 0.40 NA = 27 μm

☐   12 μm

10×, 0.25 NA = 13 μm

☐   17 μm

☐   23 μm

Figure 14-11

Comparison of pixel dimensions to diffraction spot size. Left: The pixel sizes of various CCD imagers are indicated in micrometers. Right: The diameter of the diffraction spot (Airy disk) produced by various objective lenses as it would appear on a the surface of a CCD. According to the Nyquist sampling theorm, preservation of the spatial resolution of the optics requires that a diffraction disk radius be covered by a minimum of 2 adjacent pixels on the CCD. For a 40×, 1.3 NA lens the diffraction spot radius = 40(0.61 × 0.546)/1.3 = 10 μm, so the coverage provided by a CCD with 6.8 μm pixels is just barely adequate. However, the same CCD provides excellent sampling for a 100×, 1.3 NA lens with spot radius = 100 (0.61 × 0.546)/1.3 = 26 μm, even under conditions of 2 × 2 binning (6.8 × 2 = 13.6 μm).

front-illuminated CCDs are efficient, sensitive detectors with a peak quantum efficiency of 40–50% (80% for the newest designs) at visible wavelengths ranging from 400 to 1100 nm with peak sensitivity at 550–800 nm (Fig. 14-12). With special coatings on the CCD, it is possible to extend the spectral range from 120 to >1100 nm. With special thinned, back-illuminated designs (very expensive!), quantum efficiency can be greater than 90%. However, the most recent interline CCD designs include high-transparency materials for defining pixel boundaries on the CCD surface that allow front illumination with 80% QE. Sometimes it is important to increase the sensitivity of signal detection and improve the S/N ratio in other ways, such as by decreasing the background signal, increasing the object signal, or selecting more efficient fluorescent dyes.

Noise

System noise refers to electrons in the CCD signal that have their origins in sources other than the object photons. Most noises are associated with the camera electronics (*bias noise* and *read noise*), and to a lesser extent, the *thermal noise* from the CCD. Bias signal (also called offset) and bias noise are components of the CCD signal that arise from the application of a positive bias voltage as required for proper digitization by the ADC. As already described, thermal noise refers to the generation of electrons from the kinetic vibrations of silicon atoms in the CCD substrate. It is common to refer to the bias noise and thermal noise together as the *dark noise*. As we will see later on, the photo-electron signal itself is also associated with a level of uncertainty called the *photon* or *shot noise* that always accompanies measurement of discrete quanta such as photons within a finite time or space. The amplitudes of the principal noises present in a CCD image are summarized in Table 14-1. Noise plays a major role in the determination of

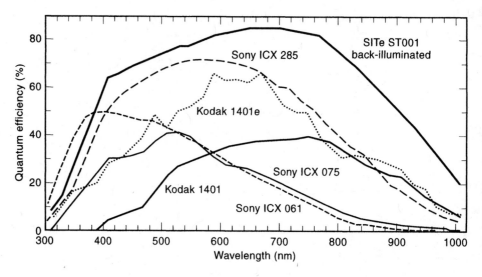

Figure 14-12

Wavelength sensitivity of some popular CCD devices. Kodak 1401: a standard CCD with limited blue light detection and peak sensitivity at ~700 nm. Kodak 1401e: The same CCD with enhanced coating for UV and blue sensitivity. Sony interline ICX-075: interline CCD with enhanced coating. Sony interline ICX-061: highly blue sensitive interline CCD with microlenses. Sony interline ICX-285: interline CCD with transparent gating structure and microlenses. SITe ST001: back illuminated CCD with increased quantum efficiency across the visible spectrum.

image quality as defined by the signal-to-noise ratio (S/N), which is discussed in detail in this chapter.

Dynamic Range

The number of resolvable steps of light intensity, gray-level steps ranging from black to white, is called the dynamic range (DR). Dynamic range is used to describe the number of gray levels in an image or the potential number capable of being recorded by a camera. In describing camera performance, DR is usually calculated as the ratio of the maximum signal electrons at saturation (the full well capacity of a pixel) to the read noise of the camera (generally 5–15 electrons). For a CCD chip with a full well capacity of 40,000 e^-/pixel and a read noise of 10 e^-, the potential DR is 4000:1. Since the full well

TABLE 14-1 Principal Noises in a CCD Image

Noise Type	Value	Source
Photon or shot noise	$\sqrt{}$ of signal e^-	Inherent noise in a photon signal
Readout (preamp) noise	~10 e^-	Preamplifier, other electronics
Thermal noise	~0.005 e^-/s	Vibration of atoms in the silicon matrix
Bias noise	~0.005 e^-/s	Noise from bias voltage during readout

capacity depends on pixel size (holding capacity is $\sim$1000 e$^-$/μm^2), CCD imagers with large pixels usually have a higher dynamic range. For example, a camera with pixels 27 μm on a side, a 580,000 e$^-$ full well capacity, and 9 e$^-$ read noise, has a DR = 65,000. The read noise of a camera is important in determining the dynamic range as well. Due to continual improvements in reducing readout noise (down to 5 e$^-$ and lower for cameras used in biological laboratories), even interline cameras with small 4.5 μm pixels can have a DR $\sim$4000.

To realize the potential dynamic range of a CCD camera, the camera must be equipped with an appropriate *digitizer*, also called an ADC or digital processor. Digitizers are described in terms of their *bit depth*. Since a computer bit can only be in one of two possible states (0 or 1), the bit depth is described as 2^x number of steps. Therefore, 8, 10, 12, and 14 bit processors can encode 2^8 . . . etc. steps, or a maximum of 256, 1024, 4096, or 65,536 gray levels. A manufacturer would typically fit a camera with a DR $\sim$4000 with a 12 bit processor to match the DR of the imaging system.

On some of the latest interline CCD cameras, the calculated dynamic range is about 2000, so the use of a 12 bit ADC seems to be more than is required. However, some of the new designs make full use of the 12 bit digitization by including a $\frac{1}{2}\times$ gain setting. This option is included because CCD chips are designed such that the pixels of the serial register hold twice as many photoelectrons as pixels in the parallel register. Thus, when a camera is operated with 2×2 binning, a common mode of operation for routine microscopy, 12-bit imaging at high quality can be obtained.

It is now quite common to see cameras fitted with processors having a much higher digitizing capacity than the inherent DR of the camera. If the read noise is designated as $1\times$ electronic gain (the conventional assignment for gain), there are potentially a large number of unused intensity levels when the bit depth of the processor exceeds the dynamic range of the camera. To fill these extra levels, manufacturers apply a (hidden) 2–$4\times$ gain to boost the signal to utilize the gray levels of the processor, but as we have seen, this operation increases the noise in the image. Therefore, by noting the full well capacity of the pixel and read noise of the camera, the dynamic range of different cameras can always be calculated and compared.

In comparison to high-bit CCD cameras, display devices such as computer monitors and dye-sublimation printers use only 8 bit processing (256 gray levels). In fact, the dynamic range of communications media such as TV/video, pictures in newsprint, and even of the eye is a meager 5–7 bits, depending on the image and illumination conditions. If the visual requirements for printers and monitors are so low, why are high-bit imaging systems necessary or even desirable? High-bit depth (DR) is required:

- For purposes of accurate quantitation of light intensities, for example, for examining ratiometric or kinetic imaging data; the larger the number of gray levels, the more accurately an intensity can be described.
- To perform multiple image processing operations without degrading the image; because of mathematical round-off errors during processing, images with more gray levels can tolerate a greater level of mathematical manipulation.
- For selecting and working with a portion of an image for display, where the region of interest covers only a narrow portion of the dynamic range of the full image. For an ROI including just 2% of the full dynamic range, this would be 82 levels for a 12 bit system, but only 5 levels for an 8 bit system. If the 5 gray levels were now stretched to fill the 256 levels for an 8 bit monitor or print, the image would look pixelated, with noticeable steps or contour lines in it.

Signal-to-Noise (S/N) Ratio

S/N ratio is used to describe the photometric accuracy of an object's signal. In qualitative terms, we use S/N ratios to describe the clarity and visibility of objects in an image. S/N ratio is calculated as the object signal (total signal minus contributing background signal) divided by the noise of the surrounding background (standard deviation of the background signal). When used to describe the imaging performance of a CCD camera, S/N is calculated in terms of a single pixel, and the calculation is always based on the number of electrons comprising the signal and the noise. The importance of S/N is easily appreciated when we examine a dim, grainy image, where the amplitude of the object signal is small and the read noise of the camera is a principal noise component in the image. In a side-by-side comparison of the imaging performance of two cameras with read noises differing by a factor of 2, the difference is clear: The camera with the lower read noise produces the clearer image. The effect of high camera noise can also be observed on the computer monitor or on a print in half-saturated, moderately exposed images. S/N characteristics are even more significant for those using a CCD camera as a photometer to monitor changes in light intensity, such as in fluorescence experiments involving FRAP, FRET, or ratio imaging of fluorescent dyes. In this case, plots of light intensity over time are smoother and more accurate when image sequences exhibit high S/N values. We will examine S/N theory and its applications in greater detail in Chapter 15.

BENEFITS OF DIGITAL CCD CAMERAS

Low-light sensitivity	Comparable film ASA $\sim$100,000 for equivalent S/N ratio; useful range of light intensity, 4–5 orders of magnitude; 3–4 orders of magnitude more sensitive than video or film
Low instrument noise	Cooling and electronics give high S/N and clear visible images; read noise/pixel as low as $3-5e^-$/pixel for biological cameras
Spatial resolution	Small (4–9 μm) pixels preserve optical resolution even at low magnifications
Time resolution	up to 10 frame/s for full-frame megapixel chip
Dynamic range	Thousands of gray levels; 10–16 bit vs. 6–8 bit for video
Less noise	Up to several hundred times less than video
Digital output	Pixels give quantitative value of light intensity; for 12 bit digitizer, up to 4096 gray levels
Linear response	<0.1% nonlinearity over 4 orders of magnitude
Flexible readout	Subarray, binning modes allow optimization of space, time, intensity

REQUIREMENTS AND DEMANDS OF DIGITAL CCD IMAGING

High costs	$10,000–$20,000 for a high-performance system
Personnel	Requires experienced user/computer technician
Additional components	Electromechanical shutter, filter wheel, computer interface card, RAM upgrade, acquisition/processing software

Image storage	Zip drives, CD burner and software
Computer maintenance	Maintenance is generally high
Slow speed	Generally not capable of video-rate image capture/display
Time requirements	Time required for image processing after acquisition; computer required for image display, processing, storage

COLOR CAMERAS

Many investigators choose to use color to represent the visual appearance of specimens seen in the microscope. Although there are some technical difficulties in displaying color in prints, the increase in information in a color image can be very significant. Some common color applications include fluorescence microscopy, stained tissue sections in histology and pathology, and beta-galactosidase and peroxidase stains in specimens viewed by bright-field or DIC microscopy.

Unlike color photographic film, a CCD device by itself is not color sensitive, so the acquisition of color images requires that red, green, and blue wavelengths be isolated with filters, acquired separately by the CCD, and then joined together to create a composite color picture. These requirements impose additional constraints that limit the resolution of space, time, and light intensity. Therefore, depending on the particular design and solution, we will see that color cameras tend to be slower, have reduced spatial resolution and dynamic range, and produce noisier images than gray-scale cameras. The designs and solutions for providing color information vary considerably:

- *External color filters or color liquid crystal device.* A motorized filter wheel rotates separate red, green, and blue filters into the light path at the location of the lamp or in front of the camera. The camera acquires separate images for each color. The full spatial resolution of the chip is maintained using this arrangement, but the speed of acquisition and display is reduced. To increase speed, a liquid crystal tunable filter can be used that is transparent and displays RGB colors in very rapid sequence.
- *Three-chip design.* The camera contains a beam-splitting prism that directs the image to three separate CCD devices each masked by a color filter. The spatial resolution of the CCD is maintained, and the frame rate (important for rapid sequences and for video output) can be very fast, because acquisition is simultaneous for each color channel; however, light intensity delivered to each detector is considerably reduced. Compared to a single-chip gray-scale camera exposed for a comparable amount of time, the color image is nearly 10-fold dimmer. Gain can be applied to brighten the color image, but at the expense of reduced S/N ratio and increased graininess.
- *Color microlenses and movable color masks.* CCDs can be manufactured with red, green, and blue microlenses applied in a specific pattern on individual pixels on the chip. These cameras are fast, and pixels are illuminated at greater intensity than in the three-chip design, but spatial resolution is reduced. Another solution for individual pixel masking features an array of red, green, and blue microlenses (a color mask) that moves rapidly and in a square pattern immediately over the surface of the CCD. Each pixel is responsible for providing RGB color information, so maximum spatial resolution is obtained. This camera design is ideal for bright-field color photography of histological specimens at low magnification.

POINTS TO CONSIDER WHEN CHOOSING A CAMERA

The criteria for reviewing the performance of a CCD camera are included in the next exercise. If you are planning to purchase a camera, you will need to examine additional issues regarding software, repair, convenience, and cost. If possible, attend an industry-sponsored camera demonstration, where a knowledgeable company representative is present; alternatively, examine camera specifications at a Web site or from product literature, or visit a neighboring laboratory that has a system up and running. It is important to gain firsthand experience by making trial exposures on the camera and to ask questions as you proceed through the steps of image acquisition, image processing, and printing. If you are preparing to buy a CCD camera, you should examine the excellent review by Spring (2000) on CCD camera performance.

Software

- Is the image acquisition software convenient and easy to use?
- Is basic image processing available (γ scaling, brightness, contrast, sharpening adjustments) and easy to use?
- Is basic image analysis available (segmentation/data extraction, measurements of shapes and geometric parameters)? Determine if you will need these operations and if are they easy to use.
- Does the software allow you to write customized scripts (computer macros) that might be needed to speed up and make more convenient special image acquisition and processing operations? Anticipate that it will take some time to master scripting, and bear in mind that the ease of scripting varies considerably among different programs.
- Can your images be saved in TIFF, PICT, JPEG, GIF, and other universal file formats for import into other programs such as Photoshop (Adobe) or Canvas (Claris) for further processing and labeling?
- Since camera operation is controlled entirely through the software, remember that there are alternative programs that will run the same camera. A camera fitted with difficult or complex imaging software is unlikely to get used. Decisions regarding the software and type of computer can be equally as important as selecting a camera. Be sure to investigate at least two software programs.

Camera Repair and Service

- Does the camera company have an established record and a good reputation for producing competitive, reliable cameras? Because of the large investment, camera reliability is important!
- What do other colleagues and friends have to say about the company?
- Does the company provide workshops or special training for customers?
- What is the turnaround time for repair? One week is considered good. By picking a responsive company, you avoid the risk of holding up a research program because of a broken camera. Remember that all cameras break down. Anticipate breakdown once every 1–2 years for a heavily used, high-end camera.

- Are loaner cameras available in times of emergency? The company should offer to help out.
- Are service contracts available? A 2-year contract at one-tenth the cost of the camera is a reasonable investment for heavy users.

Cost

- Check with microscope vendors and independent agents for the best competitive pricing.
- Ask about trade-in and upgrade policies.
- Consider cameras in the price range that is in accordance with the amount of anticipated use. Avoid the temptation to overbuy, knowing that sophisticated generally means more difficult, and that difficult means the system probably will not get used.

Convenience and Ease of Use

- Was your trial experience exciting and convenient, the camera's operation straightforward, and the potential use obvious and compelling? If not, this is telling you to get something simpler.

Exercise: Evaluating the Performance of a CCD Camera

The purpose of this exercise is to prepare you for selecting a digital CCD camera for use in your laboratory. Prepare a summary report addressing the questions, and be sure to indicate the name of the camera and its manufacturer. Conclude the report by indicating the strengths and weaknesses of the camera with respect to the intended application, and indicate if you would recommend this system for purchase by your own lab.

Considering Needs and Requirements
- Indicate the principal modes of light microscopy now in use in your laboratory.
- Make a list of the key resolution parameters: spatial resolution, time resolution, dynamic range, and S/N, and indicate the relative importance of each for your application. Keep these in mind while choosing a camera.

Spatial Resolution
- What CCD is mounted in the camera? What are the pixel dimensions?
- Indicate the magnification of the objective you typically use, and then calculate the radius (μm) of the diffraction disk in the image plane produced by this lens at 550 nm. Do the pixel dimensions on the CCD meet the Nyquist limit for preserving spatial resolution? If not, how would you solve this problem?
- Indicate the binning modes available (2×2, 3×3, 4×4, etc.). Remember that binning reduces spatial resolution, but reduces the exposure time and helps protect live cell preparations.

- Indicate the width and height of a full-frame image (in pixels). Calculate the image size of a print made using a 1200 dpi printer.

Camera Sensitivity and Noise

- What is the range of wavelength sensitivity? Does the CCD contain an enhanced coating for UV or IR wavelengths? What is the QE at 550 nm?
- What is the saturation value (full well capacity) of a pixel in electrons?
- What is the read noise per pixel (in electrons) of the camera? Determine the mean value of a bias frame to establish the contribution from bias plus read noise.
- Calculate the percent contribution of the bias signal in an image signal that reaches saturation.
- What is the operating temperature of the CCD and what is the contribution of thermal noise? For quantitative applications, thermal noise (and read noise) should be minimal. The noise levels are less important when acquiring images to prepare prints, especially if specimens are bright.

Dynamic Range

- Calculate the dynamic range (maximum number of gray levels) from the full well saturation level and the camera read noise (both values must be in units of electrons).
- Note the bit depth of the digitizer (8, 10, 12, 14 bits). This value is sometimes assigned to the camera as a whole.
- Is the bit depth of the digitizer a good match for the camera's dynamic range? The fact that the digitizer may be oversized does not condemn a camera, but you should be aware of (and calculate!) the compensating gain factor applied by the company to allow the camera to use all of the gray-level values. This factor is often overlooked, but it is important when evaluating and comparing the performance of different cameras. Remember that high gain settings (fewer electrons/ADU) give grainier images.

Temporal Resolution

- What is the minimum time required for the camera to acquire and display a full-frame, unbinned image? A one-quarter frame binned 2 × 2? To do this, use a stopwatch and adjust the software for the acquisition of 10 frames in time-lapse mode, with a minimum time interval (1 ms) between the frames.

Image quality (qualitative and quantitative aspects)

- Histogram stretch a feature spanning a limited number of pixel values and examine the quality of the displayed image in terms of noise and graininess.
- After histogram adjustment, print the image and examine its quality.
- Prepare a plot showing the ADUs of a row of pixels through an object. Examine the pixel fluctuations across a uniform feature and confirm that they are minimal.

- Determine the S/N ratio (see the text and exercise in the following chapter). This is the most important test that can be performed to judge image quality. It is imperative to perform this test for a camera that is to be used for sensitive quantitative imaging.

Image acquisition/processing/analysis software

Because the convenience in using a camera is largely determined by the computer platform and software, it is very important to examine the acquisition-processing-analysis software. The software used to demonstrate a camera may not be the most convenient. Whatever the selection, the software must have an extension called a camera driver, which is required to operate a particular camera.

- Is the image acquisition menu convenient and easy to use, and does it control the basic parameters of exposure time, binning, and gain?
- Is basic image processing available (histogram adjustment, gamma scaling, contrast, sharpening adjustments) and are these easy to use?
- Is basic image analysis available (segmentation, data extraction, particle counts, measurements of shapes and geometric parameters) and are these easy to use?
- Does the software allow you to write customized scripts that might be needed for special automated acquisition-processing-analysis operations?
- Can images be saved in TIFF, PICT, and other universal file formats so images can be imported into other programs like Photoshop for further processing and printing?

DIGITAL IMAGE PROCESSING

OVERVIEW

Taking pictures with a video or digital camera requires training and experience, but this is just the first step in obtaining a properly acquired and corrected image. Image processing is used for two purposes: (1) It is required to establish photometric accuracy so that pixel values in the image display the true values of light intensity. (2) It is also an essential tool for optimizing an image for scientific publication. In preparing an image for display or printing, it is essential that the image represent the specimen as objectively as possible, and with few exceptions, this means including all of the information (intensity values) contained in the object image. Because it is a matter of utmost importance, we discuss basic image processing operations in this chapter and guidelines for preparing images for scientific publication in Chapter 16. You should distinguish between the following terms: *image processing,* which refers to the digital manipulation of pixel values in the image, and *image analysis,* which encompasses counting and measuring operations performed on objects in the image. Many commercial software programs for camera control and image acquisition perform these functions. Image analysis, including measurements of morphological features, the use of grids in stereology, image segmentation, and other important topics, are not covered in this book. Interested readers should refer to excellent texts by Russ (1998) and Russ and Dehoff (2000). In this chapter we review four operations that are essential for understanding and using image processing, namely:

1. Adjusting the image histogram to regulate brightness and contrast
2. Performing flat-field correction to establish photometric accuracy
3. Applying spatial filters for image enhancement
4. Using S/N ratio calculations to determine confidence limits of observations

In processing images, proceed through the following steps, performing them only as necessary:

1. Save and duplicate the raw image file. Use the duplicate for processing; the original is always available in case you make a mistake.

2. Flat-field-correct the image to restore photometric accuracy of pixel values. As a bonus, this procedure removes artifacts from microscope optics, the illuminator, and the camera.

3. Adjust brightness and contrast with histogram stretching. This is important for displaying all of the meaningful content of the image.

4. Adjust the gamma (γ) to allow exponential scaling for displaying bright and dim features in the image.

5. Apply the median filter or a blurring filter to reduce speckles and noise.

6. Apply unsharp mask or other sharpening filters to increase the visibility of details.

PRELIMINARIES: IMAGE DISPLAY AND DATA TYPES

Software programs display images on the computer monitor with the help of a *look-up-table (LUT)*, a conversion function that changes pixel input values into gray-level output values ranging from 0 to 255, the 8 bit display range of a typical monitor. Manipulation of the LUT changes the brightness and contrast of the image and is required for the display and printing of most microscope images. For some programs like IPLab (Scanalytics, Inc., Fairfax, Virginia) adjusting the LUT only changes how the image is displayed; it does not alter the values of pixels in the image file. When an adjusted image is saved, the LUT settings are appended to the image file, but the pixel values remain unchanged.

Some software programs show the LUT function superimposed on or next to the *image histogram,* a display showing the number of all of the individual pixel values comprising the image. This presentation is helpful in determining optimal LUT settings. The default settings of the LUT usually assign the lowest pixel value an output value of 0 (black) and the highest pixel value an output value of 255 (white), with intermediate values receiving output values corresponding to shades of gray. This method of display guarantees that all acquired images will be displayed with visible gray values on the monitor, but the operation can be deceiving, because images taken at different exposure times can look nearly the same, even though their pixel values may be different. However, as will be pointed out later, images acquired with longer exposure times have improved quality (better signal-to-noise ratio). The default LUT function is usually linear, but you can also select exponential, logarithmic, black-white inverted, and other display functions.

Data types used for processing are organized in units of bytes (8 bits/byte), and are utilized according to the number of gray levels in an image and the number of gray levels used in image processing. The names, size, and purpose of some data types commonly encountered in microscope imaging are as follows:

- Byte files contain 8 bits (1 computer byte), giving 2^8 or 256 gray-level steps per pixel. Although byte format is used for display and printing, large format data types do not have to be saved in byte format in order to be printed. Conversion to byte format should only be performed on duplicated image files so that high-resolution intensity values in the original image are not lost.

- Short integer and unsigned 16 data types contain 16 bits/pixel or 65,536 steps. Images from 10–12 bit CCD cameras are accommodated well by these data types.
- Long integer format contains 32 bits (4.3×10^9 steps) and is used in some programs to preserve internal precision during processing.
- Floating point format with 32 bits/pixel is used to preserve a high level of arithmetic accuracy during image processing. This data type uses scientific notation, with 23 bits assigned to the mantissa, 8 bits to the exponent, and 1 bit for plus or minus, giving steps that range from 20×10^{38} to 20×10^{-38}.

Thus, an image captured by a 12 bit camera is typically contained in a 2 byte (16 bit) data type. This size is convenient because the extra bits allow for greater numeric accuracy during image processing. Be aware that changing the data type to a lower format such as byte (256 gray levels) permanently reduces the resolution of gray levels and reduces the flexibility for further processing of the image. *Once a data type is converted to a lower level, the original pixel values are permanently altered and cannot be recovered.* However, in some image processing software programs, conversions to a lower-size format such as byte might be required for certain processing operations. Most programs allow you to save image files in a variety of different data types. Finally, the data types used for processing should not be confused with common file formats such as TIFF, PICT, PIC, JPEG, GIF, and so on, commonly used for storage of files in a compressed format, import of files into other programs, or transmission to other computers.

HISTOGRAM ADJUSTMENT

Nearly all images require adjustments to the LUT, or alternatively, to the image histogram to optimize brightness, contrast, and image visibility. Image-processing programs vary in how this is performed. In some programs, such as Photoshop (Adobe), adjusting the histogram (the *Levels* command) changes the values of pixels in the image file. In Photoshop original gray-level values are lost once a file is saved and cannot be recovered. When a histogram-adjusted file is reopened, a new histogram appears, reflecting the changes made in the previous *Levels* command, not the data in the original image file. In other programs such as IPLab (Scanalytics, Inc.), adjusting the LUT (the *Normalization* command) changes the appearance of the displayed image, but does not alter pixel values in the image file. When a LUT-adjusted file is reopened, all of the original pixel values are displayed in a window showing the original histogram overlaid with the existing LUT function curve. Therefore, it is important to maintain a backup of the original image file and to understand how individual programs operate before applying brightness/contrast adjustments.

The image histogram is an invaluable tool that allows for examination of the range and relative distribution of pixel values at a glance and for adjustment of the output display of the image on the monitor. An *image histogram* is a plot showing input pixel values on the x-axis vs. the number or relative number (frequency) of pixels for any given value of x on the y-axis (Fig. 15-1). The numeric range of input values on the x-axis usually corresponds to the bit depth of the acquired image (0–255 for 8 bit images, 0–4095 for 12 bit images, etc.). For didactic reasons, Figure 15-1 and the following figures show the image histogram overlaid with the LUT function. As an image-processing tool, the interactive LUT or interactive histogram allows you to increase or decrease the white-set

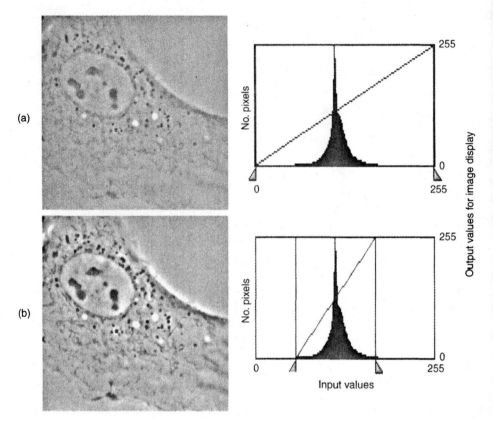

Figure 15-1

Stretching the image histogram. The y-axis (left) shows the number of pixels of value x in the histogram; the y-axis (right) shows the output values given by the LUT. Brightness and contrast are adjusted by moving the black set and white set handles at the bottom edges of the histogram. (a) Histogram display before adjustment; the darkest and brightest pixel values are normalized to span all 256 gray levels ranging from black (0) to white (255) on the monitor. (b) Histogram and image display after adjustment. The white set has been reduced to a value less than 255 and the black set increased to a value greater than 0. The pixels included between the two set points are shown in shades of gray, a procedure known as histogram stretching. All pixels with values greater than the white set point are shown as white, while all pixels with values lower than the black set point are shown as black.

and black-set values or change the value corresponding to midlevel gray. The handles at the ends of the function line are moved to define the white and black set points. Moving the handles to new locations on the histogram changes the brightness and contrast of the image (Fig. 15-1). To increase contrast, the handles are moved closer together. All pixels with values greater than the white set point are displayed as white, and all values lower than the black set point are displayed as black. All intermediate values are assigned a gray-level value according to their distribution along the LUT display function. Moving the handles closer together to increase image contrast is called *histogram stretching,* because the act of selecting new black and white set points for a middle range of gray values effectively stretches the desired portion, from black to white. In effect, the user has now defined a new LUT that the computer uses for displaying the image. To brighten or darken an image, the white and dark set handles are moved, respectively, to the left or to the right.

Before making any adjustments, the user should first get to know the image by determining the pixel values and locations on the image histogram corresponding to various regions in the object, including shadows (or dark background in the case of fluorescence images), bright features (called highlights), and other features of intermediate intensity. This is done by moving the cursor over different regions of the image and reading the corresponding numeric values from a statistics or status window. Knowledge of the intensities of different objects and their position in the histogram will allow the user to act more judiciously and conservatively in selecting the black and white set points to be used in histogram stretching. This is also a good time to check the image for indications of clipping, the appearance of saturated white or underexposed black areas in the image. Clipping should be avoided, both during acquisition and during processing. With this knowledge in mind, the operator then sets the handles on the histogram. An objective and conservative approach is to include all of the pixel values representing the specimen within the upper and lower set points. It is useful to remember that human vision cannot distinguish all of the 256 gray-scale values present in an 8 bit image, let alone a deeper one. Therefore, as the situation demands, you are justified in altering the display by histogram stretching to see important internal details. Guidelines on the use of histogram stretching are presented in Chapter 16.

There are several ways of displaying the image histogram. The standard or regular histogram has already been described and is shown in Figure 15-2a. A *logarithmic histogram* shows the input pixel value (x-axis) vs. the number of pixels having that value on a log scale on the y-axis. This type of display is useful when it is necessary to see and consider pixel values that comprise just a minority of the image and exhibit a strong response to histogram stretching (Fig. 15-2b). This should not be confused with a logarithmic LUT, which uses logarithmic scaling to display the image itself.

Another useful histogram display is the *integrated* or *cumulative histogram,* which can be convenient for adjusting the contrast and brightness of certain images (Fig. 15-2c). This histogram display is more useful for phase contrast, DIC, and bright-field images that tend to have light backgrounds than it is for dark fluorescence images, which typically include a wide range of intensities. In this histogram, the x-axis shows the input pixel values, while the y-axis shows the cumulative number of all pixels having a value of x and lower on the x-axis. For relatively bright images (DIC and phase contrast microscopy), the max-min handles (white set/black set handles) are usually moved so that they define a line that is tangent to the rapidly rising slope of the histogram.

ADJUSTING GAMMA (γ) TO CREATE EXPONENTIAL LUTs

As mentioned, LUT display functions can be linear, logarithmic, exponential, or, in the case of Adobe Photoshop software, even arbitrarily curvilinear as set by the operator. Linear functions (in byte format) have the form

$$\text{Displayed value} = 255 \ (\text{Data value} - \text{Min})/(\text{Max} - \text{Min}),$$

and are commonly used for adjusting the image display. However, linear functions present difficulties when the goal is to include all of the pixel values contained in the image.

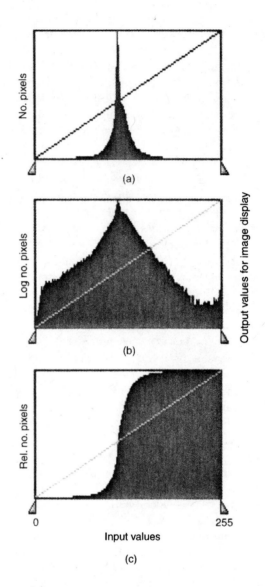

Figure 15-2

Three kinds of histogram displays. (a) The regular histogram is shown as a plot of input pixel value (x-axis) vs. the number of pixels of value x in the histogram. The LUT function is shown here as a line superimposed on top of the histogram. (b) Semi-log plot shows the input pixel value (x-axis) vs. the number of pixels having an input value of x on a log scale (y-axis). (c) Cumulative histogram showing input pixel value on the x-axis vs. the cumulative number of pixels having an input value of x or lower on the y-axis.

In some cases, such as fluorescence images, dim fluorescent features appear dark gray and seem to be lost. No linear LUT setting seems to show the full range of data satisfactorily or in the way we see the image when looking in the microscope. Partly this is because a CCD camera is a linear detector of light intensities, whereas the eye gives a logarithmic response, allowing us to see bright objects and dim, low-amplitude objects in the same visual scene. Exponential functions more closely match the nonlinear response of the eye or camera film and are created using a function with variable exponent γ such that

$$\text{Displayed value} = [(\text{Normalized value} - \text{Min})/(\text{Max} - \text{Min})]^{\gamma}.$$

The adjustment of γ and a figure showing LUTs with various γ values were presented in Figure 13-7. With γ settings <1, low pixel values are boosted relative to high values and appear a medium gray in the image; this adjustment also reduces the contrast between bright features and the darker background. A γ setting of 0.7 approximates the response of the eye, allowing the image to more closely resemble the view we perceive when looking in the microscope. Conversely, γ values >1 depress dark and medium gray pixel values and increase the visibility and contrast of bright features. The effect of adjusting γ on the image display is shown in Figure 15-3. In Photoshop, the slider used to define the midpoint of gray values performs an identical function. Photoshop gives you the choice of reading off a γ value corresponding to the slider's location or entering a numerical value of your own choosing for γ.

FLAT-FIELD CORRECTION

A single image prior to processing and adjustments is called a *raw image*. In many cases, raw images are suitable for printing directly, perhaps after first making some minor adjustments to brightness and contrast using the image histogram. However, a raw image contains two significant kinds of noises that make it unsuitable for quantitative purposes: First, it contains the bias signal and noise counts that increase pixel values beyond their true photometric values, which can cause significant errors in measuring the amplitudes of object features. Second, a raw image may contain numerous artifacts from the camera and microscope optical system (distortions due to detector irregularities, dust and scratches on lens surfaces, uneven illumination) that appear as dark shadows and specks in the image and alter the true pixel values. These artifacts are particularly visible in images with bright, uniform backgrounds (phase contrast, DIC) and in fluorescence images with medium gray or bright backgrounds. In order to restore photometric accuracy and remove the defects, the raw image must be adjusted by an operation known as *flat-field correction* (Gilliland, 1992; Howell, 1992). Only corrected images are suitable for measuring light amplitudes. Although flat-field correction is not required to display or print an image, it is very effective in removing disfiguring faults, so the procedure is widely used for cosmetic purposes as well. As shown in Figure 15-4 and described here, the correction procedure requires three frames:

- A *raw frame* contains the image signal plus signals from optical defects and electronic and thermal noises.
- A *flat-field frame* is the image of a smooth, featureless field without the object, and, if possible, is made at the same focus as the raw frame. Flat frames should be bright and utilize the full dynamic range of the camera in order to minimize the noise in the final corrected image. If both the raw and flat-field frames are dim and noisy, the final processed frame will be very noisy indeed. This situation is particularly relevant to fluorescence imaging. Therefore, the exposure time for a flat-field frame might be longer than that used for producing the raw image itself. To reduce noise

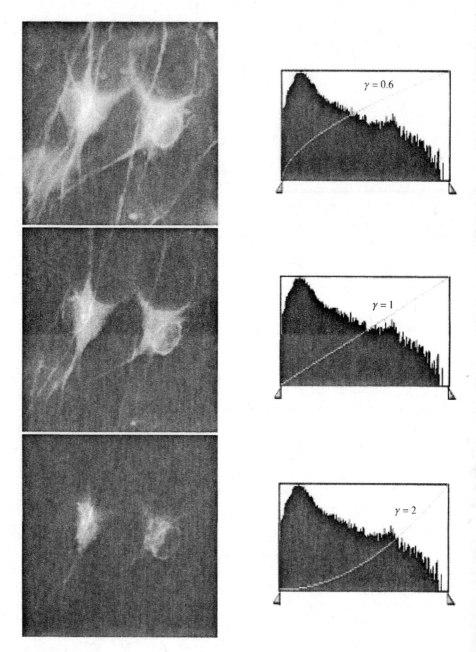

Figure 15-3

Gamma (γ) correction for differential enhancement of light and dark features. Values of $\gamma < 1$ differentially enhance dark pixels in the image, causing them to look brighter, while reducing the contrast between them and bright objects in the image; γ values > 1 differentially suppress the dark features, accentuate bright features in the image, and increase their contrast. Contrast adjustments with γ allow you to use the entire range of pixels representing the object; this is distinct from contrast adjustment by histogram stretching.

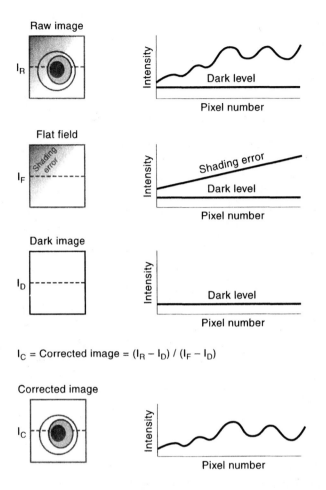

Figure 15-4

Sketch showing strategy for performing a flat-field correction of a raw image. The pairs of figures show the image and an intensity plot across its diameter. The raw image shows a central object with shading error across the background; the intensity profile shows the irregular profile of the object, an overall slope due to shading error, and an overall boost in the signal due to the dark level. The flat-field image is from a featureless background region in the specimen. The dark image, a uniform, blank frame, contains the bias and read noise of the camera. The correction equation: The dark image is subtracted from both the raw and flat images before dividing the corrected raw image by the corrected flat image. The final corrected image is shown with uneven illumination and dark-level contribution removed. (Sketch from Roper Scientific, Inc., with permission)

in this frame, a master flat-field frame should be prepared based on the average of 9–16 replicate images. In fluorescence microscopy, the flat-field frame can be the image of a uniform field of a fluorescent dye. This can be prepared by streaking a drop of fluorescein-conjugated protein or dextran across the surface of a coverslip, allowing it to dry, and then preparing a permanent reference slide using a drop of ProLong or SlowFade (Molecular Probes, Inc., Eugene, Oregon) or other antifade reagent. Although prepared for a certain dye such as fluorescein, the flat-field frame works well for a number of different filter sets and is useful for days until such time that the light source has been refocused or changed, after which a new master flat-field frame should be prepared.

- A *dark frame* contains the bias offset signal plus the electronic and thermal noise components that are present in the image. Bias counts are from the positive voltage applied to the CCD as required for proper digitization; electronic noises include the components contributing to the readout noise of the camera; thermal noise is due to the kinetic vibration of silicon atoms in the chip. At full and one-quarter saturation of a scientific-grade CCD camera, the bias count and noises may contribute roughly 5% and 20% of the apparent pixel amplitudes, respectively, and must be subtracted from the flat-field frame and the raw frame to restore photometrically accurate pixel values. Dark frames are prepared using the same exposure time as the raw image, but without opening the shutter. Since dark frames have low signals, a master dark frame should be prepared by averaging 9–16 dark frames together.

The equation for flat-field correction is

$$\text{Corr. Image} = M \frac{(\text{Raw} - \text{Dark})}{(\text{Flat} - \text{Dark})}$$

where M is the mean pixel value in the raw image. The multiplication by M simply keeps the intensity of the corrected image similar to that of the original raw image. In applying the correction, the order in which the operations are performed is important. The dark frame must be subtracted from the raw and flat frames first, followed by the division of the dark-subtracted raw frame by the dark-subtracted flat frame. The correction is performed in floating point data type to preserve numeric accuracy. To be seen optimally, the brightness and contrast of the corrected image might need to be adjusted. The visual effect of flat-field correction looks similar to that obtained by background subtraction, but performing the correction by division is more accurate. This is because light amplitudes in an image result from a multiplicative process (luminous flux × exposure time). Once corrected, the relative amplitudes of objects in the image will be photometrically accurate. Surprisingly, the corrected image lacks the optical defects that were present in the raw image. An example of a flat-field corrected image is shown in Figure 15-5. Practice in performing this correction is included as an exercise at the end of this chapter.

IMAGE PROCESSING WITH FILTERS

Filtering is used to sharpen or blur an image by *convolution,* an operation that uses the weighted intensity of neighboring pixels in the original image to compute new pixel values in a new filtered image. A matrix or kernel of numbers (the *convolution matrix*) is multiplied against each pixel covered by the kernel, the products are summed, and the resulting pixel value is placed in a new image. Only original pixel values are used to compute the new pixel values in the processed image. The kernel or mask can have different sizes and cover a variable number of pixels such as 3 × 3, 4 × 4, and so forth. Note that the sum of the numbers in the kernel always adds up to 1. As a result, the magnitude of the new computed pixel value is similar to the group of pixels covered by the kernel in the original image. After a pixel value has been computed, the kernel moves to the next pixel in the original image, and the process is repeated until all of the pixels

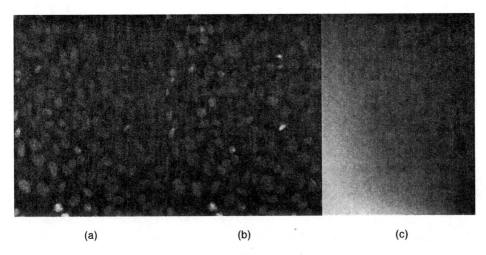

(a)   (b)   (c)

Figure 15-5
Examples of raw and flat-field-corrected CCD images. DAPI-stained nuclei of tissue culture cells. (a) Raw image. (b) Corrected image. (c) Flat-field frame. Notice that optical faults from uneven illumination are removed.

have been read, after which the new image is displayed on the screen (Fig. 15-6). For a megapixel image the process can take several seconds. Linear filters are used for smoothing, sharpening, and edge enhancement. The filters menu in the software usually provides several kernel choices for any given operation. These filters are sensitive, so there can be some difficulty in controlling the amount of filtering, with some filters giving overfiltration, and others giving a minimal effect. Therefore, most image-processing programs also allow you to create your own convolution matrix, in case the menu selections are inappropriate. With practice, you will learn to select a particular filter based on the intensity gradients and size dimensions of the detail needing adjustment in the image.

Low-Pass Filter for Blurring

This filter removes high-spatial-frequency details such as noisy pixels and sharply defined intensity transitions at the edges of objects in the image (Fig. 15-7). It blurs by partially leveling the values of pixels in a small pixel neighborhood. A low-pass filter has the effect of passing or minimally altering low-spatial-frequency components— hence its designation as a low-pass filter—and can make cosmetic improvements to grainy, low-S/N images, but at the expense of reduced resolution.

High-Pass Filter for Sharpening

High-pass filtering differentially emphasizes fine details in an image and is an effective way to sharpen soft, low-contrast features in an image. The effect of sharpening and the pattern of a strong sharpening kernel are shown in Figure 15-7. Unfortunately, this filter also emphasizes noise and can make an image look grainy.

Figure 15-6

The operation of a convolution matrix in a sharpening filter. The matrix in this example has numeric values covering a 3 × 3 pixel area (a). A kernel or mask containing these values is multiplied against the values of 9 pixels covered by the mask in the original image (b), a process called convolution. The resulting 9 products (c) are summed to obtained a value that is assigned to the central pixel location in the new filtered image (d). The kernel then moves by 1 pixel to calculate the next pixel value, and the process repeats until all of the pixels in the original image have been recalculated. Notice the values in the convolution matrix used for sharpening. The central value is emphasized (matrix value 9) relative to the 8 surrounding values. The sum of the values in the matrix is 1.

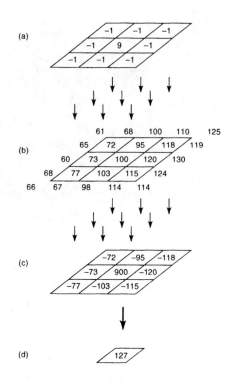

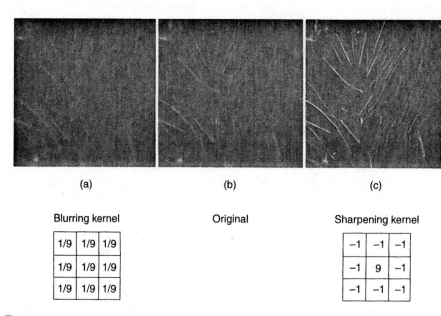

(a)   (b)   (c)

Blurring kernel			Original	Sharpening kernel		

1/9	1/9	1/9
1/9	1/9	1/9
1/9	1/9	1/9

-1	-1	-1
-1	9	-1
-1	-1	-1

Figure 15-7

Convolution filters for blurring and sharpening. The figure shows the effects of applying blurring and sharpening filters and the corresponding convolution kernels that were used. (a) Blurred. (b) Original. (c) Sharpened.

Median Filter

This filter reduces noise and evens out pixel values with respect to the values of neighboring pixels (Fig. 15-8). It is very effective at removing noise, faulty pixels, and fine scratches. A median filter can be more effective than low-pass filters in reducing noise. Like all linear filters, a median filter uses a kernel or cluster of pixels (the dimensions are determined by the operator) that moves in linear fashion, pixel by pixel and row by row, across all of the pixels in the image. For this filter, there is no convolution matrix as such. At each successive pixel location, the original pixel values covered by the kernel are rank ordered according to magnitude, and the median value is then determined and assigned to the central pixel location in a new filtered image. In the following example using a kernel size of 3 × 3, the central noisy pixel with a value of 20 in the original image becomes 7 in the new filtered image:

Pixel values covered by 3 × 3 kernel in original	Rank order of pixels	Median pixel value assigned to central pixel in new image
6 6 4		
5 20 7	4,5,6,6,7,7,8,9,20	7
7 8 9		

Histogram Equalization

Most image-processing programs have an equalize contrast function that reassigns pixel values so that each gray level is represented by the same number of pixels. The process,

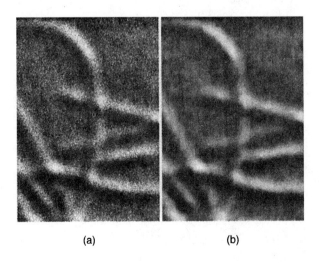

(a)   (b)

Figure 15-8

Median filter for removing noise. Highly magnified CCD image of a field of microtubules in DIC microscopy. (a) The original image was acquired at the full 12 bit dynamic range of the camera, but looks grainy after histogram stretching due to photon noise. (b) The same image after applying a median filter with a 3 × 3 pixel kernel shows that much of the graininess has been removed. The S/N ratio of the original could have been improved by averaging a number of like frames at the time of acquisition.

called histogram equalization or histogram leveling, yields a histogram that contains no peaks and has a flat, horizontal profile. In leveling, pixel values are reassigned so that each gray level is given the same number of pixels, while the rank order of the pixel values in the original picture is preserved as much as possible. This operation is used to enhance contrast in very-low-contrast (flat) images where most of the pixels have close to the same value, and where conventional methods of histogram stretching are ineffective. Equalization is an extreme measure to rescue images with low-amplitude gradients, but works well as a way to examine bias, dark, and flat-field frames, which can look nearly featureless. The effect of histogram equalization is often dramatic.

Unsharp Masking

This image-processing procedure does an excellent job of enhancing fine details in an image. Unsharp masking is well known to photographers and astronomers who used the method as a darkroom technique to enhance faint details in photographic prints. As designed by photographers, a blurred, reverse-contrast negative (or unsharp mask) is made of the original negative. The two negatives are sandwiched together in perfect registration in the enlarger and a print is made. To perform this operation on a computer, an unsharp mask is produced by blurring and reducing the amplitude of the original image; the unsharp mask is then subtracted from the original to produce a sharpened image. Extended, uniform regions are rendered a medium gray, whereas regions with sharp gradients in amplitude appear as brighter or darker intensities. The principle of unsharp masking is shown in Figure 15-9, and an example of a microscope image processed by this method is shown in Figure 15-10. If the program you are using does not perform unsharp masking, you can perform this operation manually using the following steps:

- Prepare a copy of the original and blur it with a conservative blurring filter. Images with fine details (high spatial frequencies) require more conservative blurring than images containing big blocky objects (low spatial frequencies).
- Subtract 50–95% of the amplitude of the blurred image from 100% of the original using an image math function in the program. The higher the percentage that is subtracted, the greater the sharpening effect.
- Using histogram stretching, adjust the brightness and contrast in the difference image.

Fast Fourier Transform (FFT)

This filtering operation selectively diminishes or enhances low or high spatial frequencies (extended vs. fine detailed structures) in the object image. This is a valuable operation to consider, because it reinforces concepts given in previous chapters on the location of high- and low-spatial-frequency information in the diffraction plane, which is located in the objective back aperture in the microscope. The effect is similar to that of the blurring and sharpening filters already described, but can be made to be much more specific due to an operation called spatial frequency filtering (Fig. 15-11). When an image is transformed into the so-called frequency domain through an FFT command, the information is represented in two plots (images): one containing a distribution of

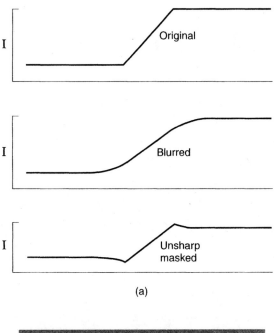

(a)

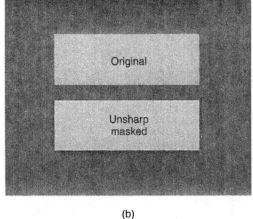

(b)

Figure 15-9

The strategy of unsharp masking for image enhancement. (a) Three intensity plots show a boundary between two regions of different light intensity as depicted by the left-hand edge of the light colored rectangles shown in drawing (b). The plots shown in (a) indicate: the original image profile, a blurred version of the original, and an unsharp mask-corrected version prepared by subtracting 50% of the blurred profile from the original profile. Notice that the small negative and positive spikes in the unsharp mask plot are seen as bright and dark edges in the bottom drawing. The amount of sharpening depends on the amount of blurring, the amount of amplitude reduction created by subtraction of the blurred image, and the amount of subsequent histogram stretching.

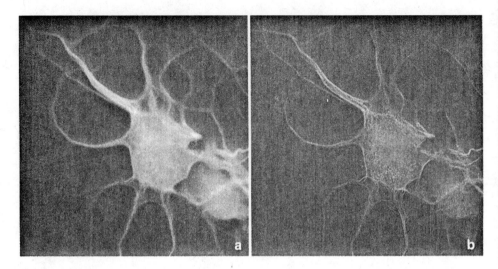

Figure 15-10

Unsharp masking emphasizes object details. The figure shows the effect of unsharp masking on an immunofluorescence image of cultured neurons labeled with a neuron-specific antibody. (a) Bright fluorescence in the cell bodies in the original image hides vesicles and filament bundles and makes it difficult to visualize small neurites in the same image. (b) Unsharp masking enhances details and differentially emphasizes faint structures.

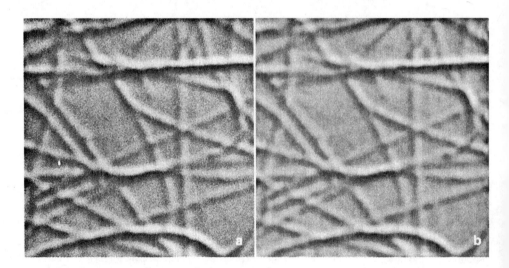

Figure 15-11

Fast Fourier Transform for processing in the spatial frequency domain. (a) The original image of microtubule filaments in DIC optics is grainy. (b) The same image after FFT processing using a mask to eliminate high spatial frequency information. The procedure reduces the graininess but does not alter the coarse structure of the filaments.

magnitudes (amplitudes) and the other of phases. The relation of the image to its Fourier transform is analogous to the relation of images in the image plane and diffraction plane (back aperture of the objective lens) of the microscope. To review these concepts, refer to Chapter 5. The FFT operation is performed in a high-resolution data type such as floating point. Usually only the amplitude (or magnitude) plot is used for display, manipulation and filtering. Most programs display the amplitude or power spectrum with a log rather than a linear scale, since the transform is otherwise too dark to see anything. Depending on the program, histogram stretching and γ scaling can be applied to improve visibility of the amplitude plot without affecting the data. In a Fourier transform:

- Frequency information is represented at different distances from the central point in the magnitude image such that information from large structural features (low spatial frequencies) is found near the center of the image, and information from small features (high spatial frequencies) is located at some distance away from the center.

- Amplitudes seen along one axis in the magnitude plot are represented in the image on an axis shifted by 90°.

- The amplitude at each location in the magnitude plot is proportional to the amount of information at that frequency and orientation in the image.

On the computer, perform an FFT transformation and obtain a pair of amplitude and phase plots. Next, apply an occluding spatial frequency mask over one of the plots (the amplitude plot) to select low, midrange, or high spatial frequencies to block out unwanted frequencies. The masked plots are then used to produce an *inverse transform:* an image that is re-created based on frequencies not occluded by the mask. Both the magnitude and phase plots are required for this operation. Processing images in the frequency domain is useful for:

- Removing noise that occurs at specific frequencies (electrical interference, raster scan lines)

- Enhancing or removing periodic structural features of the object

- Identifying spatial frequencies of defined structures in an image

- Determining the periodicity and/or orientation of indistinct features that are difficult to see in the object image

- Detecting optical aberrations such as astigmatism

- Applying convolution kernels to the magnitude plot for sharpening or blurring

SIGNAL-TO-NOISE RATIO

S/N is the ratio of a signal to the noise of the surrounding background from which the signal must be distinguished. It is the accepted parameter for describing image quality. As we shall see, S/N also has statistical meaning because it describes the confidence level (α-value) at which an object of a certain intensity can be distinguished from the background. In qualitative terms, S/N values are used to describe the visibility of an object in an image. For reference, Figure 15-12 shows a test pattern of gray squares

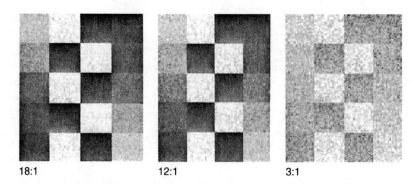

18:1   12:1   3:1

Figure 15-12

Effect of S/N ratio on the visibility of a test pattern. A checkerboard test pattern with squares of varying intensity is shown at different S/N ratios. At S/N ratios below 3, transitions between certain adjacent gray-level steps are difficult to see. Sketch from Roper Scientific, Inc., with permission.

whose visibility is shown to vary depending on S/N. Everyone using a scientific-grade camera for quantitative purposes should understand the meaning of S/N, be able to calculate it, and be familiar with its use. Using S/N theory, you can:

- Provide a quantitative descriptor of image quality. S/N is the accepted parameter for describing the visibility and clarity of an object in an image and for comparing the quality of images among themselves.

- Determine the probability that a faint signal is distinct from the background. This might occur if you were comparing changes in the fluorescence intensity in response to an experimental condition and wanted to determine if a change in fluorescence was statistically significant.

- Calculate the minimum exposure time required to obtain a signal within a certain confidence level. Your goal is to obtain smooth kinetic data to within 10% precision to accurately determine the halftime of fluorescence recovery from an image sequence; however, it is also important to keep the exposure time to a minimum to avoid photo-damage. Cosmetically attractive images require long exposures that can kill a living cell, but are not necessary for quantitative work. S/N can be used to determine the minimum exposure time required to obtain an image with the required criteria.

- Evaluate camera performance, which is important when selecting a camera for purchase.

Excellent descriptions of the application of S/N theory to quantitative image analysis have been presented by several authors, including Gilliland (1992), Howell (1992), Newberry (1991, 1994a,b, 1995a,b,c), and Rybski (1996a,b).

Definition of S/N Ratio

It is easy to determine the S/N of an object in a digital image, but before we do, we will first examine the definition of S/N and its component terms. Formally, the S/N ratio is obtained by dividing the sum of components contributing to the signal by the square root of the sums of the variances of the various noise components. If the noises are independent of each other, they are added in quadrature ($N_1^2 + N_2^2 + N_3^2 \ldots$), and the noise is shown as the square root of this term:

$$ S/N = \frac{S_1 + S_2 + S_3 \ldots}{\sqrt{(N_1^2 + N_2^2 + N_3^2 \ldots)}} . $$

In digital microscope images, the principal signal components are the photoelectrons corresponding to the object and underlying background, and bias and thermal signals; the principal noises are photon noise (see below) and the camera readout noise (for which there are several independent sources). We will recognize this general form of the equation in descriptions of practical solutions. Two additional points should be noted: (1) The statistics of photon counting are based on the number of photons converted to electrons by the detector, and correspondingly, S/N is always calculated in electrons (never in terms of analogue-to-digital units, which are the numeric values assigned to pixels in digital images). The electron equivalent of an ADU is given as ADU $\times$ gain, where gain is the number of electrons per digital unit. (2) S/N statistics can be applied on a per pixel basis for describing equipment performance, or to a group of pixels when describing an extended object in an image.

Photon Noise

Because the number of photons recorded by the camera over a discrete interval of time and space is stochastic, the accumulation of photoelectrons is described by a Poisson distribution. If the photon signal is large, as it is for most microscope images, the principal noise is *photon noise* (also called the *shot noise*), which is described as the standard deviation of the signal amplitude in photoelectrons. For a Poisson distribution, the standard deviation (and therefore the photon noise) is simply the square root of the signal. Thus, for a background signal of 10,000 electrons, the photon noise = $\sqrt{10,000}$ = 100 electrons. The fact that the photon noise is the square root of the signal explains why S/N must be calculated in electrons, not in ADUs. Several additional points are worth noting:

- An image is considered to be *photon-limited* if the photon noise of the object signal is greater than the camera read noise. For a CCD camera with 15 e$^-$ read noise, this occurs when the corrected photon count from the object reaches ~225 e$^-$ (~20 ADU/pixel), since at this level the photon noise is $\sqrt{225}$ or 15 e$^-$, the read noise of the camera. Because microscope images are relatively bright, most images are usually photon limited, and the following discussion is based on this assumption.

- Under photon-limited conditions, the S/N increases in proportion to the square root of the exposure time or the square root of the number of averaged frames. Thus, increasing the exposure time by 4 or averaging 4 like frames increases S/N 2-fold. The poor S/N of a dim image can be greatly improved by simply increasing the exposure time. For this reason, images should be acquired so that the brightest features in the image approach the saturation level of the camera (see Chapter 14). This relationship also explains why the S/N of confocal images can sometimes be dramatically improved by averaging replicate frames.

- For microscope images where the visibility of an extended object is of interest, the S/N ratio can be calculated based on dozens or even hundreds of pixels and not on a per pixel basis. In this case, an object covering a large patch of pixels can have an S/N value that is hundreds of times larger than the S/N of an individual pixel. Formulas for determining the S/N of extended objects are given in this chapter.

The fundamental principles are described in a series of excellent articles by Newberry (1991, 1994a,b, 1995a,b,c) and at the Web sites of CCD camera manufacturers.

S/N and the Precision of Numeric Values

One valuable benefit of S/N analysis is that you can state the confidence limits (in units of standard deviations) that a signal stands out from the noise of the background. This calculation is possible because S/N is the ratio of the object signal to the standard deviation of the background signal, and because standard deviations are directly related to α-values and percent confidence. Consider an image with an object signal of 100 electrons and a noise of $\sqrt{100} = 10$ e$^-$, giving an S/N ratio of 10. The inverse of this value, N/S = 0.1, is called the *relative error*—that is, the fractional component of the signal that is noise. For our example, we would say that noise is 0.1 or 10% of the signal. Since photon noise is the same as the standard deviation (SD) of the photon signal, we can say that fluctuations in the signal greater than 10% can be detected at a confidence level of 68% (1 SD). Extending this further, by multiplying by 2 for 2 standard deviations, we would say that fluctuations in the signal >20% are detected at the 95% confidence level (2 SD), and so on. Another way of interpreting this is to say that given a relative error of 0.1, we can say we detect a signal at 10% precision, or that our ability to measure the signal precisely is no better than 10%. The ability to set confidence limits has profound implications, because it allows the investigator to calculate in advance how many electrons (and therefore how many ADUs and what exposure time) are required to visualize fluctuations of a certain percent at a specified confidence level.

In summary, given S/N = 10, taking (1/S/N) $\times$ 2 SD $\times$ 100 = 20%, we say that fluctuations >20% are detected at the 95% confidence level.

Correlation of S/N with Image Quality

The relationships between S/N, % fluctuations detected at 95% confidence, and image quality are given in Table 15-1. An object image with S/N = 10–20 is poor and grainy-looking, whereas the appearance of the same object at S/N > 100 is good to excellent.

Effect of the Background Signal on S/N

In microscopy, the background signal can be large, sometimes >90% of the total signal representing an object. Figure 11-10 demonstrates the significance of the background signal in an immunofluorescence image. *In most cases, photon noise from the background is the major source of noise, not the read noise of the camera.* Thus, the S/N equation includes a term for the background noise as will be recognized in the following sections. As seen in Table 15-2, a high background signal reduces S/N considerably.

TABLE 15-1   S/N Value and Image Quality*a*

S/N ratio	% Fluctuations at 95% Confidence	Visual Image Quality
3	67	Visual limit of detection
10	20	Coarse and grainy, barely acceptable
20	10	Fine grain very obvious
30	7	
40	5	Okay to good
50	4	
100	2	Excellent
200	1	
400	0.5	
2,000	0.1	
20,000	0.01	

*a*Values for S/N are given on a per pixel basis.

Quick Estimate of the S/N Ratio of a Single Object Pixel in a Digital Image

The following procedure gives good approximate values of S/N for most images of moderate to bright magnitude where the noise is photon limited. The measurement gives S/N for a single object pixel. You will need to look up the electron gain (photoelec-trons/ADU) in the camera's operating manual, because the calculation must be performed in units of electrons, not ADUs. Since the camera's gain varies depending on MHz operating speed of the camera, be sure to pick the gain that was used at the time of image acquisition. Figure 15-13 demonstrates the concept of this approach.

- Perform flat-field correction to obtain an accurate image. Alternatively, acquire a dark frame, determine the average ADU value of the frame, and use an image math

TABLE 15-2   Effect of Background Signal on S/N*a*

Total Signal e⁻/pixel	Background/Total (%)				
	10%	25%	50%	75%	90%
43	24	20	13	7	3
171	79	65	44	22	11
685	206	171	114	48	29
2675	447	372	249	123	50
10700	921	768	512	256	102
42800	1857	1547	1031	576	206

*a*S/N values are shown for various combinations of signal strength and percent of signal contributed by the background. S/N values were calculated for a 100 pixel object and a CCD with a capacity of 42,800 e⁻/pixel using Newberry's equation.

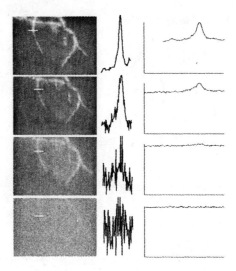

Figure 15-13
Demonstration of the S/N ratio. The CCD image depicts microtubules at the periphery of a cell labeled with fluorescent antibodies to tubulin. The white bar covers a row of pixels, whose values are shown in the intensity profile to the right. On a per pixel basis, S/N is the ratio of the amplitude of the object signal to the standard deviation of the surrounding background. Analogue-to-digital units (ADUs) must be converted to electrons using the appropriate gain factor before calculating the S/N ratio.

function in software to subtract this value from all of the pixels in the raw image (see flat-field correction previously discussed).

- Determine the ADU value of an object pixel. Since the object signal is usually the sum of the object and underlying background counts, the background must be subtracted from the total to obtain the specific object counts (ADUs). The object signal S in electrons is described as

$$S = \text{object-specific ADUs} \times \text{electron gain.}$$

- Use the ROI tool to select a group of several hundred background pixels and use the *Analyze Statistics* function to determine the standard deviation of the background sample mean in ADUs. The photon noise N of the background is given as

$$N = \sqrt{(\text{SD}^2 \times \text{gain}).}$$

- Calculate S/N.

Newberry's Analytical Equation for S/N

The following S/N equation should be used when it is necessary to obtain an accurate determination of S/N. This equation should only be applied to flat-field corrected images. See Newberry (1991) for details. The analysis equation reads:

$$S/N = \frac{\sqrt{C_o}}{\sqrt{[(1/g)+(n\sigma^2/C_o)+(n\sigma^2/pC_o)]}},$$

where

C_o = object counts in ADU where $C_{object} = C_{total} - C_{background}$

$\sqrt{C_o}$ = object noise in ADU

n = number of pixels in measured object area

p = number of pixels in measured background area

σ^2 = variance of background pixels as $(SD)^2$ in ADU

g = gain (electrons/ADU)

The C_o and $\sigma^2_{(background)}$ measurements are taken from selected object and adjacent background ROIs using the *Statistics* or *Measure ROI* command in the analysis software. The units are in ADU with the correction for electrons appearing as gain in the denominator. The background variance is the calculated statistical variance of the pixel sample (the square of the standard deviation) and is obtained from the *Measure ROI* command; it is not the Poissonian mean that was discussed earlier under photon shot noise. For convenience, set up these equations and related tables of variables in a spreadsheet.

Exercise: Flat-Field Correction and Determination of S/N Ratio

In this exercise you are to correct a raw image by flat fielding, extract numeric values from a structure in the image, and determine the signal-to-noise ratio. You will need access to Adobe Photoshop software on a computer. Although this manipulation is somewhat awkward in Photoshop, the exercise will make you think about what you are doing and reinforce the processing operations. Prepare raw, flat and dark images with a CCD camera, convert them to 8 bit data type (byte), and save them in TIFF format on the hard drive. In some cases, a supervisor may choose to do this for you. A flat-field correction is obtained by performing the following steps:

$$\text{Corrected image} = M\,[(R - D)/(F - D)]$$

a. Subtract the dark frame from the raw frame (R − D).

In PS: Image/Calculations: Source 1 = raw; source 2 = dark; blending = difference; opacity = 100%; result = new; channel = new; save the new image.

b. Subtract the dark frame from the flat frame (F − D).

In PS: same as above and save the image.

c. Determine the average pixel value M in the corrected flat frame (F − D).

In PS: Image/Histogram: Record the mean value from data table.

d. Divide the corrected raw image frame by the corrected flat-field frame (R − D)/(F − D).

In PS: You cannot divide images in PS, so take the inverse of the corrected flat frame and multiply it by the corrected raw frame. Use the menus:

Image/Map/Invert, and save the inverted flat corrected image. Then use Image/Calculations to do the multiplication.

e. Multiply each pixel element in the image from step d by M.

In PS: Prepare a new image that contains only the value of M. Make a copy of raw. Then go to Image/Adjust/Levels, and in the bottom scale bar move the upper and lower pointers to the value of M determined in step c. The image will look a uniform gray. Name and save the image. Using Image/Calculations, multiply the divided image from step d by the average image M.

That's it. Now touch up the contrast of the flat-field-corrected image. Select *Image/Adjust* and then *Brightness/Contrast*.

1. Print the raw image and the corrected images together on the same page on an inkjet or dye sublimation printer.

2. Now calculate the S/N of an object in the image.

a. Determine the average pixel value of the object (actually the object plus background together).

In PS: Use the ROI tool to draw a rectangle around the object; use Image/Histogram to obtain the mean value of the ROI. Convert to electrons by multiplying by the gain.

b. Determine the mean and standard deviation of the pixel value in the surrounding background area.

In PS: Use the ROI tool to draw a rectangle around a patch of background area. Then use Image/Histogram to determine the mean and standard deviation of the ROI. Convert to electrons by multiplying by the gain.

c. Subtract the two means to determine the specific value of the signal (S) electrons.

d. Divide the signal S by the noise N, the standard deviation of the background. The result is the S/N ratio calculated per object pixel in the image. Show your work.

3. For this object, fluctuations in the background of what percent can be distinguished at a 95% confidence level? Show your work.

4. Name two methods that could be used to improve the S/N ratio of the object.

IMAGE PROCESSING FOR SCIENTIFIC PUBLICATION

OVERVIEW

Image processing is often essential for preparing digital images for display where you want the image to look good, and publication where you want an accurate record. When these goals conflict, as they frequently do, we might ask if there are image-processing standards to guide us in making key decisions. Unfortunately for the microscope community, there is no recognized set of standards, and the usual course is for individuals to create solutions on their own. While many image-processing operations available in popular software such as NIH Image or Photoshop can be performed in a photographer's darkroom, they are more easily performed digitally on a computer, tempting us to over-process images. Figure 16-1 is processed in ways that give different interpretations regarding the intracellular distribution of a fluorescent protein. Recognizing the need to present the image in a nonarbitrary way, it would be useful to consider if image processing guidelines are feasible and if they would help assure that images remain faithful portrayals of their objects. In the author's opinion, the answer to these questions is yes. However, experts agree that attaining this goal is difficult because guidelines are constraining and inherently subjective. No single set of guidelines would be appropriate for all images, and guidelines are to some extent arbitrary. Nevertheless, as electronic imaging and image processing become indispensable in the laboratory, guidelines would help standardize the way in which we present scientific images. It should be pointed out that the main focus of this chapter deals with the processing of fluorescence microscope images, a category that poses unique problems in the display of intensity, contrast, and the use of color. However, in many places the guidelines are relevant to images obtained by other forms of light microscopy. We begin by examining the significance and history of processing in presenting microscope images.

IMAGE PROCESSING: ONE VARIABLE OUT OF MANY AFFECTING THE APPEARANCE OF THE MICROSCOPE IMAGE

Consider the large number of factors that affect the quality and appearance of a fluorescence microscope image:

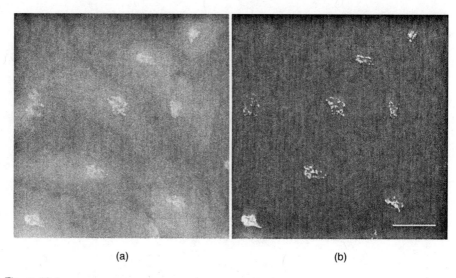

(a)   (b)

Figure 16-1

Two interpretations of an object's structure resulting from selective manipulation of image-processing tools. Bar = 10 μm. (Image courtesy of Carolyn Machamer, Johns Hopkins University.)

- Microscope optics introducing distortions to an image if not properly employed.
- Environmental conditions affecting the morphology of cells in culture and during handling
- Fixatives causing some morphological distortion and differential extraction of components
- Labeling that is uneven or unbalanced, or inappropriate choice of labels
- Different filter sets giving varying views of fluorescent signals
- Photobleaching of fluorescent dyes
- Uneven illumination
- Visual perception not matching the linear response of a CCD camera
- Instrumental distortions (fixed bias pattern noise and electrical interference)
- Variation in parameters for image acquisition (gain, offset, dynamic range)
- Image-processing operations

Any of these factors can affect the appearance of the object and bias our interpretations. So why single out image processing, the last item on the list, as a major concern? In large measure it is because the other sources of image variability share in common the fact that they are defined by physical and chemical parameters. These parameters are generally understood, or if they are variable, are specified by procedures written into the materials and methods section of the research paper. The situation regarding image processing is, at present, much less defined. With time, procedures for processing scientific images will become standardized, and image processing will be less of a concern. For now, however, it deserves attention.

THE NEED FOR IMAGE PROCESSING

Although our goal is to standardize image-processing operations, our experience tells us that image processing is an invaluable agent in scientific discovery. Video-enhanced contrast imaging of microtubules and minute cellular structures, and contrast manipulations to medical X rays and astronomical images are all examples. Here, it is very clear that nothing can be seen without substantial adjustments to the original image. Further, two important technological advances in electronic microscope imaging—video enhancement of image contrast and confocal microscopy—profoundly changed the way scientists regarded microscope images. These technologies are entirely dependent on electronic detectors, and both require extensive electronic adjustments and follow-up computer processing. The images and knowledge gained from these technologies are impressive. Nevertheless, for the typical microscope image our approach to processing and image display usually needs to be more conservative.

We would be remiss if we did not emphasize an obvious point. If processing is done to enhance the visibility of important image features that cannot otherwise be seen to good advantage, then it is appropriate and acceptable. This is no different from any other scientific data. Naturally, if you have only one picture showing some piece of information, then it is not wise to trust it. But if the same feature shows up repeatedly, then you consider it trustworthy even if it is difficult to see. In such situations, image processing might be the only way to display image details and describe an observation to others.

VARYING PROCESSING STANDARDS

Images used as data in scientific publications are distinct from the cover images of scientific journals or the enhanced images used to highlight scientific presentations. These two categories of images are both important, but serve different purposes and are prepared using different standards. Ideally, figures of scientific merit must represent a specimen objectively and completely. The fundamental tenet of scientific publishing—that the reader should be able to repeat observations based on the details given in the article—should be the goal. Images on journal covers serve a different purpose. Here the rules are relaxed, and artistic manipulation of the image may be important to emphasize a particular feature, although the cover legend should acknowledge this point. Experience in image processing suggests it is best to act conservatively. You will have the confidence that you have prepared an objective record based on defined parameters for acquisition and processing.

RECORD KEEPING DURING IMAGE ACQUISITION AND PROCESSING

Good image processing begins with keeping good records.

- *Save raw and processed image files separately.* Before processing an image, make a copy of the original raw image and save the processed image as a separate file. When preparing figures for publication, keep notes about the processing steps applied to each image. Try to standardize your operations so that images are treated similarly. Remember that it will be necessary to include details on image processing in the paper so that readers will know what you have done.

- *Keep record sheets of imaging sessions in the research notebook.* Prepare an entry describing your acquisition parameters for each experiment. For general microscopy, the sheet might include objective magnification, NA and type, and the illuminator and filters used (fluorescence filter set, ND, IR- and UV-blocking filters, and others). For confocal images, notes on the pinhole size, scan rate, black-level and gain settings, frame averaging, and step size in a z-series are important. Some image acquisition software programs save the settings in a note associated with the image file. It also helps to add notes about the amount of protein expression, labeling efficiency and signal strength, balance of fluorochromes, photobleach rate, and other observations pertinent to each exposure. One set of notes is usually adequate for each session involving the same preparation, but if settings change significantly during a session, you should write them down and indicate the image to which they pertain. It is easy to use a microscope casually without taking notes, but the value of the images will be limited, and you are more likely to repeat errors and waste time.

- *Keep folders and directories of all computer image files so that they can be easily reviewed and examined.*

Note: Guidelines for Image Acquisition and Processing

The following guidelines are listed in the order of their occurrence from specimen preparation to processing:

- *Give priority to the quality of the specimen preparation.* If the visual image does not measure up, it is always best to prepare a better specimen in the lab rather than to compensate for a poor image by image processing. For example, if fluorescence labeling is nonspecific or if there is considerable bleed-through (signal in an image from another fluorochrome), processing the image can be difficult and should not be encouraged. This is a particular concern when the goal is to determine if two signals are colocalized.

- *Optimize image acquisition with the microscope/camera system.* For fluorescence microscopy this means proper adjustment of the microscope, illuminator, and filters for the best visible image, and full dynamic range and proper gain and offset settings for the best acquired image. Of all the things affecting image quality, specimen preparation and image acquisition are the most important. All too often the offset and gain settings of the camera, and later, the histogram settings on the computer, are used without being aware of their consequences during subsequent image processing. There is little even an experienced computer operator can do with an undersaturated, low-bit image.

- *Display all of the specimen information in the image histogram.* Before beginning processing, examine the pixel values of objects and background in the image. If you are aware of where the information is first, the danger of losing information by adjusting the histogram will be reduced. Also check that the image does not show evidence of clipping at white and black values, a sign that an image has not been acquired or processed correctly. As a general rule, include all of the image data contained in the original image histogram and display it using the full range of gray-scale values from 0 to 255 using histogram stretching. For example, in a bright-field image where the information might occupy a fraction of the gray-scale range, you should bring in the white and black set sliders to encompass the values contained in the histogram display. If contrast needs to be improved, it is generally acceptable to eliminate signals corresponding to the top and bottom 1% of the data values. Beyond that, you should consider using γ scaling. These operations are described in Chapter 15. If bright saturating features (dead cells, precipitates) appear in the image, it is legitimate to determine the value of the brightest features in the cell of interest, and stretch the image histogram based on the revised range of pixel values.

If regions of an image other than the principal object show dim fluorescence, include them within the image histogram, because it is likely that there will be similar intensities in the object of interest. If you exclude the unwanted intensities by stretching the histogram, the image might look posterized, leaving extensive black (or white) areas. This is one of the most frequent errors made in image processing. It is rarely justifiable to remove visible features of the image by adjusting the histogram just to make the picture look better. This is especially true when you expect to see a signal in "unwanted" places because of the biology of the system or the properties of immunofluorescence labeling (Fig. 16-1).

- *Apply global processing operations conservatively.* For routine image processing, limit yourself to 5–10% of the estimated maximum capacity of image enhancement filters such as γ scaling, sharpening and blurring filters, and unsharp mask operations. Fortunately, there are objective criteria for thinking about this. For example, in using γ scaling on a dark fluorescence image, it is useful to remember that a γ factor of 0.7 closely matches the logarithmic response of the eye; under bright-light conditions, contrast thresholds for adjacent objects that are barely visible vs. readily distinguishable are approximately 3% and 30%, respectively. The best test for whether an image has been overprocessed is to compare the processed image, side by side, with the original unprocessed image or with the real view in the microscope. *Image processing operations should be reported in the methods section of a research article or in the figure legend* so that the reader understands how the image was manipulated.

- *Avoid using extreme black and white values in the image during printing.* Typically the dark background around a cell in an immunofluorescence image should not be set to 0. Because halftone cells blur during printing, publishers recommend that the gray scale in submitted images not extend to true black or white. The lowest value (black) should be set to ~12% above 0 (30 units on an 8 bit scale), and the highest value should be set to ~8% below 255 (235 on an 8 bit scale). The precise value depends on the kind of paper on which the image is to be published. For high-quality paper in a journal, the bottom cut-off is about 10%; for newsprint, the value is about 20%. So for newsprint, pixels with values of ~50 ($256 \times 20\%$) and below will all appear black, even though they are easily visible on the computer screen. If you follow this guideline, dark gray values will not become swallowed up by black, and light gray values will not disappear and come out looking white.

> • *Display highly manipulated features as an inset or separate figure.* Ordinarily, cellular features should not be differentially enhanced or diminished (dodge and burn operations in Photoshop) unless their very visibility is at stake. It is better to show the whole cell normally, with highly manipulated features included as an inset or as a separate picture and with processing details explained in the accompanying legend.

THE USE OF COLOR IN PRINTS AND IMAGE DISPLAYS

The eye is extremely sensitive to color, so assigning a color to a fluorochrome can be an excellent way to represent two or more different signals in the same image. Because we perceive the visual world in color and perceive color when looking in the microscope, the presence of color makes prints and monitor displays appear natural and attractive. Therefore, in fluorescence imaging involving multiple fluorochromes, a common approach is to assign a monochrome color channel to each gray-scale fluorescence image and display the information as a composite color print. Thus, color allows us to view multiple channels simultaneously and is especially effective in demonstrating the colocalization of two or more signals.

To preserve the dynamic range of original pixel values, color images should be handled and stored in as high level a data type as is practical and feasible. This is particularly important when image files are submitted directly for publication. Thus, data type is as important for color images as it is for gray-scale image files. Previously, color was limited to 8 to 12 bits per color channel, but because of improvements in computer speed and storage space, recent versions of image-processing software such as Photoshop can now support 16 bits per color channel. While monitors and printers are still limited at 8 bit resolution, color image files can be handled and stored in data types with much higher bit depth. At the present time, however, many programs still require conversion to byte (8 bits) for displaying and storing 24 bit color images.

Despite the advantages of using color and the improvements in handling and storing color images at high resolution, displaying multiple color signals in a single image presents a unique set of problems:

- Brightness and visual dynamic range, the range of light intensities, and the amount of detail that is perceived by the eye when looking at a picture (either a print or a computer monitor) are potentially much greater in a gray-scale image than in a color image.

- Some colors are perceived as being much brighter than others, owing to differences in the quantum efficiency of the eye at different wavelengths (see Chapter 2), a fact that can bias image displays.

On monitors and on prints, conversion to color can reduce brightness and visibility. While color can increase information content in an image by including multiple channels of color-encoded information, color can also reduce the brightness and visual dynamic range of light intensities perceived by the eye compared to the same image presented in full gray-scale range. Thus, on a monitor or on a print, color images appear darker because they have reduced light intensities. The reduction in intensity in going from gray scale to monochrome color is explained in the following way. On luminous color displays such as monitors and LCD projectors, distinct red, green, and blue pixels contained in the screen are excited to create the colors of the visual spectrum. For displaying white, the electron guns excite the three color phosphors at maximum, and the eye perceives the RGB pixel cluster as white. However, for a monochrome red image, the excitation of green and blue pixels is minimized, so the visual intensity of saturated red features drops to about a third of the intensity of the same feature shown as white in a gray-scale image. Therefore, for red monochrome color in an illuminated display, the brightness and visual dynamic range of the image are greatly reduced.

In color printing, red, green, and blue inks (RGB mode) or cyan, magenta, yellow, and black inks (CMYK mode) are printed as dots on a sheet of paper. For saturating red color, inks are applied with the result that only red is reflected from the page to the eye, while all other wavelengths are absorbed and are not reflected. So, again, for a monochrome red image, saturated features reflect only a fraction of the light that is capable of being reflected by white in a gray-scale print.

Blue (indigo blue at 400–450 nm) is a particularly difficult color to display (especially when shown against a black background) owing to the 40- to 100-fold reduced sensitivity of the eye to blue (430 nm) compared to its peak sensitivity to green (550 nm) (Kingslake, 1965). In color prints, blue chromosomes or nuclei against a dark background are often disappointing, because little can be perceived or resolved. A much preferred strategy, particularly for known, defined structures such as chromosomes or nuclei, is to use cyan (a blue-green color) for the blue color channel. For two-fluorochrome specimens the image display can be made brighter by choosing green and red colors to which the eye is very sensitive and avoiding blue altogether. To increase the intensity of the green-red image, you can change the color palette to show green as yellow-green and red as orange-red, since both red and green pixels are used to show each of these two colors. This raises the value of the saturated orange-red pixel by about 30–50% compared to the value for saturated red alone. Overlap of the two colors still produces yellow, the same as if we had used only pure red and green. The main point is that the intensity of a saturated feature in a gray-scale picture is significantly brighter and easier to see than the intensity of the same saturated feature in a color image. For a three-panel figure showing each of two fluorochromes separately and combined together, an effective strategy is to show the separate fluorochromes in gray scale, reserving color for just the combined image.

COLOCALIZATION OF TWO SIGNALS USING PSEUDOCOLOR

A common goal in fluorescence microscopy is to determine if two fluorescent signals are coincident in the same structure. For complex patterns, the use of a combined pseudocolor display is very useful. Two color channels are selected (red and green) so that regions of overlap appear yellow. (Other pairings, such as blue and green giving cyan and blue and red giving magenta, may also be used, but are much less effective because of the reduced sensitivity of the eye to blue.) In planning for image acquisition and display, it is best to adopt a standard that is easy to define and understand, namely, combine images having the same dynamic range for each fluorescent signal. In the case of confocal microscopy, this practice actually matches the procedure recommended for image acquisition. Once combined, overlapping bright red and green signals give an unambiguous bright yellow signal. Remember that dark yellow hues, such as colocalized dim objects and backgrounds with equal red and green contributions, look brown. Optimize the gain and offset separately for each fluorescence signal (background = 0, saturation = 255) so that each fluorochrome is displayed using a full 0–255 gray-scale range (objective, easy to do, and easy to explain and describe to others). The images are then processed if necessary and merged. This is a conventional way to acquire and display multicolor images; however, because each signal is acquired to fill the gray-scale range, this method does not allow you to determine the relative amplitudes of the two signals in the original specimen, only the relative signal strength within a given color channel.

In a perfectly aligned fluorescence imaging system, a point source in the specimen is perfectly registered, pixel for pixel, by different filter sets at the image plane in the camera. However, inadequate color correction of the objective lens or poorly aligned filters can cause misregistration of fluorescence signals during color merging. This artifact is recognizable by the uniform displacement of signals. In images with complex patterns and a mixture of bright and dim signals, this can go undetected, leading you to conclude that signal distribution in a structure is distinct or partially overlapping. Therefore, during a color merge operation, in order to obtain good registration, it is important to be able to freely move the color layers with respect to each other, an operation called *panning*. The ability to align through panning requires the presence of coincident reference points (prominent features in the object) in each color layer. If multiply stained reference points do not exist, it may be useful to add multifluorescent beads to the specimen, at a dilution giving just a few beads per field of view, prior to mounting with a coverglass. For critical applications, there is another solution: Use a single multifluorescence filter cube with a multiple-bandpass dichroic mirror and a barrier filter together with different fluorochrome-specific exciter filters so that the same dichroic mirror is used for all of the fluorescence signals; this configuration is used in confocal microscopes and other wide-field fluorescence microscopes where color alignment is critical.

More problematic is the case where the microscopist's acquisition and processing methods are uneven. Inaccurate interpretations regarding colocalization in merged color images can result from improper gain and offset settings of the camera during acquisition or from extreme histogram stretching during image processing. This is depicted in a sketch in Figure 16-2. It is important to follow the procedures outlined in this book;

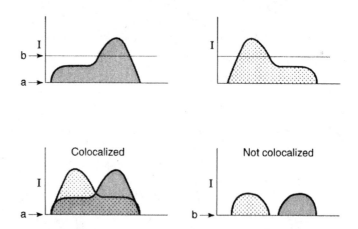

Figure 16-2

Colocalization of two fluorescent signals. The signals represented by the solid and dotted patterns are interpreted to be partially colocalized or have distinct distributions depending on the offset applied to the image. Bottom left: overlaid solid and dotted images using the offset at position a. Bottom right: overlaid solid and dotted images using the higher offset setting at position b.

otherwise, a too-high black-level setting may produce distinct segments of separate colors that under conditions of lower black-level setting and reduced contrast are observed to be contained in the same structures. For critical applications, ratio imaging methods can be used to distinguish between signals that are colocalized vs. signals that are only partially overlapping.

A CHECKLIST FOR EVALUATING IMAGE QUALITY

The following are criteria we might use in evaluating the quality of a scientific image. They contain the principles we have described along the way for specimen preparation, optical adjustments, and image acquisition, processing, and presentation.

Fixation and labeling

absence of fixation artifacts

adequate fluorochrome intensities

absence of bleed-through and nonspecific signals

supporting control experiments

The acquired image

proper optical zoom or magnification to preserve the required degree of optical resolution

full range of gray-scale values utilized

appropriate camera gain (and offset)

notes on microscope optics and camera acquisition parameters

Image processing

image studied to match pixel values with locations on the image histogram

histogram white and black sets adjusted to include all of the data in the image

no or minimal black and white saturated areas in the image

conservative use of gamma (γ) scaling and sharpening filters

insets and explanations made separately for emphasized features requiring extreme image processing

Use of color

single fluorochromes shown in gray-scale format

for colocalization of different fluorochromes, minimum and maximum values of color channels properly adjusted on the image histogram

for display of multiple fluorochromes in a montage, gray scale used for single channels, color used for combined multichannel image

The check list and guidelines presented in this chapter will allow new microscopists to use image processing knowledgeably and appropriately. Although processing needs vary depending on the application and the particular image, at a minimum, this chapter will help direct discussion on what processing operations should be performed. When a reader of your article can understand and reproduce an observation in his or her own laboratory, you will know that you have mastered the essentials of image processing and many fundamentals of light microscopy and electronic imaging.

ANSWER KEY TO PROBLEM SETS

Chapter 1

(2) For pancreatic ascinar cells, cell diameter $\sim$10 μm, nucleus diameter $\sim$5 μm; secretory granule $\sim$1 μm. (3) The field stop diaphragm is used to limit the area of illumination in the specimen to protect the specimen and reduce the amount of scattered light reaching the eye or detector. Opening the diaphragm too far causes unnecessary exposure and reduces contrast; closing the diaphragm too far masks field of view. Neither operation changes the intensity of light in the image to a significant extent. The condenser diaphragm likewise has an optimum setting, typically allowing transmission through 70% of the back aperture of the objective lens. A Bertrand lens is required to make the adjustment. If opened too far, excess light fogs the image; if stopped down, contrast is improved, but spatial resolution is compromised. The 70% position offers a good compromise between image contrast and resolution. Reduction of image brightness should be made by reducing the voltage of the power supply or by inserting a neutral density filter in the optical path.

Chapter 2

(1) The principal complementary color pairs are red-cyan; yellow-blue; green-magenta.

Chapter 4

(1a) No, the object is located slightly outside the focal length of the objective. To produce a magnified real image, the object is located between 1 and 2 focal lengths. (1b) No, the intermediate image is located slightly inside the focal length of the ocular as is required to form a magnified virtual image. (1c) The eye-brain system perceives a magnified virtual image because the linear extensions of rays, focused on the retina by the ocular and eye lens, appear to converge in a virtual image plane located about 25 cm in front of the eye. The ray diagram in Figure 1-2 helps clarify the role of perspective in perceiving a virtual image.

Chapter 6

(2) For a 40× 0.65 NA dry lens and 546 nm green light, striae or pore spacings of 0.5 µm are barely resolved. Only *Gyrosigma, Navicula,* and *Stauroneis* are clearly resolved. *Nitzschia* and *Pleurosigma* are barely resolved. (3) The spatial resolution is given as $d = 0.61\lambda/NA = (0.61 \times 0.546)/0.65 = 0.51$ µm. (5) The diffraction patterns preserve the orthogonal and hexagonal arrangements of striae and pores seen in the normal (orthoscopic) viewing mode of the microscope. The narrowest spacings in the object correspond to the widest spacings in the diffraction patterns. (6) Lowest resolution with red light and restricted aperture; best resolution with blue light and open aperture. For demonstration of the effect of condenser aperture size on resolution, try *Gyrosigma, Navicula,* or *Stauroneis* with a 20× lens in green light; alternatively, examine *Gyrosigma, Stauroneis, Nitzschia,* or *Pleurosigma* with a 40× lens in red light. (7) If a 100× oil lens is used without oil, the NA corresponds to about 1, and *Pleurosigma* is barely resolved. With oil immersion, definition and resolution are both dramatically improved and pores are evident.

Chapter 7

(3) The molarity of hemoglobin is calculated: g/L BSA at 50% match point × 1/64,000 g/mol = mol/L. The concentration of hemoglobin in erythrocytes is approximately 4–5 mM. This is 2,500 times the molarity of tubulin and 250,000 times the concentration of aldolase. (4) The range of phase densities is partly due to real differences in hemoglobin content. The ratio of blood volume to BSA solution volume should be very small and the solution should be mixed very well. Cells could be fractionated on a density gradient and the hemoglobin content of cells in various fractions determined quantitatively.

Chapter 9

(3) For the slow axis of the waveplate running NW–SE between crossed polars and a magenta-colored background: starch gains, NE–SW quadrants yellow, SE–NW quadrants blue; bordered pits, NE–SW quadrants blue, SE–NW quadrants yellow. (4) Carbon bond alignments in amylose chains are arranged radially as in a pincushion; the carbon bond alignments in lignin molecules in wood pits are arranged tangentially, like the layers of an onion.

Chapter 11

(1) Rhodamine bleeds through the long-pass filter set intended for fluorescein. (2) Rhodamine is stimulated by the 490 nm excitation wavelength for fluorescein, and since the long-pass emission filter transmits red light, the fluorescence emission of rhodamine is also transmitted by this set. (3) The fluorescein set with narrow bandpass filter is not completely effective in transmitting fluorescein-specific fluorescence. Rhodamine is weakly excited by the band of wavelengths at 490 nm used to excite fluorescein and its emission spectrum overlaps the region of peak fluorescence by fluorescein. It is unavoidable that some rhodamine fluorescence is transmitted by the fluorescein-narrow set. (4) A fluorescein set with long-pass emission filter is useful when fluorescein is the only label, fluorescence emission is weak, and it is desirable to transmit the maximum amount of emitted fluorescent light. (5) To reduce bleed-through, reduce the amount of rhodamine label or use an illuminator such as a xenon arc lamp that does not stimulate rhodamine 10 times the amount it does fluorescein (as is the case for a mercury arc lamp).

Chapter 12

(1) Removing a neutral density filter or increasing laser power provides more fluorescent light, allowing one to reduce the gain setting or scan at a faster rate; opening the pinhole diaphragm increases the thickness of the optical section in the specimen; increasing the gain on the PMT makes the image look grainy; decreasing the offset on the PMT can make the image look flat and washed out; reducing the scan speed increases the rate of photobleaching; increasing the zoom factor improves spatial resolution but increases the rate of photobleaching. (2) The relationship between pinhole size and the thickness of an optical section is nonlinear, being greater at small pinhole diameters. (3) Generally, the phenomenon of bleed-through is asymmetric—long-wavelength fluorochromes bleed more into filter sets for short-wavelength fluorochromes than occurs the other way around. This is because excitation of short-wavelength fluorochromes also excites longer-wavelength fluorochromes, and since fluorescence emission spectra begin at the excitation wavelength, some of the short-wavelength fluorescence emission of the longer-wavelength fluorochrome can pass through the filter set for the short-wavelength fluorochrome.

Chapter 13

(1) Video is suitable for time-lapse sequences lasting several minutes to hours. Membrane ruffling, organelle movements, cell migration, and mitosis are ideally suited for video time-lapse recording. (2) The horizontal cutoff frequency is calculated as $f_H = M_{objective} M_{relay} N_H/1.2D$, where M represents the magnification of the objective and relay lens, N is the lines of horizontal or vertical resolution at $N_H = 800$, and D is the diagonal measurement of the target in the video electron tube in mm. For a D of 15.875

mm and no relay lens (M_{relay} = 1), f_H = (100 × 800)/(1.2 × 15.875) = 4200 cycles/mm. This corresponds to a distance in the specimen of 1000 μm/mm × (1/4200) or 0.24 μm. (3) The addition of a VCR significantly degrades the image due to a decrease in the system bandwidth, where bandwidth BW is given: $1/BW$ of system = $\sqrt{\Sigma\ 1/(BW\ \text{each component})^2}$. Given that a bandwidth of 10 MHz is required to represent 800 lines horizontal resolution, and the bandwidths in MHz for the camera and VCR are 15 and 6 MHz, respectively, the system bandwidth is 3.9 MHz, which corresponds to 312 lines resolution, a 2.5-fold decrease in spatial resolution. (4) To avoid image degradation, the image should be magnified by employment of an intervening booster lens or "TV lens."

MATERIALS FOR DEMONSTRATIONS AND EXERCISES

Chapter 1: Calibration of Magnification

Materials

pancreas slides, H&E stain (Carolina Biological Supply)

immersion oil

lens tissue and cleaner

eyepiece reticules (Klarmann Rulings)

stage micrometers (Klarmann Rulings)

Chapter 2: Complementary Colors

Materials

3 slide projectors

red, green, blue gelatin filters, colored glass filters, or interference filters (Edmund Scientific)

aluminum foil masks with 1 cm hole in mounting for projection slide

optical bench with filter holders (Edmund Scientific)

holographic diffraction grating (Edmund Scientific)

2 × 2 inch IR cut filter (hot mirror) (Edmund Scientific)

2 × 2 inch opaque mounting with 0.5–1.0 mm diameter slit

yellow and blue food coloring

2 one-liter beakers

light box

extension cords

quartz-halogen or xenon light source (xenon is better) and power supply

4–6 inch handheld magnifying glass

spectroscope (Learning Technologies)

sheet of cardboard, white on one surface

scissors

1–2 inch test objects (paper or plastic sheet) of saturated red, green, and blue colors

Chapter 3: Spectra of Common Light Sources

Materials

optical bench

narrow slit

IR blocking filter

holographic grating

100 W quartz halogen microscope lamp

100 W mercury arc microscope lamp

100 W xenon microscope lamp

handheld spectroscopes (Learning Technologies)

aluminum foil

razor blades

scissors

Chapter 4: Optical Bench Microscope, Lens Aberrations

Materials

optical bench with lens holders and screen

50 mm lens, two 200 mm lenses for optical bench

quartz-halogen microscope lamp and power supply

lens tissue or microscope slide, marking pen, adhesive tape

aluminum foil

fine-gauge syringe needle

scissors

adhesive tape

Chapter 5: Diffraction Images in Back Focal Plane of Objective Lens

Materials

3 mW HeNe laser or laser pointer

aluminum foil

fine-gauge syringe needles

9 × 12 inch rectangular glass baking dish

small container of milk

Pasteur pipette and rubber bulb

EM grid, 400 mesh

microscope slide

adhesive tape

optical bench with lens and specimen holders

demonstration lenses, 50 and 100 mm focal length

transparent diffraction gratings, various spacings (Edmund Scientific)

transparent holographic grating

aperture mask, 0.5–1 mm slit width, made from the edges of two razor blades or
 2–3 mm diameter hole in a square of aluminum foil

IR blocking filter or IR hot mirror

tape measure

stage micrometer and eyepiece reticule

compact disk (CD)

bird feather

handheld calculators

Chapter 6: Resolution of Striae in Diatoms

Materials

immersion oil, lens cleaner, lens tissue

diatom test plates (Carolina Biological Supply)

diatom exhibition mounts (Carolina Biological Supply)

Red, green, and blue glass filters or interference filters (Edmund Scientific)

Chapter 7: Phase Contrast Microscopy

Materials

box Pasteur pipettes

rubber bulbs

Kimwipes

70% ethanol

gauze wipes

Band-Aids

lancets

microscope slides and #1.5 coverslips

sharps container

Notes on Preparing Stock Solution of Serum Albumin

1. Dialyze 50 mL of 30% (wt/vol) bovine serum albumin (Sigma liquid stock A 7284) against 2 L distilled water to remove 158 mM NaCl. Change water after 1–2 hr, then continue overnight with 2 L water containing 30 mL ion exchange resin (BioRad AG 501-X8 mixed-bed resin) on a stirring box at 4°C.

2. Transfer the albumin to a lyophilizer jar, freeze in liquid nitrogen or dry ice/ethanol bath with swirling, and lyophilize overnight until dry.

3. Resuspend the dry albumin powder in 25 mL distilled water. Break up clumps of albumin periodically during the day. It takes several hours to resuspend the albumin. The solution may need to be centrifuged to remove bubbles and foam using a benchtop centrifuge.

4. Determine the protein concentration by reading absorbance at 280 nm in a UV spectrophotometer. Dilute the stock 1:1000 in water before reading. Use care in performing the dilution, since the BSA stock is extremely thick and concentrated. The OD_{280} of a 1 mg/mL solution is 0.66. The concentration of the stock will be $\geq$ 400 mM.

5. Prepare the following 5 solutions as indicated. The final total solute concentration in each tube will be 326 mOsM. All solutions contain added sodium chloride to maintain the erythrocytes. NaCl is added to give the same value of osmolarity for each tube, but albumin contributes 95–98% of the refractive index, the balance being due to the salt, which we will ignore. (If albumin is prepared at other concentrations, the following values will be useful: from tables, the osmolarity equivalent of 100 mg/mL BSA = 13.793 mM NaCl, and 9.463 g/L NaCl = 326 mOsM.)

mg/mL BSA	g NaCl/L
200	7.863
250	7.463
300	7.063
350	6.663
400	6.263

References: Barer et al., 1953; James and Dessens, 1962.

Chapter 8: Polarized Light

Materials

2 dichroic sheet polarizers

light box or overhead projector

microscope slide

sheet of cellophane (not food wrap)

calcite crystals

Chapter 9: Polarization Microscopy

Materials

2 dichroic sheet polarizers

light box or overhead projector

sheet of cellophane

cellophane tape

red-I plate (λ-plate)

dichroic sheet polarizers, 2 × 4 cm strips, with one edge parallel to transmission axis of polarizer

red-I plates; 2 × 4 cm strips, with edges parallel to the axes of the refractive index ellipsoid of the plate (see instructions for cutting strips)

corn starch, a few grains of starch mounted on a slide under a coverslip with a drop of glycerol

leaf of *Amaryllis* or *Diffenbachia,* a drop of plant cell sap mounted on a slide under a coverslip

onion skin, a thin sliver mounted on a slide in a drop of water under a coverslip

preparation of squashed *Drosophila* thorax, mounted on slide in drop of water under a coverslip

stained section of pine wood, *Pinus strobus,* radial section (Carolina Biological Supply)

stained section of buttercup root, *Ranunculus* (Carolina Biological Supply)

stained section of striated muscle, H&E (Carolina Biological Supply)

razor blades for cutting materials

Pasteur pipettes and rubber bulbs

Preparation of Red-I Plates for Student Microscopes

On a light box, rotate two dichroic sheet polarizers to obtain extinction. Between the polars insert the red-I plate that is to be cut into 2 × 4 cm strips for the student microscopes. With the polars crossed and held in a fixed position, rotate the red-I plate until the bright interference color is observed. The axes of the refractive index ellipsoid of the plate are now oriented at 45° with respect to the transmission axes of the crossed polars. Attach a strip of cellophane tape to a microscope slide and insert the slide between the crossed polars. The long axis of the tape is parallel to the long axis of the refractive index ellipsoid (= slow axis of the wavefront ellipsoid) and will serve as a reference. Rotate the microscope slide through 360°. At two positions the tape will look bright yellow, and at two positions rotated by 90° with respect to the first, the tape will look pale yellow. In between these two axes the tape will have the same red color as the red-I plate by itself. The orientation of the tape giving the bright yellow color marks the high refractive index axis of the red-I plate. Mark this axis on the red-I plate as a short line with a marker pen. The red-I plate can now be cut into strips with a paper cutter carefully so that one edge of the strips is perfectly parallel to the axis drawn on the plate. Blacken this edge with the marker pen and draw an ellipsoid at a corner of the plate with its short axis parallel to the blackened edge.

Chapter 10: DIC Microscopy

Materials

24 coverslips of cultured cells

serum-free MEM culture medium supplemented with 10 mM HEPES, pH 7.2

filter paper

petroleum jelly

paper wipes (Kimwipes)

lens cleaner, water, immersion oil

Pasteur pipettes and rubber bulbs

plastic beakers for discarded medium

fine forceps

Chapter 11: Fluorescence Microscopy

Materials

24 coverslips of cultured cells

serum-free MEM culture medium supplemented with 10 mM HEPES, pH 7.2

filter paper

petroleum jelly

paper wipes (Kimwipes)

lens cleaner, water, immersion oil

Pasteur pipettes and rubber bulbs

plastic beakers for discarded dyes and medium

fine forceps

chlorophyll extract

fluorescein solution, 10 mg fluorescein in 1 L water

delipidated BSA (Sigma A-6003)

long-wave UV lamp (blacklight)

light box

spectrometers

cardboard sheet with 1 cm wide slit

optical bench

IR cut filter

holographic diffraction grating

1 mm slit aperture

xenon arc lamp and power supply

Stock Solutions of Fluorescent Dyes

ethidium bromide 10 mg/mL in water

JC1 dye, 10 mg/mL in DMSO

$DiOC_6$ 5 mg/mL in DMSO

bodipy-ceremide, 1 mM in ethanol

MitoTracker red

LysoTracker green

bodipy-phalloidin, 3 μM in DMSO

DAPI, 5 mg/mL in water

Preparation of Chlorophyll

Place 2 cups of fresh spinach and 1 L isobutyl alcohol in a food blender and homogenize until the leaf tissue is thoroughly dispersed. Filter the solution through several layers of cheesecloth. The same procedure can be used to prepare extracts of other tissues (50 g).

Chapter 13: Video Microscopy

Materials

culture of *Acanthamoeba, Dictyostelium,* or other protozoan

Pasteur pipettes and rubber bulbs

slides and coverslips

immersion oil

video cart with camera, image processor, VCR, and monitors

4× TV lens

diatom slide for demonstrating aliasing

SOURCES OF MATERIALS FOR DEMONSTRATIONS AND EXERCISES

3M/Optical Systems Division, Norwood, MA, tel. 781-386-6264
full-wave plate retarders
dichroic polarizing film

Carl Zeiss, Inc., Thornwood, NY, tel. 800-543-1033
Michel Levy color chart
Microscope optical pathway posters, various microscopes
booklet, *Microscopy from the Very Beginning*

Carolina Biological Supply Co., Inc., Burlington, NC, tel. 800-334-5551
Microscope slides, histological preparations:
skeletal muscle, H1310
Bordered pits, pine wood, 1.s., B520
Ranunculus root, c.s., B520
diatom test plate, 8 forms, w.m., P7-B25D
pancreas, monkey, h.-e., 98-9652

Edmund Scientific Company, Inc., Barrington, NJ, tel. 800-728-3299
diffraction grating sheet, 600 lines/inch
diffraction grating sheet, 23,000 lines/inch
holographic diffraction grating, 3600 lines/mm
calcite crystals
red HeNe laser, 0.8 mW, Uniphase class IIIa laser, K61-337, $290
hot mirror, 50 × 50 mm, reflects IR, K43-452, $52
BG-38 Schott IR-blocking filter, 50 × 50 mm, K46-434, $44
KG1 Schott IR-blocking filter, 50 × 50 mm, Y45-649, $43
GG-420 Schott UV-blocking filter, 50 × 50 mm, K46-427, $38
red, green, blue gelatin filters (Kodak Wratten filters 25, 58, 47B), 75 × 75 mm, $19
    each
polarizing film, tech spec quality, 8.5 × 15 inch sheet, K45-668, $29
polarizer pack, demo kit, 38490
lens cleaner, K53-881, $7

Hamamatsu Photonics, Inc., Bridgewater, NJ, tel. 908-231-1116
product literature on video and CCD cameras, image intensifiers

Klarmann Rulings, Inc., Manchester, NH, tel. 800-252-2401
eyepiece micrometer, 0.01 mm divisions, $58
stage micrometer, 0.01 mm divisions, $440

Learning Technologies, Inc., Cambridge, MA, tel. 800-537-8703
plastic spectrometer, PS-14, $18
holographic diffraction grating, 750 lines/mm, PS-08-A, 9 × 5 inch sheet, $8

Leica Microsystems, Inc., Allendale, PA, tel. 201-236-5900
polarization color chart
Leica booklets on modes of light microscopy

Nikon Optics, Inc., Melville, NY, tel. 516-547-8500
color brochures on fluorescence filter sets

Olympus America, Inc., Melville, NY, tel. 800-446-5967
color brochures on microscope optical pathway in upright and inverted microscopes

Roper Scientific-Photometrics, Tucson, AZ, tel. 520-889-9933
color poster of CCD arrays
product literature on CCD cameras

GLOSSARY

Aberrations of a lens. Faults in lens design that cause optical performance to deviate from that of an ideal lens. Usually attributed to materials composing the lens and the spherical curvatures of lens surfaces. Lens aberrations include chromatic and spherical aberration, astigmatism, coma, distortion, and field curvature. *50*

Absorbance or **optical density.** The amount of absorption of light by a substance as measured in a spectrophotometer and given as the log of the reciprocal of the transmittance, where transmittance is the ratio of the transmitted light intensity to incident light intensity. *38*

Achromat. A lens corrected for chromatic aberration at two wavelengths (red and blue) and for spherical aberration (green). *53*

Acousto-optical tunable filter (AOTF). A device, based on sound waves, to control the wavelength or intensity of light delivered from a laser or other light source to a specimen. Sound waves induced in a glass block set up a standing wave with alternating domains of high and low refractive index, allowing the block to act as a diffraction grating and to deflect an incident beam of light. The period of the grating is changed by altering the frequency of sound waves delivered to the block. *221*

ADC. See *Analogue-to-digital converter (ADC).*

ADU. See *Analogue-to-digital unit (ADU).*

AGCC. See *Auto-gain control circuit (AGCC).*

Airy disk. The central diffraction spot in the focused image of a point source of light. Diffraction at the front aperture of the lens disturbs the incident wavefront, causing the diffraction pattern. The Airy disk diameter is determined by the wavelength of light and the angular diameter of the lens as seen from the image plane. *65*

Alexa dyes. Registered trademark for a series of fluorescein and rhodamine derivatives by Molecular Probes, Inc., noted for their stability and resistance to photobleaching. *185*

Aliasing. In microscopy, the false pattern of spacings that is observed when periodic detail in an object is recorded by an electronic imager, which itself has periodic sampling units (pixels, raster lines), and when the number of samples per specimen period is less than 2. Aliasing is avoided by adjusting the magnification high enough so that two or more imaging elements cover a unit period in the object. See also *Nyquist criterion.* *248*

Numbers after each glossary term indicate corresponding page numbers.

Amplitude object. Objects that absorb light as opposed to those that shift the phase of light (phase objects) as the basis for image formation. *12, 91, 97*

Amplitude of an electromagnetic wave. The magnitude of the electric field vector of an electromagnetic wave. Amplitude is distinguished from intensity (irradiance), a measure of the amount of light energy or photon flux, which in visual perception is proportional to amplitude squared. *19, 22*

Analogue signal. A signal that is continuously variable—for example, the variable voltage signal sent from a video camera to a closed-circuit television. Analogue signals are distinct from digital signals, which are composed of a stream of discrete numeric values in binary computer code. *235*

Analogue-to-digital converter (ADC). A device for converting an analogue signal into the digital code of the computer. A 12 bit converter in the head of a CCD camera converts an analogue voltage signal into one of 4096 possible digital values for processing and display on a computer. ADC circuits in a CCD camera are noted for their high processing rate and low noise. *215, 269*

Analogue-to-digital unit (ADU). In digital imaging, the numeric value assigned to represent the amplitude of a photon signal. The conversion factor used in making the assignment—for example, 10 photoelectrons/ADU—is called the gain factor. *263*

Analyzer. A linear polarizer used to analyze or determine the plane of vibration of an incident ray of polarized light. *118*

Angular aperture. See *Aperture angle.*

Anisotropic. In describing the optical properties of an object or propagation medium, having dissimilar properties in different directions. *121*

Annulus. In phase contrast microscopy, the transparent ring at the front aperture of the condenser that provides illumination of the specimen. *103*

AOTF. See *Acousto-optical tunable filter (AOTF).*

Aperture angle. The angle subtended by the edges of a lens as seen from a point in the specimen plane or in the image plane. Aperture angle is included in the expression for numerical aperture ($NA = n \sin\theta$), where n is the refractive index and θ is one-half of the full aperture angle. See also *Numerical aperture (NA). 66, 69*

Aperture plane. In a microscope adjusted for Koehler illumination, the set of conjugate focal planes located at the light source, the front aperture of the condenser, the back aperture of the objective lens, and the iris of the eye. Adjustable aperture diaphragms at these locations are used to limit stray light and determine the numerical aperture, and hence the spatial resolution, of the instrument. *4*

Apochromat. A lens especially designed to correct for chromatic aberration at three or four wavelengths (red, green, blue, UV) and at two or more wavelengths (green, red) for spherical aberration. The high degree of color correction makes these lenses suitable for fluorescence microscopy and stained histological specimens in bright-field microscopy. *53*

Astigmatism. An off-axis aberration of lenses whereby rays from an off-axis object passing through the horizontal and vertical diameters of a lens are focused as a short streak at two different focal planes. The streaks appear as ellipses drawn out in horizontal and vertical directions at either side of best focus, where the point image is a small disk. Off-axis astigmatism increases with increasing displacement of the object from the optic axis. Astigmatism is also caused by asymmetric lens curvature due to mistakes in manufacture or improper mounting of a lens in its barrel. *52*

Auto-gain control circuit (AGCC). A circuit in some electronic cameras that automatically adjusts the gain to maintain a constant output signal on a display device. Auto-gain control is useful in fluorescence imaging where specimens are subject to continual photobleaching. *240*

Azimuth angle. A term used to describe the orientation of an object in a plane such as the specimen plane of the microscope. On a graph with polar coordinates marked off in 360°, the angle subtended between a fixed reference (designated 0°) and a vector rotated about the origin. *20*

Back aperture of the objective lens. An aperture plane of the light microscope located at or near the rear lens element of the objective lens and the site of formation of a diffraction image of the object. Optical devices such as phase plates, DIC prisms, and aperture masks used in forms of interference microscopy are located at or near this location. *5, 80*

Background subtraction. An operation in electronic imaging whereby an image of the featureless background near an object is subtracted from the image of the specimen to remove patterns of irregular illumination and other optical faults such as scratches and dust. *244*

Bandwidth. In video microscopy, the range of frequencies expressed as the number of cycles per second used to pick up and transmit an image signal. To resolve an object having 800 alternating black and white lines in the time of a single horizontal raster scan (the required resolution for closed-circuit TV signals), the camera electronics must be fast enough to rise and fall in response to each line at a rate that corresponds to a frequency of 10,000 cycles/s. The bandwidth of such a camera is said to be 10 MHz. *245*

Barrier filter. See *Emission filter.*

Beam splitter. An optical device for separating an incident beam of light into two or more beams. A prism beam splitter in the trinocular head of a microscope directs the imaging beam to the eyepieces and to the camera simultaneously. A polarizing beam splitter made of a crystalline birefringent material is used to produce linearly polarized light. A dichroic mirror beam splitter reflects excitation wavelengths while transmitting long-wavelength fluorescence emission. *126, 192*

Bertrand lens. A built-in telescope lens located behind the back aperture of the objective lens. When rotated into the optical path, the back aperture and diffraction plane are seen, and other planes that are conjugate to it, while looking in the oculars of the microscope. *5*

Bias noise. One of the noise components contained in a raw CCD image. Bias noise is calculated as the square root of the bias signal, and is considered in determining the S/N ratio of images. *273*

Bias retardation. In polarization microscopy, the alteration of optical path differences between O and E rays, made by adjusting a compensator. Bias retardation is introduced by the operator to change the contrast of an object or is introduced in order to measure the relative retardation Γ of the O and E wavefronts of birefringent objects. In DIC microscopy, bias retardation is introduced using a Wollaston prism to optimize image contrast. *161*

Bias signal. In digital CCD cameras, the signal resulting from the application of a bias voltage across the CCD chip, a condition required to store and read the pixels on the chip. The bias signal must be subtracted from the total image signal in order to obtain a photometrically accurate image signal. *273*

Biaxial crystal. A class of birefringent crystals having two optic axes. Mica is an example of this class of crystals. *126*

Binning. In CCD microscopy, the command given in software to combine the signal content of multiple adjacent pixels. Because the total number of photon counts required to reach saturation remains the same, binning reduces the exposure time required to reach a certain signal value by a factor equal to the number of binned pixels. However, because there are overall fewer pixels in a binned picture covering a given area, binning also reduces the spatial resolution. *270*

Bioluminescence. An oxidative process resulting in the release of energy as light emission—for example, firefly luminescence, which requires an enzyme, luciferase, to catalyze a reaction between the substrate luciferin and molecular oxygen in the presence of ATP. *181*

Birefringence. The double refraction of light in transparent, molecularly ordered materials caused by the existence of orientation-dependent differences in refractive index. Also refers to the refractive index difference experienced by a transmitted ray through such a material. Incident beams of light on a birefringent specimen are split into O and E rays that can recombine after emergence from the object, giving linearly, elliptically, or circularly polarized light. *124, 126*

Bit. In computer language, a binary digit 1 or 0. An 8 bit image therefore has 2^8 or 256 gray level steps, while a 12 bit image contains 2^{12} or 4096 steps. *262, 284*

Bleed-through. In fluorescence microscopy, the transmission of unwanted wavelengths through a filter designed to block them. Bleed-through occurs when a filter's design does not allow complete destructive interference of unwanted wavelengths, when the angle of incident rays is oblique, when the transmission and emission spectra of fluorescent dyes overlap, and for other reasons. *194, 197*

Brewster's angle. The unique angle formed by an incident ray and the perpendicular to a reflective dielectric substance such as water or glass (the transmitting medium) at which the reflected ray is totally linearly polarized. Brewster's angle θ is given as $\tan\theta = n_i/n_t$, where n_i and n_t are the refractive indices of the medium of the incident beam and the transmitting medium, respectively. *121*

Bright-field microscopy. A mode of optics employing the basic components of objective and condenser lenses for the examination of amplitude objects such as stained histological specimens. *12*

Brightness. A qualitative expression for the intensity of light. *55*

Byte. In computer language, a unit of information containing 8 bits. *284*

Camera control unit (CCU) or **camera electronics unit (CEU).** The main external electronics control unit of an electronic imaging system. For video cameras, there is often a stand-alone unit that is used to adjust gain, offset, and other functions; in CCD cameras, the controlling circuits are often contained in the camera head and are operated by software on a computer. *269*

CCD. See **Charge-coupled device (CCD).**

CEU. See **Camera control unit (CCU).**

Charge-coupled device (CCD). A slab of silicon semiconductor that is divided into an array of pixels that function as photodiodes in a light-sensitive photodetector. In the presence of an applied voltage and incident light, the pixels generate and store "photoelectrons" (alternatively, electron holes) resulting from the disruption of silicon bonds from incident photons. The number of stored "photoelectrons" determines the amplitude of pixel signals in the displayed image. *236, 260*

Chromatic aberration. An aberration of lenses, whereby light waves of different wavelength are brought to focus at different locations along the optic axis. In the typical case of a simple thin lens, the focal length is shorter for blue wavelengths than it is for red ones. *51*

Circularly polarized light. A form of polarized light whereby the E vector of the wave rotates about the axis of propagation of the wave, thus sweeping out a spiral. If the wave could be viewed end-on, the movement of the E vector would appear to trace the form of a circle. *131*

CLSM. See *Confocal laser scanning microscope (CLSM).*

CMYK. A popular format for color printing, whereby colors are reproduced using four different inks: cyan, magenta, yellow, and black. *313*

Coherent light. A beam of light defined by waves vibrating in the same phase, although not necessarily in the same plane of vibration. To maintain the same phase over long distances, coherent waves must be monochromatic (have the same wavelength). Laser light is monochromatic, linearly polarized, and highly coherent. *20, 82*

Collector lens. A focusable lens of the illuminator capable of collecting light over a wide area and directing it toward the specimen. In Koehler illumination, the collector lens is used to focus a magnified real image of the filament or arc of the bulb in the front aperture of the condenser. *7*

Collimated beam. A beam in which rays proceed in the same direction and follow trajectories that are parallel to one another. Collimated light need not be monochromatic, polarized, or coherent. *20*

Colored-glass filter. A slab of glass containing colloids or metals that absorb certain wavelengths while freely transmitting others. In microscopy, colored-glass filters are commonly employed in fluorescence filter sets and as effective blockers of UV and IR light. *39*

Coma. An off-axis aberration of lenses, whereby rays from an off-axis point passing through the edge of the lens are focused closer to the optic axis than are rays that pass through the center of the lens, causing a point object to look like a comet with the tail extending toward the periphery of the field. Coma is the most prominent off-axis aberration. For lenses having the same focal length, coma is greater for lenses with wider apertures. *52*

Compensator. In polarization microscopy, a birefringent slab that is positioned between the polarizer and analyzer and can be tilted or rotated. This action varies the optical path difference between the O and E rays emergent from a birefringent object and is performed to make quantitative measurements of relative O and E wave retardations, or for qualitative purposes in changing image contrast and brightness. *136, 140*

Composite view or **projection view.** In confocal microscopy, an image created by adding together multiple optical sections acquired along the z-axis. The images of three-dimensional objects, although blurry in conventional wide-field fluorescence mode, are often remarkably clear in confocal composite view. *213*

Compound light microscope. An optical instrument that forms a magnified image of an object through a two-step series of magnifications: The objective forms a magnified real image of the object, and the eyepiece forms a magnified virtual image of the real image made by the objective. This basic design forms the basis of all modern light microscopes. *1*

Condenser annulus. In phase contrast and dark-field microscopy, a transparent annulus in an opaque black disk located in the front aperture of the condenser that serves as the source for illuminating the specimen. *103*

Condenser lens. A lens assembly located near the specimen and specimen stage that collects light from the illuminator and focuses it on the specimen. Proper optical performance requires that the condenser be highly corrected to minimize chromatic and spherical aberration. *2, 56*

Cone cell photoreceptors. Retinal cells responsible for color vision and visual acuity. See also *Fovea. 24*

Confocal laser scanning microscope (CLSM). A mode of light microscopy whereby a focused laser beam scans the specimen in a raster and the emitted fluorescent light or reflected light signal, sensed by a photomultiplier tube, is displayed in pixels on a computer monitor. The dimensions of the pixel display depend on the sampling rate of the electronics and the dimensions of the raster. A variable pinhole aperture, located in a plane confocal with the specimen, rejects out-of-focus signals and allows for optical sectioning. *208*

Conjugate focal planes. In light microscopy, two sets of field and aperture planes whose precise geometrical positioning in the microscope is assured by adjusting the focus of the objective, condenser, and lamp collector lenses as required for Koehler illumination. The two sets of focal planes are conjugate with each other but not with the focal planes belonging to the other set; as a consequence, looking from one focal plane along the optic axis simultaneously reveals the images of the other conjugate focal planes. *4*

Constructive interference. In wave optics and image formation, the condition where the summation of the E vectors of the constituent waves results in an amplitude greater than that of the constituents. For interference to occur, a component of one wave must vibrate in the plane of the other. *63*

Contrast. Optical contrast is the perceived difference in the brightness (intensity or irradiance) between an object and its surround, and is usually given as the ratio of the light intensity of an object I_o to the light intensity of the object's background I_b, thus: $C = (I_o - I_b)/I_b$, or alternatively as $C = (I_o - I_b)/(I_o + I_b)$. *22*

Contrast threshold. The minimal contrast required for visual detection. The contrast threshold is strongly dependent on the angular size, shape, and brightness of the specimen, the brightness of the viewing environment, the region of the retina used for detection, and other factors. For extended objects, the contrast threshold is usually given as 2–3% in bright light and 30–100% or even greater in dim light. *22*

Convolution. In image processing, the procedure of combining the values of image pixels with a 3×3, 5×5, etc., matrix of numbers through a defined function for purposes of image sharpening and blurring. The processing values in the convolution matrix or kernel are applied to each image pixel and to neighboring pixels covered by the mask in order to calculate new pixel values in a processed image that takes into account the values of neighboring pixels. *292*

Curvature of field. An aberration of a lens that causes the focal plane to be curved instead of flat. *52*

Cyanine dyes. Fluorescent dyes produced by Amersham, Inc., that are known for their photostability, solubility, and relatively high quantum efficiency. *185*

Dark count and **dark noise.** In electronic cameras, the photon-independent signal in an image. The major constituents of the dark count are the bias signal, thermal noise, and camera read noises from the detector and processing circuits. The dark count must be subtracted from a raw image in order to obtain a corrected image based solely on photon-dependent counts. The dark noise is defined as the square root of the dark count and remains a part of the corrected image. The dark count is determined by acquiring an image (a dark frame) without opening the camera shutter. *273*

Dark-field microscopy. A mode of light microscopy in which 0th order undeviated light is excluded from the objective, and image formation is based solely on the interference of waves of diffracted light. Typically, dark-field optics are obtained by illuminating the object with a steeply pitched cone of light produced with a transparent annulus in an otherwise opaque mask at the condenser front aperture. A relatively small NA objective is used so that undeviated light does not enter the objective, allowing only diffracted waves to enter the objective and form an image. *112*

Dark frame. In image processing, a picture containing the dark count (thermal and bias counts) and showing a featureless, low-amplitude background that is used to prepare a flat-field-corrected image of the object. A dark frame is prepared using the same exposure time as for the raw image but without opening the shutter. *292*

Deconvolution microscopy. A method that applies computer deconvolution algorithms to a through-focus stack of images along the z-axis to enhance photon signals specific for a given image plane or multiple focal planes in an image stack. A stepper motor attached to the microscope focus drive guarantees image acquisition at regular intervals through the specimen. In typical applications, deconvolution methods are used to deblur and remove out-of-focus light from a particular focal plane of interest. In more sophisticated applications, image frames of an entire stack can be deconvolved to allow clear views of a specimen displayed in projection view or in 3D viewing mode. *205*

Depth of field. The thickness of the optical slice through a specimen that is in focus in the real intermediate image. The thickness measurement is dependent on geometric and wave-optical parameters. For a high NA objective, the thickness of the optical slice in the specimen Z is given as $Z = n\lambda/NA^2$, where n is the refractive index of the medium between the lens and the object, λ is the wavelength of light in air, and NA is the numerical aperture. *90*

Depth of focus. The thickness of the image at the real intermediate image plane in the microscope. Like depth of field, the focus thickness depends on geometric and wave-optical parameters. The depth of focus is given approximately as $[1000\ M_{objective}/(7\ NA\ M_{total})] + [\lambda M_{objective}^2/2\ NA_{objective}^2]$, where M is magnification, λ is wavelength, and NA is the numerical aperture. *90*

Descanning. In confocal microscopy, the optical design of allowing the fluorescent light emitted at the specimen upon excitation by the scanning laser spot to retrace its path back through the objective lens and scanner mirrors to the dichroic mirror. With descanning, the fluorescent image spot at the detector pinhole remains steady and does not wobble. *210*

De Sénarmont method of compensation. In polarization and DIC microscopy, the use of a fixed quarter-waveplate retarder together with a rotating analyzer as a method for measuring optical path differences (relative retardations) between O and E rays and for introducing compensating retardation to adjust image contrast. Retardations of $\lambda/20$ to 1λ can be measured with an accuracy of ± 0.15 nm. *145*

Destructive interference. In wave optics and image formation, the condition where the summation of the E vectors of constituent waves results in an amplitude less than that of the constituents. For interference to occur, a component of one wave must vibrate in the plane of the other. *63*

Dichroic mirror. In fluorescence microscopy, an interference filter that exhibits a sharply defined transition between transmitted and reflected wavelengths. When inclined at a 45° angle with respect to incident light beams, the mirror reflects short excitation wavelengths through 90° onto the specimen and transmits long fluorescent wavelengths to the real intermediate image plane. A dichroic mirror is one of the three filters contained in a fluorescence filter set. *191*

Dichroism. The property exhibited by linear polarizing films and certain naturally occurring minerals, whereby incident wavelengths are differentially absorbed, causing the object to appear in two different colors depending on the angle of view and the orientation of incident waves. The phenomenon reflects the difference between the absorption curves for chromophores oriented in different directions in the dichroic object. *123*

DIC microscopy. See *Differential interference contrast (DIC) microscopy.*

DIC prism. See *Wollaston prism.*

Dielectric constant. A parameter describing the electrical permittivity of a material (the degree to which a material is permeated by the electric field in which it is immersed). Substances with low electrical conductivity (and low permittivity) such as glass, plastic, and water are called insulators or dielectrics. The dielectric constant ϵ is related to the refractive index n such that $n = \sqrt{\epsilon}$. *129*

Differential interference contrast (DIC) microscopy. A mode of light microscopy employing dual-beam interference optics that transforms local gradients in optical path length in an object into regions of contrast in the object image. Also referred to as Normarski optics after the name of its inventor, George Nomarski. The specimen is illuminated by myriad pairs of closely spaced coherent rays that are generated by a crystalline beam splitter called a Wollaston prism. Members of a ray pair experience different optical path lengths if they traverse a gradient in refractive index in a phase object. Optical path differences become translated into amplitude differences (contrast) upon interference in the image plane. DIC images have a distinctive relief-like, shadow-cast appearance. *155*

Diffracted wave. In phase contrast and other modes of interference microscopy, waves that become deviated from the path of 0th-order (background) waves at the object. Diffracted waves can be shown to be retarded in phase by $\sim \frac{1}{4}$ wavelength from the background wave by vector analysis. Diffracted waves combine with background waves through interference in the image plane to generate resultant particle (P) waves of altered amplitude that are perceived by the eye. See also *Particle wave* and *Surround wave*. *101*

Diffraction. The bending or breaking up of light that occurs when waves interact with objects, much in the way that waves of water bend around the edge of a log or jetty. Light waves that become scattered upon interacting with an object (diffracted waves) follow paths that deviate from the direction followed by waves that do not interact with the specimen (nondiffracted or undeviated waves). *20, 61*

Diffraction grating. A transparent or reflective substrate containing an array of parallel lines having the form of alternating grooves and ridges with spacings close to the wavelength of light. Light that is reflected by or transmitted through such a grating becomes strongly diffracted. Depending on the geometry of illumination and wavelength, a grating can generate color spectra and patterns of diffraction spots. *71, 75*

Diffraction plane. One of the aperture planes of the light microscope containing the focused diffraction image of the object. Under conditions of Koehler illumination, the diffraction plane is located in or near the back focal plane of the objective lens. *4*

Digital image processor. In video imaging, a signal processing device that converts analogue video signals to digital format for rapid image processing operations such as frame averaging, background subtraction, and contrast adjustment. *249*

Digitizer. See *Analogue-to-digital converter (ADC).*

Distortion. An aberration of lenses, where the magnification factor describing an image varies continuously between the central and peripheral portions of the image. Depending on whether the magnification is greater at the center or at the periphery, the distortion can be of the barrel or the pincushion type, respectively. *52*

Double refraction. In polarization optics, the splitting of light into distinct O and E rays in a birefringent material. When a birefringent crystal of calcite is placed on a page of printed words, the effects of double refraction are clearly observed as an overlapping, double image of the text. *124, 126*

DR. See *Dynamic range.*

Dwell time. In confocal microscopy, the length of time that the scanning laser beam remains at a unit of space corresponding to a single pixel in the image. In a laser scanning microscope, dwell time is typically 0.1–1.0 μs or more. Long dwell times increase the rate of photobleaching and decrease the viability of living cells. *218*

Dynamic range (DR). The term describing the resolution of light intensity, the number of steps in light amplitude in an image that can be resolved by an imaging device. This number can range from several hundred to a thousand for video and low-end CCD cameras to greater than 65,000 for the 16 bit CCD cameras used in astronomy. For CCD cameras, the DR is defined as the full well capacity of a pixel (the number of photoelectrons at saturation) divided by the camera's read noise. *216, 246, 274*

Elliptically polarized light. A form of polarized light whereby the E vector of a polarized wave rotates about the axis of propagation of the wave, thus sweeping out a right- or left-handed spiral. If you could view the wave end-on, the movement of the E vector would appear to trace the form of an ellipse. *131*

Emission filter. In fluorescence microscopy, the final element in a fluorescence filter cube, which transmits fluorescence emission wavelengths while blocking residual excitation wavelengths. Commonly called a barrier filter. Emission filters are colored-glass or interference filters and have the transmission properties of a bandpass or long-pass filter. *191*

Emission spectrum. In fluorescence, the spectrum of wavelengths emitted by an atom or molecule after excitation by a light or other radiation source. Typically, the emission spectrum of a dye covers a spectrum of wavelengths longer than the corresponding excitation spectrum. *181*

Epi-illumination. A common method of illumination in fluorescence microscopy, where the illuminator is placed on the same side of the specimen as the objective lens, and the objective performs a dual role as both a condenser and an objective. A dichroic mirror is placed in the light path to reflect excitatory light from the lamp toward the specimen and transmit emitted fluorescent wavelengths to the eye or camera. *190*

Equalize contrast function. See *Histogram equalization.*

Excitation filter. In fluorescence microscopy, the first element in a fluorescence filter cube and the filter that produces the exciting band of wavelengths from a broadband light source such as a mercury or xenon arc lamp. Commonly the excitation filter is a high-quality bandpass interference filter. *191*

Excitation spectrum. In fluorescence, the spectrum of wavelengths capable of exciting an atom or a molecule to exhibit fluorescence. Typically the excitation spectrum covers a range of wavelengths shorter than the corresponding fluorescence emission spectrum. *181*

Extinction and **extinction factor.** In polarization optics, the blockage of light transmission through an analyzer. This condition occurs when the vibrational plane of the E vector of a linearly polarized beam is oriented perpendicularly with respect to the transmission axis of the analyzer. If two Polaroid filters are held so that their transmission axes are crossed, extinction is said to occur when the magnitude of light transmission drops to a sharp minimum. The extinction factor is the ratio of amplitudes of transmitted light obtained when (a) the E vector of a linearly polarized beam and the transmission axis of the analyzer are parallel (maximum transmission) and (b) they are crossed (extinction). *120, 123, 159*

Extraordinary ray or **E ray.** In polarization optics, the member of a ray pair whose velocity varies with the direction of transmission through a birefringent medium. The surface wavefront of E waves emanating from an imaginary point source in a birefringent medium can be described as the surface of a three-dimensional ellipsoid. See also *Ordinary ray. 124*

Eyepiece or **ocular.** The second magnifying lens of the microscope used to focus a real magnified image on the retina of the real intermediate image produced by the objective. The added magnification provided by the eyepiece increases the angular magnification of the virtual image perceived by the eye. The typical range of eyepiece magnifications is 5–25×. *56*

Eyepiece telescope. See *Bertrand lens.*

Fast axis. In polarization optics, the long axis of the wavefront ellipsoid, a construction used to describe the surface of an emergent wavefront from a point source of light in a birefringent material. The fast axis indicates the direction of low refractive index in the specimen. See also *Refractive index ellipsoid. 128*

Fast Fourier transform (FFT). A filtering operation used to selectively diminish or enhance low or high spatial frequencies (extended vs. fine detailed structures) in the object image. In generating a transform, image details are separated into sinusoidal frequency components to create a map of spatial frequencies. The transform can be covered with a mask to enhance or diminish spatial frequencies of interest. Reversing the procedure produces an image (the inverse transform) in which contrast of spatial details is modified. *296*

FFT. See *Fast Fourier transform (FFT).*

Field diaphragm. A variable diaphragm located in or near the aperture plane of the light source that is used to reduce the amount of stray light in the object image. Since the edge of the diaphragm is conjugate with the object plane under conditions of Koehler illumination, the field diaphragm is used as an aid in centering and focusing the condenser lens. *10*

Field planes. That set of conjugate focal planes representing the field diaphragm, the object, the real intermediate image, and the retina. *4*

Flat-field correction. In image processing, the procedure used to obtain a photometrically accurate image from a raw image. A so-called dark frame containing bias and thermal counts is subtracted from the raw image and from a "flat" or "background" image. The dark-subtracted raw image is then divided by the dark-subtracted flat-field image to produce the corrected image. With operation, all optical faults are removed. The photometric relation of pixel values to photoelectron count is also lost during division, although the relative amplitudes of pixel values within an image are retained. See also *Dark frame* and *Flat-field frame. 289*

Flat-field frame. In image processing, a picture of featureless background that is used to prepare a flat-field-corrected image of the object. A flat-field frame is obtained by photographing a featureless region, close to and in the same focal plane as the object. *289*

Fluorescence. The process by which a suitable molecule, transiently excited by absorption of external radiation (including light) of the proper energy, releases the energy as a longer-wavelength photon. This process usually takes less than a nanosecond. *179, 181*

Fluorescence filter set. On a fluorescence microscope, an assembly of filters used to transmit excitation and emission wavelengths in an epi-illuminator. A filter set typically includes an exciter filter, a dichroic mirror, and an emission (or barrier) filter. *190*

Fluorescence microscopy. A mode of light microscopy whereby the wavelengths of fluo-rescence emission from an excited fluorescent specimen are used to form an image. *177*

Fluorite or **semiapochromat lens.** Objective lenses made of fluorite or Ca_2F, a highly transparent material of low color dispersion. The excellent color correction afforded by simple fluorite elements accounts for their alternative designation as semiapochromats. The maximum numerical aperture is usually limited at 1.3. *53*

Fluorochrome. A dye or molecule capable of exhibiting fluorescence. *181*

Fluorophore. The specific region or structural domain of a molecule capable of exhibiting fluorescence. Examples include the fluorescein moiety in a fluorescein-conjugated protein and the tetrapyrrole ring in chlorophyll. *181*

Focal length. The distance along the optic axis between the principal plane of a lens and its focal plane. For a simple converging (positive) lens illuminated by an infinitely distant point source of light, the image of the point lies precisely one focal length away from the principal plane. *43, 45*

Focal ratio or **f-number.** The ratio of the focal length of a lens to the diameter of its aperture. *65*

Fovea. A 0.2–0.3 mm diameter spot in the center of the macula on the retina that lies on the optic axis of the eye and contains a high concentration of cone cell photoreceptors for color vision and visual acuity in bright light conditions. *24*

Frame accumulation. In electronic imaging, the method of adding together a number of image frames to create a brighter image with improved signal-to-noise ratio. *249*

Frame averaging or **Kalman averaging.** In electronic imaging, the method of averaging a number of raw image frames to reduce noise and improve the signal-to-noise ratio. The signal-to-noise ratio varies as the square root of the number of frames averaged. *217*

Frame grabber board. In electronic imaging, the computer board that determines the frame memory, a remote access memory for storing the pixel values comprising an image. Grabbing is the process of acquiring and storing an image into a frame buffer. *235*

Frame-transfer CCD. A CCD whose pixel array is divided into two equal halves. One section is uncovered and contains active photosites; the other area is masked with a light-shielding metal film and acts as a storage site. This design of CCD speeds image capture, because readout of the storage site and acquisition at the image site occur simultaneously and because the camera can work without a shutter. *267*

Frequency of vibration. The number of periods per unit time or unit distance of an electromagnetic wave. *16*

Full-frame CCD camera. A CCD design in which all of the pixels function as active photosites. *267*

Full wave plate. In polarization optics, a birefringent plate (retarder) capable of introducing an optical path length difference between O and E rays equivalent to a full wavelength of light. Since the unique wavelength emerges with a linearly polarized waveform, positioning the plate diagonally between two crossed polars under conditions of white light illumination extinguishes the linearly polarized wavelength, resulting in the visual perception of an interference color. A first-order red plate shows a first-order red interference color because 551 nm green wavelengths have been removed. *141*

Full width at half maximum (FWHM). A parameter describing the spectral range of transmitted wavelengths of a bandpass filter. The cut-on and cut-off boundaries are defined as the wavelengths giving 50% of maximum transmittance of the filter. A FWHM of 20 indicates that the transmitted bandwidth spans 20 nm. Also called the half bandwidth (HBW). *37*

FWHM. See *Full width at half maximum (FWHM).*

Gain. In electronic imaging, the amplification factor applied to an input signal to adjust the amplitude of the signal output. In video, an input voltage signal might be multiplied by a positive gain factor to increase the voltage of the signal output and hence the brightness of the signal on the TV. In digital CCD microscopy the number of photoelectrons is divided by a gain factor (photoelectrons/ADU) to determine ADUs assigned to an image pixel. *221, 238, 270*

Gamma (γ). The exponent in the mathematical expression relating an output signal to its corresponding input signal for a camera, monitor, or other display device. In image processing, changing gamma increases or decreases the contrast between high and low pixel values according to the function, displayed value = [(normalized value − min)/(max − min)]$^\gamma$. For a video camera, the display signal and input signal are related through the relationship $i/i_D = (I/I_D)^\gamma$, where i and I are the output current and corresponding light intensity for a given signal; i_D and I_D are the dark current and its corresponding intensity. In both cases, γ is the slope of the curve in a log-log plot of this function. For $\gamma = 1$, the relationship between signal strength and display intensity is linear. *240, 287*

GFP. See *Green fluorescent protein (GFP)*.

Green fluorescent protein (GFP). A fluorescent 30 kDa protein from the jellyfish *Aequorea victoria,* which is commonly used as a fluorescent marker to determine the location, concentration, and dynamics of a protein of interest in cells and tissues. The excitation and emission maxima of enhanced GFP occur at 489 and 508 nm, respectively. The DNA sequence of GFP is ligated to the DNA encoding a protein of interest. Cells transfected with the modified DNA subsequently express fluorescent chimeric proteins in situ. *185*

Half bandwidth (HBW). See *Full width at half maximum (FWHM)*.

Halo. In phase contrast microscopy, characteristic contrast patterns of light or dark gradients flanking the edges of objects in a phase contrast image. Halos are caused by the phase contrast optical design that requires that the image of the condenser annulus and objective phase plate annulus have slightly different dimensions in the back focal plane of the objective. *108*

High-pass filter. In image processing, a filter that transmits high spatial frequencies and blocks low spatial frequencies. After filtering, edges and fine details are strongly enhanced, whereas large extended features are made less distinct. *293*

Histogram equalization or **histogram leveling.** In image processing, an operation that reassigns pixel values so that each step of the gray scale is represented by an approximately equal number of pixels. In equalization, the cumulative histogram distribution is scaled to map the entire range of the gray scale, the rank order of the original pixel values being maintained as closely as possible. This operation is used to increase the contrast of flat images in which the pixel values differ by only a small amount compared to the potential gray scale range of the image. *295*

Histogram stretching. In image processing, the operation whereby a limited range of gray-level values is displayed on a scale having a greater range of amplitudes, frequently from black to white. The effect of this operation is to increase the contrast of the selected range of gray levels. *286*

HMC. Hoffman modulation contrast. See *Modulation contrast microscopy (MCM)*.

Huygens' principle. A geometrical method used to show the successive locations occupied by an advancing wavefront. An initial source or wavefront is treated as a point source or a collection of point sources of light, each of which emits a spherical wave known as a Huygens' wavelet. The surface of an imaginary envelope encompassing an entire group of wavelet profiles describes the location of the wavefront at a later time, *t*. Huygens' principle is commonly used to describe the distribution of light energy in multiple interacting wavefronts as occurs during diffraction and interference. *72*

Image analysis. Any number of measurement operations of objects contained in an image including particle counts; geometrical measurements of length, width, area, centroid location, and so forth; intensity measurements; contour measurements; stereology; and many other operations. *283*

Image distance and **object distance.** With respect to the principal planes of a lens, the image-to-lens and object-to-lens distances, as predicted by the lens equation in geometrical optics. See also *Lens equation. 45*

Image histogram. A frequency histogram of pixel values comprising an image, with pixel values shown on the x-axis and the frequency of occurrence on the y-axis. The histogram allows one to see the relative contributions of dark, gray, and bright pixels in an image at a glance and is useful for adjusting image brightness and contrast. *284*

Image processing. The adjustment of an image's pixel values for purposes of image correction and measurement (flat-field correction) or display (adjustment of image brightness and contrast). *283*

Immunofluorescence microscopy. A mode of fluorescence microscopy in which a certain molecular species in a specimen is labeled with a specific fluorescent antibody. Fluorescence emission from excited antibodies is collected by the objective lens to form an image of the specimen. Antibodies can be made fluorescent by labeling them directly with a fluorescent dye (direct immunofluorescence) or with a second fluorescent antibody that recognizes epitopes on the primary antibody (indirect immunofluorescence). *179*

Incandescent lamp. A bulb containing an inert gas and metal filament that emits photons as the filament becomes excited during passage of electric current. The spectrum of visible wavelengths emitted by the filament shifts to increasingly shorter wavelengths as the amount of excitation is increased. The output of incandescent lamps is very high at red and infrared wavelengths. *29*

Index ellipsoid. See *Refractive index ellipsoid.*

Infinity corrected optics. The latest optical design for microscope objective lenses in which the specimen is placed at the focal length of the lens. Used by itself, the image rays emerge from the lens parallel to the optic axis and the image plane is located at infinity. In practice, a tube lens or Telan lens located in the body of the microscope acts together with the objective to form an image in the real intermediate image plane. This optical design relaxes constraints on the manufacture of the objective lens itself and allows for placement of bulky accessory equipment such as fluorescence filter cubes in the space between the objective and the tube lens. *50*

Integrated histogram. A modified form of an image histogram in which the x-axis indicates the pixel value and the y-axis indicates the cumulative number of pixels having a value of *x* and lower on the x-axis. The edge of the histogram defines a display function from which one can determine the rate of change in light intensity at any value of *x* along the gray scale. Useful for determining the gray-level midpoint and for determining if LUTs should be linear, exponential, etc. *287*

Intensifier silicon-intensifier target (ISIT) camera. A video camera tube used for low-light imaging applications. An ISIT tube is essentially a SIT tube modified by the addition of an image intensifier coupled by fiber optics as a first stage of light amplification. *250*

Intensity of light. Qualitatively, the brightness or flux of light energy perceived by the eye. By universal agreement, the term *intensity,* meaning the flow of energy per unit area per unit time, is being replaced by the word *irradiance,* a radiometric term indicating the average energy (photon flux) per unit area per unit time, or watts/meter^2. As a term describing the strength of light, intensity is proportional to the square of the amplitude of an electromagnetic wave. *22*

Interference. The sum of two or more interacting electromagnetic waves. Two waves can interfere only if a component of the E vector of one wave vibrates in the plane of the other wave. Resultant waves with amplitudes greater or less than the constituent waves are said to represent constructive and destructive interference, respectively. *63*

Interference color. The color that results from removal of a band of visible wavelengths from a source of white light. *39, 141*

Interference filter. A filter made from alternating layers of different dielectric materials or layers of a dielectric material and thin metal film that transmits a specific band of wavelengths. The spacings between the layers of one-quarter or one-half wavelength allow constructive interference and reinforce propagation through the filter of a particular wavelength λ. All other wavelengths give destructive interference and are absorbed or reflected and do not propagate through the filter. *39*

Interline transfer CCD. A form of CCD having alternate columns of pixels that function as exposed photosites and masked storage sites. During operation, the masked pixels are read out and digitized while exposed sites simultaneously capture photoelectrons. This double-duty action speeds up camera operation. Newer interline CCDs contain microlenses that cover storage- and photosite-pixel pairs to increase light-gathering efficiency. Upon completion of an exposure, the transfer of signal charges from a photosite pixel to an adjacent storage pixel is so fast that a camera shutter is not required. *268*

Inverse transform. In image processing, the image created by a Fourier transform operation used primarily to blur and sharpen images. A Fourier transform looks like a diffraction pattern and represents spatial details as spatial frequencies. After applying a mask to select low- or high-spatial frequency information, the transform is converted back into an image (the inverse transform). Spatial frequency masking with a Fourier transform gives similar results as blurring and sharpening convolution filters, but can give a more even result. *81, 299*

Ion arc lamp. Lamps containing an ionized gas or plasma between two electrodes that radiates visible wavelengths when excited by an electric current. Arc lamps used in light microscopy usually contain mercury vapor or xenon gas. *30*

Irradiance of light. The radiometrically correct term for light intensity. Irradiance is the radiant flux incident per surface unit area and is given as watts/meter^2. Irradiance is a measure of the concentration of power.

ISIT camera. See *Intensifier silicon-intensifier target (ISIT) camera.*

Isotropic. In describing the optical properties of an object or propagation medium, having identical properties in different directions. *121*

Jablonski diagram. A diagram showing the energy levels occupied by an excited electron in an atom or molecule as steps on a vertical ladder. Singlet and triplet excited states are shown separately as ladders standing next to each other. *180*

Kernel. See *Convolution.*

Koehler illumination. The principal method for illuminating specimens in the light microscope, whereby a collector lens near the light source is used to focus an image of the light source in the front aperture of the condenser. The microscope condenser is focused to position the conjugate image of the light source in the back focal plane (diffraction plane) of the objective lens. The method provides bright, even illumination across the diameter of the specimen. *7*

Lens equation. In geometrical optics, the equation $1/f = 1/a + 1/b$ describing the relationship between the object distance a and the image distance b for a lens of focal length f. *46*

Light microscope. A microscope employing light as an analytic probe and optics based on glass lenses to produce a magnified image of an object specimen. *1*

Linearly polarized light. A beam of light in which the E vectors of the constituent waves vibrate in planes that are mutually parallel. Linearly polarized light need not be coherent or monochromatic. *117*

Long-pass filter. A colored glass or interference filter that transmits (passes) long wavelengths and blocks short ones. *37*

Long working distance lens. An objective lens having a working distance many times greater than that of a conventional objective lens of the same magnification. A long working distance lens is sometimes easier to employ and focus, can look deeper into transparent specimens, and allows the operator greater working space for employing micropipettes or other equipment in the vicinity of the object. However, the NA and resolution are less than those for conventional lenses of comparable magnification. *54*

Look-up-table (LUT). In image processing, a mathematical function that converts input values of the signal into output values for display. *284*

Lumen. A unit of luminous flux equal to the flux through a unit solid angle (steradian) from a uniform point source of 1 candle intensity. *284*

LUT. See *Look-up-table (LUT)*.

Lux. A unit of illumination equal to 1 lumen per square meter.

Magnitude. See *Amplitude of an electromagnetic wave*.

Median filter. In image processing, a nonlinear filter for blurring and removing grain. The filter uses a kernel that is applied to each pixel in the original image to calculate the median value of a group of pixels covered by the kernel. The median values computed at each pixel location are assigned to the corresponding pixel locations in the new filtered image. *295*

Microchannel plate. A device for amplifying the photoelectron signal in Gen II (and higher-generation) image intensifiers. The plate consists of a compact array of capillary tubes with metalized walls. Photoelectrons from the intensifier target are accelerated onto the plate where they undergo multiple collision and electron amplification events with the tube wall, which results in a large electron cascade and amplification of the original photon signal. *251*

Modulation contrast microscopy (MCM). A mode of light microscope optics in which a transparent phase object is made visible by providing unilateral oblique illumination and employing a mask in the back aperture of the objective lens that blocks one sideband of diffracted light and partially attenuates the 0th-order undeviated rays. In both MCM and DIC optics, brightly illuminated and shadowed edges in the three-dimensional relief-like image correspond to optical path gradients (phase gradients) in the specimen. Although resolution and detection sensitivity are somewhat reduced compared with DIC, the MCM system produces superior images at low magnifications, allows optical sectioning, and lets you examine cells on birefringent plastic dishes. *171*

Modulation transfer function (MTF). A function showing percent modulation in contrast vs. spatial frequency. MTF is used to describe the change in contrast between input signals and output signals for optical and electronic signal-handling devices. To obtain the MTF for an optical system, a test pattern of alternating black and white bars can be used, whose amplitudes are shown as sinusoidal curves and whose interline spacing changes from large to small across the diameter of the pattern. At large spacings where the contrast is maximal, the percent modulation is 1; at the limit of spatial resolution for a device, the so-called cutoff frequency, the percent modulation drops to 0. *252*

Molar extinction coefficient, ϵ. In spectrophotometry, the factor used to convert units of absorbance into units of molar concentration for a variety of chemical substances. ϵ is given as the absorbance for a 1 M concentration and a 1 cm path length at a reference wavelength, usually the wavelength giving maximum absorbance in its absorption spectrum. *182*

Monochromatic. In theory, light composed of just one wavelength, but in practice, light that is composed of a narrow band of wavelengths. Owing to Heisenberg's uncertainty principle, true monochromatic light does not exist in nature. Even the monochromatic emission from a laser or an excited atomic source has a measurable bandwidth. Therefore, while the light produced by a narrow bandpass interference filter is called monochromatic, this is just an approximation. *20*

MTF. See *Modulation transfer function (MTF).*

Multi-immersion objective lens. An objective lens whose spherical aberration is corrected for use by immersion in media of various refractive indices, including water, glycerin, and oil. A focusable lens element used to minimize spherical aberration is adjusted by rotating a focus ring on the barrel of the objective. *54*

Multiple fluorescence filter set. A filter set for simultaneous viewing or photography of multiple fluorescent signals. The transmission profile of each filter in the set contains multiple peaks and troughs for the reflection and transmission of the appropriate excitation and emission wavelengths as in a conventional single-fluorochrome filter set. Because of constraints on the widths of bandwidths, the steepness of transmission profiles, and the inability to reject certain wavelengths, the performance is somewhat less than that of individual filter sets for specific fluorochromes. *194*

NA. See *Numerical aperture (NA).*

Negative colors. Colors resulting from the removal of a certain band of visible wavelengths. Thus, white *light* minus blue gives the negative color yellow, because simultaneous stimulation of red and green cone cells results in this color perception. Similarly, the mixture of cyan *pigment* (absorbs red wavelengths) and yellow pigment (absorbs blue wavelengths) gives green, because green is the only reflected wavelength in the pigment mixture. *25*

Negative lens. A lens that diverges a beam of parallel incident rays. A simple negative lens is thinner in the middle than at the periphery and has at least one concave surface. It does not form a real image, and when held in front of the eye, it reduces or demagnifies. *43*

Negative phase contrast. In phase contrast optics, the term applies to systems employing a negative phase plate that retards the background 0th-order light by $\lambda/4$ relative to the diffracted waves. Since the diffracted light from an object is retarded $\sim\lambda/4$ relative to the phase of the incident light, the total amount of phase shift between background and diffracted waves is 0 and interference is constructive, causing objects to appear bright against a gray background. *106*

Neutral density (ND) filter. A light-attenuating filter that reduces equally the amplitudes of all wavelengths across the visible spectrum. The glass substrate contains light-absorbing colloids or is coated on one surface with a thin metal film to reduce transmission. Neutral density filters are labeled according to their absorbance or fractional transmission. *38*

Nipkow disk. In confocal microscopy, a thin opaque disk with thousands of minute pinholes, which when rotated at high speed provides parallel scanning of the specimen with thousands of minute diffraction-limited spots. The return fluorescence emission is refocused at the same pinhole in the disk, which provides the same function in rejecting out-of-focus light as does a single pinhole in a conventional confocal microscope. Nipkow disk confocal microscopes produce a real image that can be inspected visually or recorded on a high-resolution CCD camera, whereas images of single-spot scanning microscopes are reconstructed from signals from a PMT and are displayed on a computer monitor. *229*

NTSC (National Television Systems Committee). The format for color television broadcasting, based on a 525-line, 60-field/s (30 frames/s) format, in use in the United States, Japan, and other countries. *236*

Numerical aperture (NA). The parameter describing the angular aperture of objective and condenser lenses. NA is defined as $n \sin\theta$, where n is the refractive index of the medium between the object and the lens, and θ, the angle of light collection, is the apparent half-angle subtended by the front aperture of the lens as seen from a point in the specimen plane. *85*

Nyquist criterion. With respect to preservation of spatial resolution in electronic imaging, the requirement that a resolution unit (the spacing between adjacent features in a periodic specimen) be registered by at least two contiguous sampling units (pixels, raster lines) of the detector in order to be represented correctly in the image. Thus, an object consisting of a pattern of alternating black and white lines will be represented faithfully by a detector capable of providing two or more image samples per black and white line. *249*

Object distance. See *Image distance.*

Objective lens. The image-forming lens of the microscope responsible for forming the real intermediate image located in the front apertures of the eyepieces. *2*

Offset. In electronic imaging, the electronic adjustment that is made to set the black level in the image. Adjusting the offset adds a voltage of positive or negative sign sufficient to give the desired feature or background a display intensity of 0. *221, 240*

Optical path length. In wave optics, a measure of the time or distance (measured in wavelengths) defining the path taken by a wave between two points. Optical path length is defined as $n \times t$, where n is the refractive index and t indicates the thickness or geometrical distance. A complex optical path composed of multiple domains of different refractive index and thickness is given as $\Sigma\, n_1 t_1 + n_2 t_2 + \ldots n_i t_i$. *68, 103*

Optical path length difference. The difference in the optical path lengths of two waves that experience refractive index domains of different value and thickness. In interference optics, differences in optical path length determine the relative phase shift and thus the degree of interference between 0th-order and higher-order diffracted waves that have their origins in a point in the object. *69, 103, 108, 127*

Optic axis. In an aligned optical system, a straight line joining the centers of curvature of lens surfaces contained in the system. In polarized light optics, the path followed by the ordinary or O ray in a birefringent material. *7, 125*

Optovar. A built-in magnification booster lens that can be rotated into the optical path to further increase the magnification provided by the objective by a small amount. *57*

Ordinary ray or **O ray.** In polarization optics, the member of a ray pair that obeys normal laws of refraction and whose velocity remains constant in different directions during transmission through a birefringent medium. See also *Extraordinary ray. 124*

Panning. In electronic imaging, the movement of the camera to bring into view portions of an object field that cannot be included in a single image frame; in image processing, the relative displacement of one image over another for purposes of alignment. *314*

Paraboloid condenser. A high numerical aperture condenser for dark-field microscopy having a reflective surface that is a segment of a figure of revolution of a parabola. The steeply pitched illumination cone produced by the condenser is suitable for dark-field examination with high-power oil immersion objectives. *114*

Parallel register. In CCD cameras, the extended array of imaging (and storage) pixels of the imaging area of a CCD device. Columns of pixels in the parallel register deliver their charge packets (photoelectrons) to a separate serial register from which the image signal is read and digitized by the camera electronics. *261*

Parfocal. The property of having the same distance between the specimen and the objective turret of the microscope. With parfocal lenses, one can focus an object with one lens and then switch to another lens without having to readjust the focus dial of the microscope. *9*

Particle wave. In phase contrast and other modes of interference microscopy, the wave (P wave) that results from interference between diffracted and surround waves in the image plane, and whose amplitude is different from that of the surrounding background, allowing it to be perceived by the eye. See also *Diffracted wave* and *Surround wave*. *101*

Peltier thermoelectric device. A compact bimetallic strip that becomes hot on one surface and cold on the other during the application of a current. Peltier devices are commonly used in CCD cameras where it is necessary to quickly and efficiently cool the CCD 50–60°C below ambient temperature in a compact space. *267*

Phase. Relative position in a cyclical or wave motion. Since one wavelength is described as 2π radians or 360°, the phase of a wave is given in radians or degrees or fractions of a wavelength. *63*

Phase contrast microscopy. A form of interference microscopy that transforms differences in optical path in an object to differences in amplitude in the image, making transparent phase objects appear as though they had been stained. Surround and diffracted rays from the specimen occupy different locations in the diffraction plane at the back aperture of the objective lens where their phases are differentially manipulated in order to generate a contrast image. Two special pieces of equipment are required: a condenser annulus and a modified objective lens containing a phase plate. Because the method is dependent on diffraction and scattering, phase contrast optics differentially enhance the visibility of small particles, filaments, and the edges of extended objects. The technique allows for examination of fine details in transparent specimens such as live cells. *97*

Phase gradient. In interference microscopy, the gradient of phase shifts in an image corresponding to optical path differences in the object. *154*

Phase object. Objects that shift the phase of light as opposed to those that absorb light (amplitude objects) as the basis for image formation. See also *Amplitude object*. *97*

Phase plate. In phase contrast microscopy, a transparent plate with a semitransparent raised or depressed circular annulus located at the rear focal plane of a phase contrast objective. The annulus reduces the amplitude of background (0th order) waves and advances or retards the phase of the 0th-order component relative to diffracted waves. Its action is responsible for the phase contrast interference image. *105*

Phase or **centering telescope.** See *Bertrand lens*.

Phosphorescence. The relatively slow ($>10^{-9}$ s) emission of photons after excitation of a material by light or other radiation source. *181*

Photobleaching. The diminution of fluorescence emission due to the chemical modification of a fluorochrome upon continued exposure to excitation wavelengths. Photon-induced damage resulting in photobleaching is largely due to the generation of free oxygen radicals that attack and permanently destroy the light-emitting properties of the fluorochrome. The rate of photobleaching can be reduced by including anti–free radical reagents such as ascorbate or by reducing the concentration of oxygen in the mounting medium. Bleaching is usually irreversible. *183, 223*

Photodiode. A semiconductor device such as a CCD for detecting and measuring light by means of its conversion into an electric current. Incident photons generate charge carriers (electrons and electron holes) in the CCD matrix that support the conduction of a current in the presence of an applied voltage. *261*

Photomultiplier tube (PMT). An electrical device for amplifying photon signals. Photons striking a target at the face of the PMT liberate free electrons, which are accelerated onto a dynode that in turn liberates a shower of electrons. By arranging several dynodes in a series, a great amplification of the original photon is achieved, which is then transmitted to other signal-processing circuits. PMTs amplify photon signals but do not form an image as does an image intensifier. *209*

Photon limited. In image processing, a signal is said to be photon limited when the photon noise is greater than the electronic noises associated with the camera (bias, thermal, and read noises). *301*

Photon noise or **shot noise.** In image processing, the noise associated with the photon signal itself. Since the accumulation of photons per unit area or time is random (Poissonian) by nature, the photon signal is typically described as the mean of a collection of measurements, and the noise is described as the square root of the photon signal. Since the signal is based on "photoelectron counts," the photon noise is calculated in terms of electrons, not in the digitized signal units (analogue-to-digital units or ADUs) used by the computer. *273, 301*

Photopic vision. The mode of vision based on cone cell photoreceptors in the retina that provides visual acuity and color perception under bright light conditions. *24*

Pinhole aperture. In confocal microscopy, the variable diaphragm in the real intermediate image plane that is adjusted to receive the focused spot of fluorescent light that is emitted by an excited laser-scanned spot in a fluorescent specimen. A photomultiplier tube located behind the pinhole generates an electrical signal that is proportional to the number of photons passing through the pinhole. *208*

Pixel. A "picture element" in electronic cameras. In confocal microscopes, pixels correspond to the photon signal received and digitized per unit of sampling time during laser scanning. In CCD cameras, pixels are square or rectangular areas of silicon substrate that are delimited by conducting strips on the CCD surface. *236*

Plane parallel. In wave optics, the term applies to waves vibrating in a plane that is parallel to some reference plane, but not necessarily in the reference plane itself. See also *Linearly polarized light. 117*

PMT. See *Photomultiplier tube (PMT).*

Polar. The common term applied to a sheet of linear polarizing film (dichroic filter or Polaroid filter) and particularly to its use as a polarizer or analyzer in producing and analyzing polarized light. *119*

Polarizability. In polarization optics, a property describing the strength of interaction of light with molecules in a manner that depends on the orientation of atomic bonds. Light waves interact more strongly with molecules when their E vectors are oriented parallel to the axis defining light-deformable (polarizable) covalent bonds such as the axes of long-chain hydrocarbon polymers like polyvinyl alcohol, cellulose, and collagen. This geometry is supported when an incident light ray is perpendicular to the long axis of the polymer. Interaction of light with molecules along their polarizable axis retards wave propagation and accounts for the direction-dependent variability in their refractive index, a property known as birefringence. *130*

Polarization cross. In polarization microscopy, the appearance of a dark upright cross in the back aperture of the objective lens under conditions of extinction with two crossed polars. Ideally, the back aperture is uniformly dark under this condition, but the depolarization of light by the curved lens surfaces of the condenser and objective lenses causes brightenings in four quadrants and hence the appearance of a cross. *138*

Polarization microscopy. A mode of light microscopy based on the unique ability of polarized light to interact with polarizable bonds of ordered molecules in a direction-sensitive manner. Perturbations to waves of polarized light from aligned molecules in an object result in phase retardations between sampling beams, which in turn allow interference-dependent changes in amplitude in the image plane. Typically the microscope contains a polarizer and analyzer, and a retardation plate or compensator. Image formation depends critically on the existence of ordered molecular arrangements and a property known as double refraction or birefringence. *135*

Polarized light. Light waves whose E vectors vibrate in plane-parallel orientation at any point along the axis of propagation. Polarized light can be linearly polarized (vibrations at all locations are plane parallel) or elliptically or circularly polarized (vibration axis varies depending on location along the propagation axis). Polarized light need not be monochromatic or coherent. *20*

Polarizer. A device that receives random light and transmits linearly polarized light. In microscopy, polarizers are made from sheets of oriented dichroic molecules (Polaroid filter) or from slabs of birefringent crystalline materials. *118*

Polaroid sheet or **polar.** A sheet of aligned long-chain polyvinyl alcohol molecules impregnated with aligned microcrystals of polyiodide. The E vectors of incident waves vibrating along the axis parallel to the crystal axes are absorbed and removed, resulting in the transmission of waves that are linearly polarized. *119*

Positive colors. Colors that result from mixing different wavelengths of light. The equal mixture of red and green wavelengths results in the perception of yellow, a positive color. *24*

Positive lens. A lens that converges a beam of parallel incident rays. A simple positive lens is thicker in the middle than at the periphery, and has at least one convex surface. A positive lens forms a real image and enlarges or magnifies when held in front of the eye. *43*

Positive phase contrast. In phase contrast optics, the term applies to systems employing a positive phase plate that advances the background wave by $\lambda/4$ relative to the diffracted wave. Since the diffracted light from an object is retarded $\sim\lambda/4$ relative to the phase of the incident light, the total phase shift between background and diffracted waves is $\lambda/2$ and interference is destructive, causing objects to appear dark against a gray background. *105*

Principal plane. For a simple thin lens, the plane within the lens and perpendicular to the optic axis from which the focal length is determined. Thick simple lenses have two principal planes separated by an intervening distance. Complex compound lenses may have multiple principal planes. *43*

Processed image. A raw image after it has been subjected to image processing. *244*

Progressive scan cameras. A reference to video cameras and camcorders with interline CCD detectors. The designation "progressive scan" indicates that the entire image signal is read off the chip at one time and that images are displayed as a series of complete frames without interleaving as in the case of conventional broadcast video signals. *268*

QE. See *Quantum efficiency.*

Quantum efficiency (QE). The fraction of input photons that are recorded as signal counts by a photon detector or imaging device. *182, 272*

Quenching. The reduction in fluorescence emission by a fluorochrome due to environmental conditions (solvent type, pH, ionic strength) or to a locally high concentration of fluorochromes that reduces the efficiency of fluorescence emission. *183*

Raster. A zigzag pattern of straight line segments, driven by oscillators, used to scan a point source over the area covered by a specimen or an image. *208, 236*

Raw image or **raw frame.** In image processing, the name of an image prior to any modifications or processing. *244, 289*

Rayleigh criterion for spatial resolution. The criterion commonly used to define spatial resolution in a lens-based imaging device. Two point sources of light are considered to be just barely resolved when the diffraction spot image of one point lies in the first-order minimum of the diffraction pattern of the second point. In microscopy, the resolution limit d is defined, $d = 1.22 \, \lambda/(NA_{objective} + NA_{condenser})$, where λ is the wavelength of light and NA is the numerical aperture of the objective lens and of the condenser. *88*

Readout noise or **read noise.** In digital imaging, read noise refers to the noise background in an image due to the reading of the image and therefore the CCD noise associated with the transfer of charge packets between pixels, preamplifier noise, and the digitizer noise from the analogue-to-digital converter. For scientific CCD cameras, the read noise is usually 2–20 electrons/pixel. *273*

Readout rate. The rate in frames/s required for pixel transfer, digitization, and storage or display of an image. *269*

Real image. An image that can be viewed when projected on a screen or recorded on a piece of film. *45*

Real intermediate image. The real image focused by the objective lens in the vicinity of the oculars of the microscope. *2, 3*

Red fluorescent protein (RFP). A fluorescent protein from a sea anemone of the genus *Discosoma* used as a fluorescent marker to determine the location, concentration, and dynamics of a protein of interest in cells and tissues. The DNA sequence of RFP is ligated to the DNA encoding a protein of interest. Cells transfected with the modified DNA subsequently express fluorescent chimeric proteins in situ. *185*

Refraction. The change in direction of propagation (bending) experienced by a beam of light that passes from a medium of one refractive index into another medium of different refractive index when the direction of propagation is not perpendicular to the interface of the second medium. *20*

Refractive index ellipsoid and **wavefront ellipsoid.** An ellipsoid is the figure of revolution of an ellipse. When rotated about its major axis, the surface of the ellipsoid is used to describe the surface wavefront locations of E waves propagating outward from a central point through a birefringent material. The same kind of figure is used to describe the orientation and magnitude of the two extreme refractive index values that exist in birefringent uniaxial crystals and ordered biological materials. *128, 130*

Region of interest or **ROI.** In image processing, the subset of pixels defining the "active region" of an image, determined by defining xy coordinates or by drawing with a mouse over the image. The ability to define an ROI is useful during image processing, and when saving a part of an image as a new image. *269*

Relative error. The reciprocal of the signal-to-noise ratio and therefore the ratio of the noise to the signal. *302*

Relative retardation. In polarization optics, the relative shift in phase between two waves expressed in fractions of a wavelength. *127, 139*

Relay lens. An intermediate magnifying lens in an imaging system placed between the objective and the real intermediate image. In video, so-called TV lenses increase the magnification of the image projected on the camera 2- to 8-fold. *246*

Resolving power. See *Spatial resolution.*

Retardation plate or **retarder.** In polarization optics, a birefringent plate positioned between the polarizer and analyzer that introduces a relative retardation between the O and E rays in addition to that produced by a birefringent object to change image contrast or render path differences in color. When used in a device to measure the amount of relative retardation in the object, the plate is called a compensator. See also *Compensator. 136, 139*

RGB. Red, green, and blue. A mode of color image display that is based on the red-green-blue tricolor stimulus system for the visual perception of color. Commonly used in television, computer monitors, and other image display devices. *313*

Rod cell photoreceptors. Retinal cells located outside the fovea and in the periphery of the retina responsible for vision in dim light. Rod cell vision, or scotopic or night vision, is absent of color. *23*

ROI. See *Region of interest.*

RS-170 broadcast format. The format used for signal generation and display in commercial television broadcasting, based on the 2:1 interlace of two raster fields producing an image frame of 485 display lines at a rate of 30 frames/s. *236*

RS-330 broadcast format. A variant of the RS-170 format frequently used for closed-circuit television, differing in how raster lines are encoded, synchronized, and displayed on the monitor. The format is based on the 2:1 interlace of two raster fields producing an image frame of 485 display lines at a rate of 30 frames/s. *236*

Scotopic vision. Night vision based on rod cell photoreceptors. See *Rod cell photoreceptors. 23*

Serial register. In CCD cameras, a single row of pixels that receives charge packets from all of the adjacent columns of pixels in the parallel register. The serial register transmits the charge packets pixel by pixel to the on-chip preamplifier and analogue-to-digital converter until the register is emptied. *262*

Shade-off. In phase contrast microscopy, the gradient in light intensity that is observed from the edge to the center of extended phase objects of uniform thickness and refractive index. In central regions of uniform objects where diffraction is minimal, the corresponding amplitude in the image can approach that of the background. *109*

Shading correction. In video microscopy, the electronic controls of an in-line digital image processor that are used to remove light gradients across the image field, such as those arising from uneven illumination. *241*

Shear axis. In DIC microscopy, the axis of displacement of the O and E rays caused by the DIC prisms. In a DIC image, the axis is identified as the line connecting the bright and dark shaded edges of refractile objects. *159*

Short-pass filter. A colored-glass or interference filter that transmits (passes) short wavelengths and blocks long ones. *37*

Shot noise. See *Photon noise.*

Signal-to-noise (S/N) ratio. The ratio of the signal of an object to the noise of the surrounding background, where noise is the square root of the sum of the variances of contributing noise components. In the case that noise is photon limited and background noise may be approximated by the square root of the background signal, S/N gives the number of standard deviations that distinguish the object signal from the mean signal of the background. *217, 247, 276, 299*

Sign of birefringence. In polarization optics, a reference to the sign of birefringence b, where $b = n_e - n_o$. For the case that the refractive index describing the trajectory of the E ray is greater than that describing the trajectory taken by the O ray, the sign of birefringence of the material is positive. *127*

Silicon-intensifier target (SIT) camera. A video camera tube for imaging under low-light conditions. The camera tube contains a photocathode that accelerates photoelectrons onto a silicon diode target plate, which greatly amplifies the signal. A scanning electron beam neutralizes the target while generating a beam current containing the signal. *250*

Simple lens. A lens consisting of a single lens element and distinct from a compound lens having multiple lens elements. *45*

SIT camera. See *Silicon-intensifier target (SIT) camera.*

Slow axis. In polarization optics, the short axis of the wavefront ellipsoid, a construction used to describe the surface of an emergent wavefront from a point source of light in a birefringent material. The slow axis indicates the direction of high refractive index in the specimen. See also *Refractive index ellipsoid. 128*

Slow-scan CCD camera. Another designation for a full-frame CCD design. The time required for the serial readout of the parallel register of the CCD is slow compared to the time to read an interline CCD of comparable size and dimensions. *259*

S/N ratio. See *Signal-to-noise (S/N) ratio.*

Spatial filter. A filter that selectively manipulates a location in an image such as an aperture in a field plane of a microscope or a sharpening or blurring filter in image processing. *208, 250*

Spatial frequency. The reciprocal of the distance between two objects (periods/distance). *245*

Spatial frequency filter. A filter that selectively manipulates a location in the diffraction plane in a microscope (aperture plane masks in modulation contrast microscopy) or a mask applied to Fourier transforms to manipulate low and high spatial frequency information in image processing. *171, 296*

Spatial resolution. The resolution of component features in an image. In optical systems, resolution is directly proportional to the wavelength and inversely proportional to the angular aperture. The practical limits on wavelength and angular aperture determine the limit of spatial resolution, which is approximately one-half the wavelength of light. *87, 215, 245, 272*

Spectral range. The range of wavelengths, or bandwidth, under consideration. *272*

Spectroscope. A device for determining the wavelength of a certain emission line, bandwidth, or color. A diffraction grating is positioned between the eye and a narrow slit, and the eye-slit axis is directed at a target light source. The grating produces a spectrum of the constituent wavelengths admitted by the slit, and the spectrum is superimposed on a ruling so that you can determine the wavelength of a particular spectral color by simple inspection. *25*

Spherical aberration. A lens aberration typical of lenses with spherical surfaces that causes paraxial rays incident on the center and periphery of a lens to be focused at different locations in the image plane. The degree of aberration increases with the decreasing focal ratio of the lens. The aberration can be corrected in simple lenses by creating aspherical surfaces. *52*

Stepper motor. A motor whose drive shaft does not rotate continuously, but advances in discrete intervals or steps. *205*

Stokes shift. The distance in nanometers between the peak excitation and peak emission wavelengths of a fluorescent dye. *182*

Subarray readout. An option for image acquisition with a CCD camera whereby a portion of the total available imaging area of the CCD is selected as the active area for acquiring an image. Selection of the subarray region is made in the image acquisition software. In subarray readout mode, the acquisition rate is fast, and images take up less storage space on the hard drive. *269*

Super-resolution. In electronic imaging, the increase in spatial resolution made possible by adjusting the gain and offset of a camera. In confocal microscopy, super-resolution is obtained by constricting the confocal pinhole to about one-quarter of the Airy disk diameter. *216*

Surround wave or **background wave.** In phase contrast and other modes of interference microscopy, waves that traverse an object but do not interact with it. Surround waves are not deviated by the object and do not become altered in phase. For purposes of describing diffraction and interference, such waves are called the 0th-order component. Surround (S) waves combine with diffracted (D) waves through interference in the image plane to generate resultant particle (P) waves of altered amplitude that are perceived by the eye. See also *Diffracted wave* and *Particle wave*. *99*

System MTF. A function describing the percent modulation (percent reduction in the peak to trough amplitude difference for a signal) of a signal resulting from transit through a series of signal-handling devices. For a cascaded series of devices, the system MTF for a given frequency f is the product of the values of percent modulation for each individual component in the system so that % modulation of the system = $a\% \times b\% \times c\% \dots$ *252*

Thermal noise. In CCD imaging, the noise of the thermal signal in an image caused by the kinetic vibration of silicon atoms in the matrix of a CCD device. Thermal noise is considerably reduced by cooling the CCD to $-20°C$. Low-light-level cameras used in astronomy are sometimes cooled to the temperature of liquid nitrogen to effectively eliminate thermal noise. *264*

Thin lens. A lens whose thickness is small compared to its focal length. A line through the center of the lens (a plane representing the two coincident principal planes of the lens) provides a reasonably accurate reference plane for refraction and object and lens distance measurements. Lenses are assumed to be thin when demonstrating the principles of graphical ray tracing. *45*

Tube lens or **Telan lens.** An auxiliary lens in the body of the microscope, which in conjunction with an infinity focus objective lens forms the real intermediate image. The Telan lens provides some of the correction for chromatic aberration, which lessens constraints on the manufacture of the objective lens. *50*

Two-photon and **multi-photon laser scanning microscopy.** In laser scanning confocal microscopy, a method of fluorochrome excitation based on an infrared laser beam whose energy density is adjusted to allow frequency doubling or tripling at the point of beam focus in the specimen. Thus, molecules that simultaneously absorb two or three photons of 900 nm fluoresce the same as if excited by a single higher-energy photon of 450 or 300 nm, respectively. The method allows deep penetration into thick tissues. Because fluorescence emission is contained within a single focal plane, a variable pinhole aperture is not required before the detector. *226*

Uniaxial crystal. A birefringent crystal characterized by having a single optic axis. *126*

Unsharp masking. An image sharpening procedure, in which a blurred version of the original image (called an unsharp mask) is subtracted from the original to generate a difference image in which fine structural features are emphasized. Edges in the difference image are sharpened, and the contrast between bright and faint objects is reduced. *296*

Video electron tube. An electron tube that functions as a video pickup tube for recording an image for subsequent display on television. The tube contains a photosensitive target, magnetic coils for deflecting an electron beam in a raster over the target, and electronics for generating an analogue voltage signal from an electric current generated during scanning of the target. *236*

Video-enhanced contrast microscopy. In video microscopy, a method of image enhancement (contrast enhancement) in which the offset and gain positions of the camera and/or image processor are adjusted close together to include the gray-level values of an object of interest. As a result, the object image is displayed at very high contrast, making visible features that can be difficult to distinguish when the image is displayed at a larger dynamic range. See *Histogram stretching.* *242*

Virtual image. An image that can be perceived by the eye or imaged by a converging lens, but that cannot be focused on screen or recorded on film as can be done for a real image. The image perceived by the eye when looking in a microscope is a virtual image. *45*

Wavefront ellipsoid. See *Refractive index ellipsoid.*

Wavelength. The distance of one beat cycle of an electromagnetic wave. Also, the distance between two successive points at which the phase is the same on a periodic wave. The wavelength of light is designated λ and is given in nanometers. *16*

Wollaston prism. In interference microscopy, a beam splitter made of two wedge-shaped slabs of birefringent crystal such as quartz. In differential interference contrast (DIC) microscopy, specimens are probed by pairs of closely spaced rays of linearly polarized light that are generated by a Wollaston prism acting as a beam splitter. An important feature of the prism is its interference plane, which lies inside the prism (outside the prism in the case of modified Wollaston prism designs). *157*

Working distance. The space between the front lens surface of the objective lens and the coverslip. Lenses with high NAs typically have short working distances (60–100 μm). Lenses with longer working distances allow you to obtain focused views deep within a specimen. *9, 11*

Zone-of-action effect. See *Shade-off.*

Zoom factor. In confocal microscopy, an electronically set magnification factor that is used to provide modest adjustments in magnification and to optimize conditions of spatial resolution during imaging. Since the spatial interval of sampling is small at higher zoom settings, higher zoom increases the spatial resolution. However, since the same laser energy is delivered in a raster of smaller footprint, increasing the zoom factor also increases the rate of photobleaching. *219*

REFERENCES

Allen, R. D. (1985). New observations on cell architecture and dynamics by video-enhanced contrast optical microscopy. *Ann. Rev. Biophys. Biophysical Chem.* 14, 265–290.

Allen, R. D., David, G. B., and Nomarski, G. (1969). The Zeiss-Nomarski differential interference equipment for transmitted-light microscopy. *Z. Wiss. Mikroskopie u. Mikrotechnologie* 69, 193–221.

Allen, R. D., Allen, N. S., and Travis, J. L. (1981a). Video-enhanced contrast: differential interference contrast (AVEC-DIC) microscopy: a new method capable of analyzing microtubule-related motility in the reticulopodial network of *Allogromia laticollaris. Cell Motility* 1, 291–302.

Allen, R. D., Travis, J. L., Allen, N. S., and Yilmaz, H. (1981b). Video-enhanced contrast polarization (AVEC-POL) microscopy: a new method applied to the detection of birefringence in the motile reticulopodial network of *Allogromia laticollaris. Cell Motility* 1, 275–289.

Barer, R., Ross, K. F. A., and Tkaczyk, S. (1953). Refractometry of living cells. Nature 171, 720–724.

Bennett, A. H., Jupnik, H., Osterberg, H., and Richards, O. W. (1951). *Phase Microscopy: Principles and Applications* (New York: Wiley).

Berek, V. M. (1927). Grundlagen der Tiefenwahrnehmung im Mikroskop. Marburg Sitz. Ber. 62, 189–223.

Brenner, M. (1994). Imaging dynamic events in living tissue using water immersion objectives. *Am. Laboratory,* April 14–19.

Buil, C. (1991). *CCD Astronomy* (Richmond: Willmann-Bell).

Ellis, G. W. (1978). Advances in visualization of mitosis in vivo. In: *Cell Reproduction: In Honor of Daniel Mazia* (E. Dirksen, D. Prescott, and C. F. Fox, eds.), pp. 465–476 (New York: Academic Press).

Françon, M. (1961). *Progress in Microscopy* (Evanston, IL: Row, Peterson).

Galbraith, W., and David, G. B. (1976). An aid to understanding differential interference contrast microscopy: computer simulation. *J. Microscopy* 108, 147–176.

Gall, J. G. (1967). The light microscope as an optical diffractometer. *J. Cell Sci.* 2, 163–168.

Gilliland, R. L. (1992). Details of noise sources and reduction processes. In: *Astronomical CCD Observing and Reduction Techniques,* Astronomical Society of the Pacific Conference Series, vol. 23. San Francisco pp. 68–89.

Haugland, R. P. (1996). *Handbook of Fluorescent Probes and Research Chemicals,* 6th. ed. (Eugene, OR: Molecular Probes, Inc.)

Hecht, E. (1998). *Optics,* 3rd. ed. (Reading, MA: Addison Wesley Longman).

357

Hiraoka, Y., Sedat, J. W., and Agard, D. A. (1987). The use of a charge-coupled device for quantitative optical microscopy of biological structures. *Science* 238, 36–41.

Hoffman, R. (1977). The modulation contrast microscope: principles and performance. *J. Microscopy* 110, 205–222.

Hoffman, R., and Gross, L. (1975). Modulation contrast microscopy. *Appl. Opt.* 14, 1169–1176.

Holst, G. C. (1996). *CCD Arrays, Cameras, and Displays* (Bellingham, WA: SPIE Optical Engineering Press).

Howell, S. B. (1992). Introduction to differential time-series astronomical photometry using charge-coupled devices. In: *Astronomical CCD Observing and Reduction Techniques,* Astronomical Society of the Pacific Conference Series, vol. 23. San Francisco pp. 105–129.

Inoué, S. (1981). Video image processing greatly enhances contrast, quality, and speed in polarization-based microscopy. *J. Cell Biol.* 89, 346–356.

Inoué, S. (1989). Imaging of unresolved objects, superresolution, and precision of distance measurement with video microscopy. *Methods Cell Biol.* 30, 85–112.

Inoué, S. (2001). Polarization Microscopy, in: Current Protocols in Cell Biology, in press.

Inoué, S., and Inoué, T. (2000). Direct-view high-speed confocal scanner—the CSU-10. In: *Cell Biological Applications of Confocal Microscopy* (B. Matsumoto, ed.), 2nd. ed. (New York: Academic Press).

Inoué, S., and Oldenbourg, R. (1998). Microtubule dynamics in mitotic spindle displayed by polarized light microscopy. *Mol. Biol. Cell* 9, 1603–1607.

Inoué, S., and Spring, K. R. (1997). *Video Microscopy: The Fundamentals,* 2nd. ed. (New York: Plenum).

James, J. (1976). *Light Microscopic Techniques in Biology and Medicine* (Netherlands: Martinus Nijhof).

James, J. and Dessens, H. (1963). Immersion-refractometric observations on the solid concentration of erythrocytes. J. Cell Comp. Physiol. 60, 235–41.

Kingslake, R. (1965). Applied Optics and Optical Engineering, Volume 1. Light: Its Generation and Modification, (New York: Academic Press).

Koehler, A. (1893). A new system of illumination for photomicrographic purposes. *Z. Wiss. Mikroskopie* 10, 433–440. Translated in *Royal Microscopical Society—Koehler Illumination Centenary,* 1994.

Lang, W. (1970). Nomarski differential-interference contrast system. *Am. Laboratory,* April pp. 45–51.

Lang, W. (1975). Nomarski differential-interference contrast microscopy. I. Fundamentals and experimental designs; II. Formation of the interference image; III. Comparison with phase contrast; IV. Applications. *Carl Zeiss Publication 41-210.2-5-e.*

Minnaert, M. (1954). *The Nature of Light and Color in the Open Air* (New York: Dover).

Nathans, J. (1984). In the eye of the beholder: visual pigments and inherited variation in human vision. *Cell* 78, 357–360.

Newberry, M. V. (1991). Signal-to-noise considerations for sky-subtracted data. *Publ. Ast. Soc. Pacific* 103, 122–130.

Newberry, M. V. (1994a). The signal to noise connection. *CCD Astronomy,* summer 1994, pp. 34–39.

Newberry, M. V. (1994b). The signal to noise connection, Part II. *CCD Astronomy,* fall 1994, p. 3 (corr.) and 12–15.

Newberry, M. V. (1995a). Recovering the signal. *CCD Astronomy,* spring 1995, pp. 18–21.

Newberry, M. V. (1995b). Dark frames. *CCD Astronomy,* summer 1995, pp. 12–14.

Newberry, M. V. (1995c). Pursuing the ideal flat field. *CCD Astronomy,* winter 1996, pp. 18–21.

Oldenbourg, R. (1996). A new view on polarization microscopy. *Nature* 381, 811–812.

Oldenbourg, R. (1999). Polarized light microscopy of spindles. *Methods Cell Biol.* 61, 175–208.

Oldenbourg, R., Terada, H., Tiberio, R., and Inoué, S. (1993). Image sharpness and contrast transfer in coherent confocal microscopy. J. Microsc. 172, 31–39.

Padawer, J. (1968). The Nomarski interference-contrast microscope. An experimental basis for image interpretation. *J. Roy. Micr. Soc.* 88, 305–349.

Pawley, J. B. (1995). *Handbook of Biological Confocal Microscopy,* 2nd. ed. (New York: Plenum).

Ploem, J. S. (1967). The use of a vertical illuminator with interchangeable dielectric mirrors for fluorescence microscopy with incident light. Z. wiss. Mikrosk. 68, 129–142.

Pluta, M. (1988). *Advanced Light Microscopy,* vol. 1, *Principles and Basic Optics* (Amsterdam: Elsevier).

Pluta, M. (1989). *Advanced Light Microscopy,* vol. 2, *Specialized Methods* (Amsterdam: Elsevier).

Pluta, M. (1993). *Advanced Light Microscopy,* vol. 3, *Measuring Techniques* (Amsterdam: Elsevier).

Ross, K. A. F. (1967). Phase Contrast and Interference Microscopy for Cell Biologists (London: Edw. Arnold, Ltd).

Russ, J. C. (1998). *The Image Processing Handbook,* 3rd. ed. (Boca Raton, Florida: CRC Press).

Russ, J. C., and Dehoff, R. T. (2000). *Practical Stereology,* 2nd. ed. (New York: Plenum).

Rybski, P. M. (1996a). What can you really get from your CCD camera? *CCD Astronomy,* summer 1996, pp. 17–19.

Rybski, P. M. (1996b). What can you really get from your CCD camera? (Part II). *CCD Astronomy,* fall 1996, pp. 14–16.

Shotton, D. (1993). *Electronic Light Microscopy: Techniques in Modern Biomedical Microscopy* (New York: Wiley).

Slayter, E. M. (1976). *Optical Methods in Biology* (Huntington, NY: Krieger).

Slayter, E. M., and Slayter, H. S. (1992). *Light and Electron Microscopy* (New York: Cambridge University Press).

Sluder, G., and Wolf, D. E. (1998). *Methods in Cell Biology,* vol. 56, *Video Microscopy* (San Diego: Academic Press).

Spencer, M. (1982). *Fundamentals of Light Microscopy* (New York: Cambridge).

Spring, K. (1990). Quantitative imaging at low light levels: differential interference contrast and fluorescence microscopy without significant light loss. In: *Optical Microscopy for Biology;* (Proceedings of the International Conference on Video Microscopy held in Chapel Hill, North Carolina, June 4–7, 1989.) Herman, B and Jacobson, K., editors. (New York: Wiley).

Spring, K. R. (2000). Scientific imaging with digital cameras. *BioTechniques* 29, 70–76.

Strong, J. (1958). *Concepts of Classical Optics* (San Francisco: W. H. Freeman).

Sullivan, K. F., and Kay, S. A. (1999). *Methods in Cell Biology,* vol. 58, *Green Fluorescent Proteins* (San Diego: Academic Press).

Texereau, J. (1963). How to Make a Telescope (Garden City, New York: Doubleday). pp. 5–6.

White, J. G., Amos, W. B., and Fordham, M. (1987). An evaluation of confocal versus conventional imaging of biological structures by fluorescence light microscopy. *J. Cell Biol.* 105, 41–48.

Wilhelm, S., Gröbler, B., Gluch, M., and Heinz, H. (2000). *Confocal Laser Scanning Microscopy: Principles.* Carl Zeiss Publication 40-617e.

Wood, E. A. (1964). *Crystals and Light: An Introduction to Optical Crystallography* 2nd ed. (New York: Dover), pp. 84–87.

Yuste, R., Lanni, F., and Konnerth, A. (2000). *Imaging Neurons: A Laboratory Manual* (Cold Spring Harbor, NY: Cold Spring Harbor Laboratory Press).

Zernike, F. (1946). Phase contrast, a new method for the microscope observation of transparent objects; in: Achievements in Optics, Bouwers, A., ed. (New York, Amsterdam: Elsevier).

Zernike, F. (1955). How I discovered phase contrast. *Science* 121, 345–349.

INDEX